Most people believe that women's entry into political life is a recent phenomenon. This book shatters that myth. It restores to history the career of one remarkable woman and reveals the vital role that women in the early part of the twentieth century played in American politics—despite their absence from elective or appointive office.

Belle Moskowitz was Alfred E. Smith's closest political adviser during his four terms as Governor of New York, and, as a result, the most powerful woman in American political life during his ascendancy in the Democratic Party in the late 1920s. She served as Smith's political strategist and campaign manager and was a major force in shaping the social welfare programs for which his administration is best known today; many of these programs were precursors of programs later developed on the national level by Franklin D. Roosevelt under the New Deal.

Moskowitz achieved her prominence without benefit of either family connections or social position resulting from birth or marriage. This book explains how she managed this astonishing feat. Her role as Governor Smith's adviser followed a career of almost twenty years in settlement work, social reform, and industrial and labor relations. When Smith first ran for governor in 1918, Moskowitz organized the women's vote for him. She later masterminded his nomination for President at the Democratic convention in 1928. Throughout Smith's years in office, Moskowitz worked as his public relations counselor, a new profession in which she was a pioneer.

Moskowitz's career has not received close attention until now. Since she worked at a secondary level of politics in which her own career never seemed as important as that of the man she served, she discarded many of her private papers; others were lost. Elisabeth Perry, who is Moskowitz's granddaughter, has dug deeply into the public record and the records of Moskowitz's associates, as well as drawing on the reminiscences of Moskowitz's daughter, Perry's own aunt. The result is this riveting portrait.

BELLE MOSKOWITZ

BELLE MOSKOWITZ

Feminine Politics and the Exercise of Power in the Age of Alfred E. Smith

Elisabeth Israels Perry

New York • Oxford
OXFORD UNIVERSITY PRESS
1987

Oxford University Press

Oxford New York Toronto
Delhi Bombay Calcutta Madras Karachi
Petaling Jaya Singapore Hong Kong Tokyo
Nairobi Dar es Salaam Cape Town
Melbourne Auckland

and associated companies in
Beirut Berlin Ibadan Nicosia

Published by Oxford University Press, Inc.,
200 Madison Avenue, New York, New York 10016

Oxford is a registered trademark of Oxford University Press

Library of Congress Cataloging-in-Publication Data
Perry, Elisabeth Israels.
Belle Moskowitz.
Bibliography: p.
Includes index.
1. Moskowitz, Belle, 1877–1933. 2. Smith, Alfred Emanuel, 1873–1944.
3. Politicians—New York (State)—Biography.
4. Social reformers—New York (State)—Biography.
5. New York (State)—Politics and government—1865–1950. I. Title.
F124.M885P47 1987 974.7'042'0924 [B] 87-11318
ISBN 0-19-504426-6

Portions of this book are reprinted from *Labor History,* 23 (1982), 5–31,
and *American Quarterly* 37 (1985), 719–33, with the kind permission of the publishers.

10 9 8 7 6 5 4 3 2 1

Printed in the United States of America
on acid-free paper

For David, Susanna, and Lewis

Contents

Introduction

Let it not be imagined that she was an ambitious woman, that hers was the ambition to lead. It was the will to serve and not the desire to lead that ruled her life.

Rabbi Stephen S. Wise
Eulogy for Belle Moskowitz, 4 January 1933

The news that Belle Moskowitz had died sent a wave of shock and disbelief through the crowd assembled in Albany for the inauguration of New York Governor Herbert H. Lehman. "It's a disaster!" former Governor Alfred E. Smith cried. Moskowitz had been his closest adviser. A few weeks earlier she had fallen down the front steps of her home and broken her arms. Since then, she had been laid up, presumably recovering. Few had known that more serious complications had set in. Smith and his entourage, including Moskowitz's oldest son, Carlos Israels, left Albany for New York. Three days later, three thousand mourners poured into Temple Emanu-El, and more gathered outside, jamming Fifth Avenue.

The day after her death on January 2, 1933, at the age of fifty-five, the *New York Times* published a long obituary that began on the first page. Its author said that, during Al Smith's ascendancy within the Democratic party, Moskowitz had been the most powerful woman in American politics. The author did not need to add that, throughout the 1920s, Moskowitz had been the most powerful woman in New York State politics. After assisting in Smith's first election as governor in 1918, she served as his political strategist and campaign manager during his four terms in office. A social reformer whose efforts dated from the Progressive Era, she developed many of the programs for which Smith later became famous. In 1928 she was the first woman to direct publicity in a major political party's national campaign, and was the only woman on the Democratic National Committee's executive committee. When Smith lost the election, he retained Moskowitz as his publicity and literary agent, and chose her to organize his bid for renomination in 1932. Her death meant the loss of a major prop in his battle to remain a power in American politics.

Today Belle Moskowitz's name is no longer widely known. Further, the story of how she achieved her position in an age when few women played major political roles has never been told. One reason is that Moskowitz did not save most of her political papers. We will never know why. Perhaps she discarded them when she moved residences or offices, or when she finished one project and moved on to the next. Perhaps she did not think they were important. To her, what mattered was what she accomplished, not the record of her role. Moreover, since she died fairly young and suddenly, after only a brief illness, she never had a chance to reflect upon the past. Had she lived longer, she might have made an effort to collect and preserve her traces.

Another reason her papers disappeared may lie with her family's treatment of them. After she died, her youngest son Josef Israels cleaned out the attic. Although he recognized that his mother had been "a great woman," he lacked the perspective to recognize the importance of the material she had saved, especially from her early life. He preserved a remnant, which her family later donated to Connecticut College for Women, but threw out the rest. On January 13, 1933, he wrote his sister Miriam Franklin, "Most of mother's things are pretty well cleaned up. I've been doing that personally because it would give Dad the Willies and Carlos just couldn't be much help at it." He complained, "Mother never threw anything away. I've just mercilessly thrown out masses of ancient junk rather than pass their meaningless dust on to new generations." On August 11 he reported that he was still moving "useless junk" out of the house, a "vast accumulation" that included "Grandma's wedding lace, torn and yellow, pictures of unheard of relatives and old magazines and newspapers from 1905. I sort out and throw out a carload a day, and it seems an endless job." He did save locks of his own and Carlos's baby hair.

Joe, a writer, later proposed to a publisher that he write his mother's biography, a proposal that he either never pursued or that was never accepted. One wonders on what documents he intended to base his story. Documentation does exist. It lies buried, often unindexed, in the archives and private papers of others, especially those of Al Smith, and can also be found in contemporary newspapers and journals, and in the archives of the various organizations with which she was involved.

The initial reason I took on the task of digging up this information was that Belle Moskowitz was my paternal grandmother. She died before my parents, Carlos Israels and Irma Commanday, had even met. After they divorced, I spent relatively little time with my father, the only person who could have told me more about her. As it was, I recall no one mentioning her to me until one day, when I was about sixteen, my mother's father, Frank Commanday, said, "When you grow up you must write a book about your Grandma Belle." Who was she? I asked. "Ask your father," he replied, and I did. I remember his exact words: "She was Al Smith's campaign manager, and you can't write a book about her. No one can because she wasn't a 'saver.' She threw away her papers." I knew nothing of Al Smith, of campaign managing, of the uniqueness of a woman playing such a role in politics, or of "papers." My father did not explain, and the subject never came up again.

Years later, the women's movement led to the creation of a new field, women's history. I had become a historian, not of women but of seventeenth-century France. Meanwhile, in the late 1960s and early 1970s, Belle Moskowitz's name appeared in several books about Smith,

Franklin D. Roosevelt, and Robert Moses. While reading an obituary of another historian, J. Salwyn Schapiro, who had been a boyhood friend and settlement work colleague of Henry Moskowitz, my husband Lewis Perry suggested I write a book about my grandmother. My grandfather Frank's imperative of twenty odd years earlier came back to me. So did my father's warning.

But I was older, and an experienced researcher. By that time, my father had died. Others close to Belle were also gone—Frances Perkins, Joseph Proskauer, Samuel Rosenman. Robert Moses was still alive, but (as I soon learned) loath to be interviewed. James Farley died two weeks after our first phone contact. So it went. Oral sources in the end provided the least of my documentation. Oscar Handlin's biographical sketch in *Notable American Women,* her detailed obituary in the *New York Times,* memorial notices sent in from organizations to which she had belonged, and her personal papers at Connecticut College offered the first clues. I received some gems from colleagues who heard of my work and came across uncatalogued "finds" in collections they were studying. The rest I uncovered over many years of often interrupted, but fruitful digging in libraries and archives. The process took longer than I had anticipated. Nonetheless, it has been worthwhile, not simply because I can now document Belle Moskowitz's career and thus complete the picture of her historical role, but for its historical significance.

Moskowitz's career helps explain an enigma of American women's political history. When women won the vote in 1920, they also won their first chance to reach for the tangible prizes of political power—elected or appointed office and positions of influence within the national parties. But only an exceptional few won major bureaucratic, elected, or party posts. Parties expanded their organizations to include women's divisions or to make the number of women committee members equal to that of men, but for the most part women failed to penetrate inner decision-making ranks or to receive nominations for major offices.

Historians have suggested several explanations. Women, for one, lacked political experience. Those who ran for office necessarily opposed male incumbents who were hard to displace; when they lost, they became identified as "losers" and could not raise money for subsequent races. In addition, as the older generation of women suffragists aged and died, younger women took for granted what their elders had achieved and began to pursue other, mostly non-political goals. Finally, most male politicians were not ready to grant women predominant roles in government and in political parties. They paid lip service to equality, but when it came time to advance candidates, they chose men.

To these reasons, I can now add another: the few models of successful political women available to younger women in the 1920s were women involved in politics less for the purpose of advancing their own careers than to promote ideals and programs. Eleanor Roosevelt, Frances

Perkins, and Belle Moskowitz—three of the most influential political women in the 1920s—entered politics primarily because they knew that their social programs could be achieved only through legislation. None were careerists. Both Moskowitz and Roosevelt took back seats to men, Belle to Al Smith, Eleanor to her husband. Perkins achieved high public office in Franklin Roosevelt's cabinet, but never sought power; it came to her because she needed to earn a living, and because her experience in social reform positioned her to receive an appointment.

Belle Moskowitz made choices that no personally ambitious woman would make. She refused to take credit for what she did for Smith; she turned down his offers of government posts in the belief that she would be more effective out of office than in it. Had she accepted a post, her duties would have been so circumscribed that she could not have acted as a broad influence on Smith's programs. She was right to have declined appointment, but as a result the record of her achievement nearly died with her.

Moreover, in order not to threaten the pride of the political men with whom she worked, Moskowitz adopted a strategy of what I call "feminine" (as distinct from "feminist") politics. Women were gaining a new place in party politics, but Moskowitz adhered to the precedents and role models of previous generations. She looked backward, first, to the Victorian ideology of "true womanhood" which persisted into the early twentieth century and which told women that their chief priority in life was to serve, not lead, men. Second, she looked to the "new women" of the Progressive Era, who, as volunteers in welfare agencies, social workers in settlement houses, or executive secretaries of reform organizations, struggled selflessly to achieve specific social reforms. These were the women who trained her and after whom she modeled her behavior.

In addition to her executive abilities and creativity, Moskowitz's adoption of a feminine strategy was the key to her success in the male world of politics. Because she never threatened the establishment she sought to influence, she achieved the reality, but never the appearance (except to insiders), of power. But because she left no public record of her achievement, and because her old-fashioned strategy held little appeal to a younger generation, she failed to provide a model that encouraged other women.

Moskowitz's life is historically significant in another way. Political observers in the 1920s thought she was the author of many of Smith's social programs. In a memorable phrase, Molly Dewson, a Democratic women's leader in the 1930s, called her "Al Smith's tent pole." This critically supportive role can now be documented, as can the extent to which the sources of those programs lay within the traditions of the Progressive Era social reform movements in which women were a dominant influence. In the mid-1930s Franklin D. Roosevelt, lamenting Smith's latter-day conservatism, commented that the groundwork for many New Deal policies had been laid by the programs that Roosevelt, as New York's governor after 1928, had inherited from Smith, his predecessor. By establishing Moskowitz's role in those programs, this book not only helps to explain Smith's progressivism in the 1920s but sheds further light on the nature of the connection between the progressivism fostered by women reformers and the programs of the New Deal.[1]

It is important that Moskowitz was Jewish. Up to the time Smith ran for president, Jewish professionals, most of them connected either to the Ethical Culture or Free Synagogue movements, constituted a powerful element in his inner circle of advisers—men like Abram Elkus, Joseph Proskauer, Bernard Shientag, Julius Henry Cohen, Samuel Untermyer, and, of course, Robert Moses. Elkus and Proskauer had brought Belle Moskowitz into the circle; she in turn brought in Moses, her husband Henry, Herbert and Irving Lehman, Aaron Rabinowitz, and Clarence Stein. The role this group of Jewish liberals played in formulating Smith's reforms as well as in preparing the ground for the New Deal has not yet received sufficient recognition.

I have tried to avoid a pitfall, common to biographers, of attributing all acts of importance to my subject alone. Many others influenced Smith, as Robert Caro's work on Robert Moses and Louis Hacker's and Mark Hirsch's on Joseph Proskauer have shown. I have documented everything I have attributed to Moskowitz, and have based all speculations about her influence on strong circumstantial evidence. The need to establish Moskowitz's influence has required substantial documentation, so, in the interest of limiting the book's length, I have tried to avoid telling more of Al Smith's story than was essential to the telling of Belle Moskowitz's. There are plenty of sources on Smith's career, and will be more in the future.

If there is a larger lesson to be learned from this book, it might be this: the American woman's interest in politics did not die after she won the vote. True, in the 1920s women activists divided bitterly during the first congressional debates on the Equal Rights Amendment, and this division probably hurt their political effectiveness. But even the loss of a common goal did not kill the desire of women activists to continue influencing political decisions. The absence of their names on legislation or treaties does not mean they were not working for them. Women have not been as powerless as their lack of public office would lead us to think. Although excluded from the power elites that men have always dominated, they nonetheless have found creative, if unrecorded, ways to exercise power.

Many individuals and institutions helped with this project. First and foremost are Irma Commanday Bauman and Mordecai Bauman, who gave unstintingly of their time to help me find documents and track down interviewees. Throughout the project's long gestation, they gave me their love and support, reading many drafts of the book's chapters and encouraging me to keep pressing forward. This book owes a great debt to them.

A deepened relationship with my aunt Miriam Israels Gabo was an unanticipated reward of this project. I will treasure the memory of the many hours we spent talking about her mother and other family matters. Thanks to Miriam, also, her friend Sophia Mumford gave me priceless editorial advice during the manuscript's final stages. Two other family friends, Aline MacMahon and Dorothy Rosenman, persuaded Robert Moses to see me at a time when he was refusing other interviews about the past.

In 1975 the Yale University History Department appointed me Visiting Fellow, which gave me access to Sterling Library during my first year of research. The New School for Social Research, the Wellesley Center for Research on Women in Higher Education and the Professions, the American Council of Learned Societies, the Eleanor Roosevelt Institute, and the Graduate Schools of the State University of New York at Buffalo and of Vanderbilt University have all awarded me research grants at various times. At a critical juncture in my career, I won a fellowship from the National Endowment for the Humanities. I am grateful to all these institutions for their confidence and support.

I have relied heavily on the energetic efforts of many reference librarians and archivists. Thanks go especially to James MacDonald of Connecticut College; Abraham Peck and Fannie Zelcer of the American Jewish Archives; and Martha Katz-Hyman of the American Jewish Historical Society. Also deserving thanks are Jim Corsaro of the Manuscripts Division of the New York State Library and former New York State Archivist Edward Weldon for the help and encouragement they have given me over the years. David Wigdor at the Library of Congress provided many leads, as did archivists at the Franklin Delano Roosevelt and Herbert Hoover presidential libraries. Librarians and archivists at the following institutions also deserve credit for their extra efforts: the Manuscripts Division of the Library of Congress, the National Archives, Yale University, the State University of New York at Buffalo, Indiana University, Columbia University, the New York Public Library newspaper and manuscript divisions, the Arthur and Elizabeth Schlesinger Library on the History of Women, the University of Louisville, the New York State School of Industrial and Labor Relations, Cornell University, and Vanderbilt University.

Thanks to the cooperation of many individuals, I gained access to documents not usually open to the public. Leon Stein, Lazar Teper, and Walter Mankoff of the International Ladies' Garment Workers Union allowed me to study the union's transcripts of grievance and arbitration procedures from Moskowitz's years as an official in the dress and shirtwaist trade. The New York Society for Ethical Culture, the Council of Jewish Women-New York Section, and the Women's City Club of New York also gave me access to their records.

Scholars who helped with research problems and read portions or all of the text at various stages include Tom Archdeacon, Martha Banta, Blanche Cook, Nancy Cott, Ellen Dwyer, Paula Eldot, Sue Elwell, Anita Feldman, Frank Freidel, Ralph Janis, George Juergens, Linda Kerber, Richard Kirkendall, Richard Lowitt, Samuel McSeveney, David Montgomery, Maurice Neufeld, James Patterson, Kathryn Sklar, Eugene Tobin, and Robert Wesser. Other friends offered me hospitality during research trips or listened tolerantly over the years to my many attempts to clarify my ideas about Moskowitz: Eleanor and John Baker, Judith and Roger Daniels, Joan and David Hollinger, Toni and David Davis, Judith and Todd Endelman, Ronald Grele, Barbara Hanawalt, Phyllis and Melvyn Leffler, Susan and Martin Sherwin, and Jan Shipps. Together they have formed a supportive network on which I have leaned for many years. I am glad of this opportunity to thank them all.

Thanks finally to Gerard McCauley and Sheldon Meyer, whose enthusiasm for the book was a vital sustaining factor as my work drew to an end.

To my son David, whose lifespan almost coincides with the life of the project and whose loving cheers as I neared the finish line I will never forget; to my daughter Susanna, whose pride in and acceptance of her mother as a friend grew along with the project; and to my husband Lewis, who remained throughout the source of love, inspiration, professional counsel, and savory gourmet meals, I dedicate this book. Words are inadequate to express my appreciation for their patience with what threatened to become a lifetime work. As I entered the home stretch, all three picked up more than their share of the family pieces. The book is as much theirs as it is mine.

Nashville, Tennessee E. I. P.
Spring, 1987

BELLE MOSKOWITZ

CHAPTER 1

Guardian Angel of the Lower East Side

I was born and raised in New York, and I haven't lost all of my ideals, either.

Belle Lindner Israels
New York Times, November 10, 1912

Belle Lindner grew up in a storefront house on Third Avenue and 127th Street, not far from the Harlem River. In the late nineteenth century, Harlem was a loosely defined area north of Central Park bounded by the Harlem River on the east and Morningside and St. Nicholas Parks on the west. Much of the area was still rural, but around the Lindners' home a busy residential and shopkeeping community thrived, serving those immigrants to the city who could afford to escape the congested tenements downtown.

Belle's parents, Isidor and Esther Freyer Lindner, arrived in America from East Prussia in 1869, and thus belonged to a generation later known as "old immigrants," a term that distinguished them from the "new immigrants" who poured into New York over the following decades. Because they arrived early, the Lindners avoided the crush of population on housing and jobs that afflicted later immigrants. Settling first on Canal Street on Manhattan's Lower East Side, Isidor, a watchmaker, opened a shop that did well. In a few years the Lindners were able to move uptown to the three-story clapboard house in Harlem where Belle was born on October 5, 1877. Her birth certificate lists her as Esther's sixth child. Unless this is an error, three earlier births must have ended in the infants' deaths, for when Belle came there were only two older siblings, Walter, twelve, and Max, four. A later child, Clara, lived until school age. Running across Third Avenue to greet her father, she was caught between two passing trolleys and killed.

The Lindner house is gone now, replaced by a modern brick structure. Only a few houses like theirs still survive on side streets. To us they seem minimally habitable, but to the Lindners they offered a welcome, though modest, standard of comfort. A large display window advertised the ground-floor shop, behind which a parlor provided space for company. Here Belle often listened to the neighbors discussing politics, and later recalled the gatherings as part of her education. Behind the parlor was a combination kitchen-dining area, and behind that a shed used for storage and food preparation, weather permitting. The privy was in the backyard. Years passed before the family could afford a bathroom, which was installed in an upstairs bedroom behind curtains hung on poles.

Like most Jewish wives, Esther Lindner catered to her husband. Each day before dinner, like clockwork, she gave him a glass of schnapps, and then a meal of soup and meat. While Isidor napped, she cleaned the pearls in the jewelry cases and took care of the shop. A family joke reflects how hard she worked to please him: when Belle was about four or five, Isidor took her to Europe to meet her grandparents; the family said he had taken the trip for Esther's health, not his own. For Esther, subservience to her husband seemed proper. She respected tradition. Her marriage had been arranged, and when the matchmaker told her that her intended neither drank nor smoked, her only comment was, "That's the boy for me!" After her marriage she asserted herself more, refusing to wear the same kind of "shaitel" (wig) and cap her mother, a rabbi's wife, had worn, and throwing her skirt hoops overboard when she and Isidor sailed to America. And later, when a thief threatened her in the Third Avenue shop, she chased him down the street and caught him, despite her plump and diminutive form. Esther's grandchildren remember her as "smart" and "practical." She may have catered to the family patriarch, but it was she who "ran the family."

Isidor, who always wore a beard and dressed in a Prince Albert frock coat, was a scholar who knew seven languages, including Hebrew. He was also an accomplished cantor. On Sabbath evenings he sang in Temple Israel, and for a time, after the resignation of a rabbi who never got paid, led the services. In 1878, he served as the Temple's president, its fourth. The Lindners' religious practices gave Belle a strong Jewish identity and a lifelong commitment to the Judaic commandment to serve others. But their orthodoxy did not prevent her from deviating later from traditional ways of expressing their faith.

It is hard to say which parent had a greater influence on the young Belle. That Isidor took her alone to Europe indicates a certain favoritism. His scholarly example instilled in her a love of reading. But both parents made sacrifices to foster her education, and Belle surely noticed her mother's ability to cater to a man while remaining in control of the household. This model proved useful when, as an adult, Belle entered the male-dominated worlds of social reform and politics.[1]

Belle began her education at local public schools. When she reached high school age, her parents enrolled her in a private school for reasons that are not known but can be guessed. She was a bright, articulate child whose parents were proud of her and felt she deserved special attention which New York public schools could not provide. Since the late 1880s, the city had been engaged in a bitter struggle over the centralization and professionalization of the school system. By the nineties uptown merchants in particular vigorously protested centralization, fearing it would diminish their control over their children's education. The Lindners may have shared this sentiment. Moreover, the city's classrooms in general were crowded and poorly supplied. Generally untrained teachers relied heavily on rote repetition, singsong drill, and strict discipline to instill a curriculum that had not been changed for many years. These failures were criticized by professional educators who, following the lead of Horace Mann, founder of the Massachusetts common school system, began to try new approaches toward the end of the nineteenth century. Mann had argued that basic skills and a common value system could be taught without destroying the child's individuality. This view laid the groundwork for the progressive education movement.

At the forefront of this movement was a laboratory school founded in 1887 by faculty at New York's Teachers College. Named after Horace Mann, it opened with sixty-four kindergarten and elementary pupils, adding a high school of nine pupils two years later. By the time Belle Lindner began to attend, over two hundred fifty were enrolled. Her own class never held more than twenty. For an annual tuition of $100, Belle was thus guaranteed a small class and a solid foundation in English literature, research methods, rhetoric and public speaking, ancient and modern history, government, mathematics, and free hand drawing. She also went on weekly excursions to museums and other places of cultural value, and was taught to observe critically and apply her observations in practice. Her teachers used unrepressive methods designed to inspire a love of learning.[2]

After graduating from Horace Mann in 1894, Belle enrolled as a non-matriculated "Special Student" at Teachers College. Not that she envisioned a teaching career. Endowed with a voice that could "fill a room," she had experienced success as a dramatic reader and planned to study the art with Professor Ida Benfey, whose public readings were enjoying a minor vogue on the New York theatrical scene. She dramatized portions of prose masterpieces such as *The Mill on the Floss, Les Misérables, A Tale of Two Cities,* and the Book of Job, or gave readings of short stories such as Mary Wilkins's "Young Lucretia" and "A Gala Dress." Instead of impersonating characters she "revealed" them through selected passages declaimed almost from memory. Critics praised her "graphic power" and interpretative skills. Benfey adopted Belle as a favorite, teaching her voice placement, enunciation, emphasis, and phrasing.

But Belle stayed only a year at Teachers College. She happened to hear a lecture at Columbia College given by Heinrich Conried, well-known director of the German repertory company at the Irving Place Theatre, and later of the Metropolitan Opera. She asked him for help. He usually did not coach individuals, but her talent impressed him and he accepted her as a private student, prompting Belle to leave Teachers College. Under Conried's tutelage she developed a dramatic *persona* called "James, the Tailor-Made Girl," and for the next few years delivered monologues in private homes and taught elocution and dramatic reading to children.

These years were a period of exploration during which Belle Lindner discovered much about herself. While working on her dramatic reading, she lived at home and helped out in the shop. She read a great deal, wrote letters to friends, and attended lectures and political speeches. During the summers she swam at Brighton Beach with her brother Walter and groups of friends. Costume parties with "tableaux vivants" and recitations were favorite pastimes. She also liked dancing and socializing. A later friend, Judith Bernays, described Belle as "beautiful." Her trim and petite figure, her handsome features and long chestnut hair, which she drew back in a bun, and her "intelligent, unostentatiously self-assured" manner made her "unusually attractive in a very quiet way." Like most young women of her age, she was interested in men, but her most intense relationships were with women.[3]

In Belle's day, such relationships were not unusual. Victorian society accepted them more readily than premarital relationships between men and women. Young women frequently developed crushes on other girls or older women, and some crushes grew into deep love relationships. One partner might call the other by a male name. When apart, the pair longed for one another, exchanging letters daily; reunited, they might sleep together without its being thought wrong. Some of these relationships survived the marriage of one or both partners, one woman attending the other at childbirth or adopting the offspring in the event of death.[4] As an adolescent in the late 1890s, Belle developed several relationships of this type. One was with Anna George, the daughter of economist Henry George. Anna was one year Belle's junior at Horace Mann, but they remained close after Belle graduated, exchanging letters daily for many years. It was Anna who named Belle's monologue character "James," a name that eventually came to be applied to Belle herself. Another Horace Mann friend, Fannie Sax, lived in Wilkes-Barre, Pennsylvania; she and Belle wrote one another three times a week. Years later, when Belle's first husband died and she was desperate for income to support her family, she turned to Fannie for ideas for articles on household hints which she hoped to sell to newspapers. There was also a Miss Stiurr, "sweet and strong and good," who gave her "strength and courage and example" to survive a crisis with "the article known as man."

Not all Belle's friends came from school. She met Abbie Seldner Fridenberg through the Temple Israel Sisterhood and a "Wednesday Club" of women interested in literature. Abbie wrote poems, one of which was published in the *American Jewess,* a short-lived and moderately feminist publication that grew out of the National Council of Jewish Women. She also shared Belle's taste for the Romantics, especially Browning and Tennyson. In the summer of 1896, when Abbie, who was married to a doctor, traveled abroad to "calm her nerves," Belle, almost nineteen at the time, wrote two longing letters to her. "*Why* aren't you within reach when I want you so?" she lamented. She needed her older and more experienced friend to help her "see all sides" of the profound questions that disturbed her. These concerned an unspecified but stressful experience with a man. She had emerged "heart-whole," much further along in "experience" and "education." The whole business had been "veil-dropping," she wrote, and caused her to see herself from an entirely new standpoint. "Can this be I?" she asked in adolescent self-query. She emerged from the ordeal no more "optimistic" about "the other sex," but "broadened and deepened and saddened." One could read and think about many things, she avowed, but only feel and know by undergoing something oneself. She then announced: "I have been testing myself, finding out just what my vaunted self-control can do. So far, I have been successful; but the real test is yet to come and you will know who is victor."

Quoting Tennyson's dedication to his *Idylls of the King,* she used the line "shadowing Sense at war with Soul" to reflect upon her inner struggles:

> . . . that is after all, most of the battle in life; there seems the root of the matter. Now comes the question, which is the one that *should* be victor? Of course, you answer at once, soul: but are you so sure? May not the life of the *sense* be pleasant and right too? Right, because after all, pleasure and not pain is the ultimate end.

This halting, inconclusive attempt to reconcile reason and passion led ultimately to the triumph of Belle's soul, but not without pain. She had been raised in the prevailing morality of her day: sensuality was permitted only in marriage; any other indulgence led to quick ruin. This letter may refer to a strong temptation that she seemed capable of resisting at the cost, however, of hurtful tension.

Belle's letters to Abbie reflect another tension. They contain terms that suggest considerable intimacy between the two women. For example, Belle referred to the "memory of the kiss [that] lingers yet," and the "delightful surprise of the letter from the steamer, and in that letter the hint of another kiss to come—verily you may fear to spoil me." Such expressions of physical endearment, though common among women friends, seem unusually strong here. Belle wondered if Abbie had thought of her at Browning's tomb only because Belle loved Browning. "Didn't you think of me, just a wee bit, because I loved *you?* Well, you should have." Belle ended the letter cryptically, in German: "Be affectionately greeted and—embraced (after all, it's only on paper or in the imagination)." Perhaps an unqualified embrace would have pushed their relationship too far. We do not know how Abbie's husband reacted to his wife's and Belle's mutual infatuation. Belle asked to be remembered "kindly to your lord and—companion," once again using the long dash to convey hesitation before substituting a word more genial than "master."[5]

Belle's closest and most lasting female friendship was with Grace Harriet Goodale, a classics student at Barnard. Had Belle not found a man she could marry, she and Grace might ultimately have lived together in "manless marriage," Grace's own term for her lifelong relationship with Margaret Roys, a librarian with whom she later lived. Grace and "Peggy," perceived by outsiders as lesbians, denied there were "serious abnormalities" between them and described their relations as "fine and wholesome."[6]

Grace met Belle through Amelia Wohlfarth, a former Horace Mann student and later a classmate of Grace's, who brought Belle to a Barnard tea. "We knew at once," Grace later wrote, that it was "friendship at first sight." At the start, they shared mostly literary and dramatic interests. They put on a revue together at the Fidelio Club, a German-Jewish social club, and Belle helped Grace edit her class's edition of the Barnard yearbook, the *Mortarboard.* Prepared in the spring of 1898, this was Barnard's second yearbook. It offered witty characterizations of classmates written in prose and verse. The doggerel about Grace makes fun of her narrow interests:

Two things alone
To mankind known
This Barnard girl was pat in;
All Kipling's verse
She could rehearse,
Likewise a little Latin.

The Classicist
Upon our list
Appeared one autumn morning.
She loved her Greek,
Could Latin speak,
All other subjects scorning.

Grace's talents were not confined to poetry and language. She managed the business affairs of the yearbook so well that it made a profit. As editor, Grace received $25 from this bounty and spent it on the Cambridge edition of Browning, two balcony seats for *Cyrano de Bergerac,* and two weeks' board and lodging. Her copy of the yearbook, now in Barnard's archives, contains a cryptic tribute to a non-Barnard friend, probably Belle. The handwritten line, "It must be Funny but not Rot!" was followed by a short paragraph typed by Grace: "This standard of criticism, carelessly tossed out by a friend in no way connected with Barnard, did stick in our editorial mind, and was quoted, half seriously at least, in making editorial decisions."[7]

Their love of literature bound the two women. Although Belle's ethnic roots gave her fluency in German, she identified culturally with the English Victorians, especially Kipling, who provided rich examples of moral uplift: love of work, loyalty between friends, courage in adversity, self-sacrifice for higher ends. The women did not think to criticize his glorification of the British empire, but in those days few did. Nor were they put off by the masculine world he depicted. They took from him what they wanted, selectively ignoring the rest. Grace composed poems "after Kipling," and both women quoted him to each other when they felt troubled or were in trying circumstances.

Browning was another favorite. One summer Belle plowed through the entire *Ring and the Book,* an immense verse novel that retold a seventeenth-century Italian blood tragedy. It was heavy, and even Belle admitted to Abbie Fridenberg—calling it "high treason"—that August was "really too warm" for Browning. The young women in Belle's circle also read George Meredith, Thomas Hardy, Robert Louis Stevenson, and Charlotte Brontë, whose novel *Shirley* Belle found "unnatural, high-flown, and stilted," too much "of the old school." In addition to the Victorians, Shakespeare was her "food," his comedies "just about right." In contrast, Molière seemed "wearing," always "the same set of characters, the same coarse wit." As for American literature, the popular stories of Mary Wilkins, whom Ida Benfey also loved, pleased her for their "exquisite simplicity and beauty." Set in New England villages, Wilkins's stories studied from a woman's perspective the interaction of "will" with crises of personal dignity or ethics. Like Kipling, Wilkins provided Belle and Grace with models of self-control and pluck.[8]

Belle Lindner and her friends did not spend all their time lamenting the absence of loved ones, pondering inner conflicts, or analyzing literature. In some ways, they were typical adolescents, caught up in the fads of their day. One was the mid-nineties "poster" craze, which began when magazine publishers started using lithographed poster art to advertise their weekly or monthly issues. In 1893, *Harper's Magazine* hired Edward Penfield to design their covers; soon other magazines hired artists, and posters became collectors' items. Costing from twenty-five to fifty cents, not only were they a cheap way to decorate a room, but their innovative designs—flat planes, bright colors used unnaturalistically, semi-nude androgynous figures, mirror illusions, and fantasies—evoked strong emotional responses. Some critics panned poster art for flouting traditional aesthetics, but Belle Lindner covered the walls of her room with them. She and Anna George, who dubbed Belle's room "the Poster Room," also organized "poster parties," another fad. Guests came dressed in costumes, struck poses, and asked others to identify which poster they represented. An existing photograph of a costumed group at Anna's possibly was taken at one of these events. Belle wears a dress with magazine covers all over it and, on her head, what looks like a straw wastebasket. Anna's amused glance at her friend seems to suggest she is making fun of her own taste.[9]

In any case, Belle's cultural and personal tastes reflect her continuing battle between sense and soul. She collected poster art, which critics derided and which she herself mocked. Romantic heroes moved her, but not if they were too "high-flown." Molière's earthiness fatigued her. Her letters reveal her as a person of deep emotion, prepared even for a life on the stage; but convention and propriety held her back. Women, to whom she felt attracted, were safe up to a point, but when she felt herself going too far she withdrew. At the same time she was also attracted to men, but since they endangered her "vaunted self-control" they too were kept at a distance. Buffeted by such conflicts, she naturally prized self-control, at once her ideal and her bane.

Sometimes her efforts to achieve self-control resulted in misunderstandings. Grace Goodale said that Belle's ability to meet "trouble and danger and discomfort with steadiness and calm" led some to call her "cold." Even Grace, herself reserved and seldom "explosive," thought this of Belle, and for a time shied away from the "hackneyed signs" of girl friendships with her. Then, in the spring of 1901, when Belle was terribly ill, they went through "days of anxiety" together that shook them from their mutual shyness. It was only then—three years after their first meeting—that they openly declared their mutual dependence. They remained close, even after they chose to follow different life paths. As early as 1898, Grace jested that she was a "pre-destined, fore-ordained, natural-born, dyed-in-the-wool old maid."[10] The jest came true, at least in the traditional sense, since she became a professor of classics at Barnard and settled into her "manless marriage" with Margaret Roys.

Whatever gender confusion Belle experienced in the late 1890s, by the early 1900s she had emerged with a clearer sense of how she wanted to live. Her formal photographs chronicle this passage. The society in which Belle grew up offered limited options to intelligent, ambitious women, who must either marry or take up a career "like a man." The two were seldom combined. And, even if one chose the latter, the possibilities were few. Most careers, unless one chose to be a teacher, librarian, or social worker, were closed to women. Truly exceptional women of the nineties overcame the mutual exclusivity and limitations of the standard choices. Most chafed under these conditions, but accepted them nonetheless as part of a social reality they could not change. At the same time, they found ways to express dissatisfaction. Belle, and many young women of her generation, fell into this category. College girls, for instance, often dressed up as men in their theatrical productions, acting out forbidden roles without risking gender identity. The revue Grace and Belle mounted at the Fidelio Club may have featured this kind of transvestism. The club's very name, recalling the image of Beethoven's heroine Leonora, who dresses as a man in order to rescue her husband from prison, lends plausibility to this assumption.

In 1896 Belle took up the character of "James, the Tailor-Made Girl," and had herself photographed in the role. The combination of the name "James" with "Girl," the clothes she wore, and pose she took, convey a mixed gender message. In the photo a masculine fedora and sporty gloves complement her trim, plain-bodiced, worsted suit. Puffed sleeves and skirt are concessions to femininity, but with her arms held akimbo and her shoulders thrust forward, she seems to be saying, "I am ready for anything, just like a man." Even so, the portrait is definitely of a girl. Like Fidelio, Belle's *persona* of 1896 partook of the best of both worlds: a quasi-male exterior that enabled her to play-act forbidden roles, with sufficient reminders that, when the play was over, she would return to being a young woman.

A year later she posed again for the same photographer, to much different effect. This time she dressed with more femininity, in white, with a lacy, finely pleated bodice and puffed sleeves. The high neck of the dress suggests the restraint and decorum of the proper Victorian woman. Still, the quest for gender definition had not ended. Instead of sitting in a chair, like a "lady," perhaps framed by draperies or holding a flower—poses typical of women's photographs in the 1890s—Belle perches jauntily on a table top and leans on one arm. Her slight smile and the glint in her eye promise mischief. While a little bit of "James" remains, she is neither totally "James" nor "his" opposite, a charmer waiting for a dashing suitor to sweep her away. This photograph is transitional. It conveys, in ways similar to "James," a lack of conventionality and readiness for surprise, yet the womanly dress signals that Belle was moving toward a clearer sense of how she wished to present herself to the world.

The final resolution of the gender identity quest appears in a photograph taken when she married. "James" is gone, probably for good, and her femininity is in full bloom. She is exquisitely attired in a full-skirted white dress with an elaborately decorated bodice. Sleeves drape softly down her arm, replacing the puffed sleeves of her youth, no longer in fashion. In a pose typical of the "pensive" woman romanticized in the early twentieth century, Belle stretches out one arm rather unnaturally behind her across the arm of a chair, and leans her body in the opposite direction, toward a table. On it sits a book that engages her in deep concentration. As "James/Fidelio," Belle was male on the outside and female on the inside. Here is the verso: enveloped in womanly attire, the interior of her soul shines through, all intellect and power, a reversal of the roles she had play-acted as a late adolescent.[11]

Belle Lindner Israels would be a traditional woman, marrying twice, bearing children, and changing her name with each marriage (no Lucy Stoner, she). Reveling in maternal duties, she would never resent being publicly portrayed, no matter how noted a public figure she became, as first and foremost a devoted "mother." But with this photograph she reminded posterity that domesticity was not inconsistent with the life of the mind. Her choice was different from that of her friend Grace and of other exceptional women of her age. It would both define and limit her feminism.

As the child of "old immigrants," Belle Lindner enjoyed privileges denied the children of the next generation of Jewish immigrants. Her parents, who had their own business, earned their living under conditions entirely different from the garment industry sweatshops where most new immigrants worked. New immigrant families shared overcrowded, dark tenements; even their beds might belong to others at different times of the day. Belle had her own room on the third floor of her own house, away from others. She could entertain friends there, decorate its walls as she wished, write letters, and read. She was also privileged in her education. Few daughters of immigrant Jews finished high school, much less went to a private one. Most took low-paying jobs to support the family while their male siblings went to school. Even for those Jewish girls who could afford not to work, higher education was usually out of reach.[12] The successes of Belle's siblings—Walter became a corporation lawyer, Max a department store executive—did not come at her expense.

Still, Belle Lindner's years of personal growth lacked a goal. The century was drawing to a close, and she had reached adulthood. Heinrich Conried, her acting teacher, urged her to pursue a theatrical career. But declamations in a private home were one thing, the public stage another. Her parents drew the line, and Belle acquiesced. Perhaps she secretly agreed with them that the theater put a girl's virtue at risk. Perhaps she sensed her own temperamental incompatibility with the stage. Unlike dramatic reading, which sheltered the performer in a safe, predictable environment, stage acting demanded uninhibited, emotional interaction with others. In late adolescence, Belle had sought control over her emotions. Abandoning her hard-won equanimity for the volatile life in the theater was out of the question. Nonetheless, her creative energies, her intelligence, and expressive gifts needed an outlet. The challenge was to find one that was safe.

An institution fairly new in American urban life, the "social" or "neighborhood settlement," offered her a haven. Settlements were of English origin. In 1884, two Oxford University men took up residence in an East London slum in the hope of becoming a part of it and of helping the poor improve their lot. Two years later, Stanton Coit, a young American scholar influenced by Felix Adler and the Society for Ethical Culture, visited the English settlement and brought the idea home. In August 1886, Coit moved to the Lower East Side, where a group of reformers joined him in establishing the "Neighborhood Guild." They hoped the residence would become the nucleus of a network of guilds and the basis of a "civic renaissance" in New York City. Their ambition was never realized, but the idea of living in a ghetto to create a center for communality and activism remained. In 1889 a group of college women founded the College Settlement, and at about the same time Jane Addams founded Chicago's Hull House. In 1891 two of Coit's original associates established the University Settlement. John Lovejoy Elliot's Hudson Guild and Lillian Wald's Nurses or Henry Street Settlement followed shortly after. By the late 1890s there were seventy-four settlements in the United States.[13]

Some were distinctly Jewish in their programs, personnel, and their purpose—to serve the rising numbers of Jewish immigrants. In the 1870s, approximately two hundred thousand Jews came to America. During the 1880s, another three hundred thousand arrived. The largest flood came between 1900 and World War I, when another one and a half million joined their coreligionists. Most of these Jews came not from Central Europe, like the Lindners, but from Eastern Europe and Russia, where they had lived in isolated villages or "shtetls," and suffered from a devastating combination of economic and religious persecution. Most who arrived in America settled on Manhattan's Lower East Side, where they found family and friends, cheap housing, familiar religious and dietary customs, and jobs. But they suffered considerable hardship: they came without money, lived in dark, unhealthy, overcrowded tenements, and worked for subsistence wages in unsafe and unsanitary sweatshops owned mostly by German Jews. Impoverished upon arrival, most immigrants stayed that way. The vision of a brighter future for their children kept them going, and to make that vision a reality, many young American-born adults took up social work and came down to the "district" to help.

Some historians suggest that these social workers were not altruists. In the 1890s, exposés on slum conditions such as Jacob Riis's *How the Other Half Lives* (1890) and *The Children of the Poor* (1892) had begun to appear. Meant to inspire sympathy, the books achieved a rather different purpose, some historians argue: middle- and upper-class citizens began to fear that ghetto crime and disease might spread into their own neighborhoods. Social workers may thus have been motivated by self-protection. Other historians see the work of German-Jewish social workers, in particular, as motivated by the fear that a large, poor, and unassimilated Eastern European Jewish population would incite American anti-Semitism and thus threaten the social and financial status of German Jews.[14] In any case, it is impossible to know the precise impetus behind the settlement movement. Some social workers may have been inspired by fear, conscious or unconscious; others may have been interested more in self-fulfillment than in the welfare of the poor. Still others, sincere altruists, patronized their beneficiaries, or tried to make them over in their own images. The situation on New York's Lower East Side was mixed. There were tensions between benefactors and beneficiaries, but a great deal of good was accomplished.

By 1900, Belle Lindner knew she had to find something constructive to do. Helping her coreligionists was not only a worthy cause, but acceptable in her circle. As a young girl, she had observed and probably participated in the Temple Israel Sisterhood. Founded in 1891, the Sisterhood collected money, organized sewing for the poor, and cooperated with the United Hebrew Charities in helping the needy who lived north of 116th Street. They also organized a "Working Girls' Vacation Fund" and "Working Girls' Club" to help young women workers improve the conditions of their lives in the city. As an adult, Belle helped found similar institutions, testimony to the Sisterhood's influence on her view of the Jewish woman's duty. As she later wrote, "To think of others is as natural to the Jewish woman as to breathe. She gives and has been trained for generations to give with all her heart going with it."[15]

Yet for Belle Lindner, charitable giving was not enough. She and many of her contemporaries believed that only preventive work resulted in permanent social change. The settlements, where most of the advanced social work of the time was done, seemed to be the best place to do something "useful." Accordingly, in January 1900, reversing the migratory path of her parents, Belle took up social work downtown at the Educational Alliance, one of New York's Jewish settlements. Originally called the Hebrew Institute, the Alliance had been formed in 1889 by the united effort of several Jewish organizations. At the time Belle started work there, it occupied a spacious, five-story building on the corner of East Broadway and Jefferson Street.

The work of the Alliance required cooperation among three different constituencies—the settlement's trustees, its staff, and residents of the Lower East Side. Since each group had its own sense of the Alliance's purpose, they sometimes clashed over policies and programs.

The institution's trustees included some of New York's wealthiest German Jews: businessmen and financiers such as Isidor Straus, Marcus M. Marks, Jacob H. Schiff, and Felix M. Warburg; educators Edwin R. A. Seligman and Julia Richman; and lawyer Louis Marshall. Their chief aim was to help integrate new immigrants into American life by eradicating their shtetl ways. "If you will bear in mind," Straus said in his President's report for 1896,

> that thousands of these immigrants are unwilling refugees to our shores, that they come mostly from a benighted country where law was looked upon as inimical to the individual, instead of his protector; where oppression and persecution have for generations been the only attributes of government of which they had any knowledge, it must cease to surprise you that these people are what they are.

Straus sympathized with the immigrants but scorned their cultural baggage and hoped they would abandon it quickly. Then they would no longer need Jewish charity, or continue to frighten the trustees with their political radicalism. Hull House and Henry Street allowed political meetings, but the Alliance forbade them and restricted the use of its hall to patriotic assemblies. The Alliance was to be, as Straus hyperbolized, a "People's Palace," a "temple of pleasure, of education and of refinement," an oasis in the desert of poverty that surrounded the immigrants. But it was to be an American oasis, with few traces of what the immigrants had left behind.[16]

Because the Alliance staff were in daily touch with the immigrants, they applied trustee policies more flexibly. By 1895, they had developed four programs: a social program of game rooms and reading rooms, clubs for all ages, music and drama, and art exhibits; an educational program of three lectures a week, one in German, and a reading room with Russian, Hebrew, and Yiddish newspapers; a physical program of medical examinations, with a gymnasium, baths, summer camps, a roof garden, outdoor excursions, cheap lunches and free milk, and health lectures; and a moral program of lectures, "Penny Provident Thrift Fund" for children, legal aid and desertion bureaus, a "people's" synagogue, and religious education. There were also special programs for women in domestic science and literature, and instruction for both sexes in white-collar skills. Higher moral purposes informed the entire structure. The social program, a trustee committee reported in 1894, should be kept on a "high plane of excellence," serving as a "means of elevation as well as of mere recreation." The educational program should Americanize the downtowners, spreading "distinctively American ideas on government, policy and civil life." Since wholesome leisure was essential to health, even the physical program was not for fun alone but saved lives. Finally, to develop moral character and arouse good impulses, the Alliance's moral program should be its "crowning work."

These higher purposes appealed to Alliance staff. Many of them, like Belle Lindner, were former college students or young professionals who wanted to serve others, but through an approach based in what Henry Moskowitz's friend Paul Abelson called "rational altruism." Charity and temporary relief were not enough. Settlement work promised not only to provide immigrants with skills and attitudes needed for survival, but to replace the immigrants' lost community with another they would come to love. Accepting the basic tenets of this new community, tenets embodied in the American gospel of success, Alliance staff encouraged total assimilation into American life yet did not scorn the immigrants' past. To them, the newcomers did not threaten the future of American Jews. In daily contact with the immigrants in clubs, on the roof garden, in the assembly hall and the kitchen, Alliance staff kept "higher purposes" in mind but also saw the faces in the crowd.[17]

If assimilated Jews had done nothing to help the new arrivals, society would have condemned them. As it was, some members of the third Alliance constituency, the immigrants themselves, condemned Alliance benefactors and accused them of misguided altruism. In a satire called *The Benefactors of the East Side* (1903), Yiddish playwright Jacob Gordin had financier Jacob Schiff call Russian Jews "schnorrers and beggars." He also showed lawyer Louis Marshall suggesting that the elimination of flabby muscles through physical culture would stop crime on the Lower East Side. Other characters included ladies whose panaceas ranged from cookbooks to baths. Resentful of German condescension, Gordin indicted assimilated Jews for their palliative approach. Three years earlier, Gordin had established a "counter-settlement," the Educational League, the core of which was the radical discussion and activism forbidden at the Alliance. This organization attracted intellectuals and would-be revolutionaries, but lacking the financial backing the Alliance enjoyed it could not provide programs for those seeking upward mobility.[18]

Some immigrants defended the Alliance. Joseph Gollomb became a club leader there and later wrote a novel about his experiences. *Unquiet* (1935) describes how a young immigrant at first resents the Alliance for treating him like a "slum product to be 'uplifted' by someone from the 'silk stocking' upper reaches of Society." Eventually, the Alliance's social activities help him cope with adolescence, and he comes to depend on them. So it was with others. In later years, many Alliance members looked back on its clubs, plays, classes, and outings as high points in lives of drudgery and despair, and on its practical courses as crucial to their success. Thousands of immigrants passed through Alliance programs. In 1902 alone, thirty-five thousand attended lectures, one hundred thousand went both to religious services and indoor entertainments, and three hundred thousand used the reading room and the roof garden. Still, the benefactors' patronizing attitudes remained a sore point. In 1902, after some bitter exchanges surfaced in the Jewish press, the Alliance set up a formal Advisory Committee of "influential men of the District" which it promised to consult on all "matters relating to innovation." The Alliance staff simultaneously undertook to show greater respect for the dignity of their charges.[19]

Belle Lindner, who joined the Alliance staff at the peak of the controversies, set herself the goal of fulfilling this commitment. She held a variety of posts at the Alliance, for several summers supervising the roof garden, and once, when the "Directress of Women's Work" proved to have "insufficient force," temporarily taking over her functions. But her main interest was the settlement's program of exhibits and entertainments. This activity, which linked her earlier dramatic career with her quest for "useful" work, provided her with great opportunities for service.

During Belle's three and a half years at the Alliance, entertainments and exhibits increased in importance. In 1896, the trustees' committee on social work had noted with dismay the unelevated nature of Lower East Side amusements. "Furnishing diversion is their sole aim," the report complained. "What is really needed is an organized effort to place among the masses the facilities which are so plentifully afforded those who have the advantage of residing in better neighborhoods." The trustees hired an organizer to coordinate the entertainment efforts of various downtown groups. Before Belle joined the staff, a music director performed this service. Later, a special committee chaired by trustee Sidney Blumenthal assigned Belle the task. By the fall of 1902, monthly exhibits of varying kinds were held. There was also a full schedule of orchestral, choral, and dramatic events that alternated weekly throughout the season, Sunday afternoon children's shows, and plays and tableaux staged by children to celebrate national and Jewish holidays. In addition to organizing all this, Belle also taught elocution to youngsters and coached dramatic groups, one year writing a dramatic version of Mark Twain's *Pudd'nhead Wilson* which she staged with a boys' club.[20]

When Belle started to work on exhibits and entertainments, the Alliance was already noted for its annual spring and autumn flower and nature shows. To these offerings Belle added others more pertinent to the Lower East Side. For example, in 1902, in addition to creating opportunities for East Side artists to show their work, she mounted exhibits for uptown artists inspired by downtown subjects. She treated as art the Jewish ceremonial objects immigrants had brought with them from Europe and assembled a show featuring utensils used in synagogues and homes, such as brass samovars, candelabras, and menorahs, special cups for Sabbath hand-washing, and vessels for inhaling spices while the labor of the coming week was blessed. Exhibits of Alliance student work accompanied every fund-raiser. Students won prizes not only for their exhibits but for designing the best poster for advertising the exhibit.[21] In all of this work, Belle displayed sensitivity to the needs of Lower East Siders as well as respect for their culture, old and new.

Her efforts to raise the standards of Alliance entertainments reflected her refusal to patronize. Before she had joined the Alliance, the trustees had provided only amateur performances for the district. These had not been well received. In the fall of 1902, illustrating a change in trustee policy, she announced that Franklin H. Sargent, president of the American Academy of Dramatic Arts, was helping stage "first class plays" at a "popular price" at the Alliance on alternate Saturday evenings. These would include two modern plays, one from the eighteenth century, Shakespeare's *Twelfth Night,* a German play acted by Heinrich Conried's Irving Place Theatre, and one of Jacob Gordin's Yiddish plays. In addition to enlisting the services of her old teacher Conried, Belle also lined up reader Ida Benfey, billed as "one of the best story tellers in America." For music, the Halevy Singing Society and members of the American Symphony Orchestra, directed by Sam Franko, performed on alternate Sundays. Downtown audiences were enthusiastic.

A downtown writer calling himself "One of the Submerged" discussed the experience in the *American Hebrew,* the organ of the German-Jewish community. Alliance concerts used to be "simply awful," he wrote. Audiences hissed, or left the hall in disgust. But everything had suddenly changed. "Some guardian angel stepped in and declared that the 'Ghetto' actually liked good music. 'Let's try to give them good shows, and see what will happen,' this wise person suggested." After praising Sam Franko for a program of string quintets, the writer continued, "No better demonstration of the kind of 'work' that we need down here, has ever been given, than the striking changes of expression which appeared on the faces of the audience. . . . To see the expression in those strong Jewish faces change from the gay and light to the serious and solemn, as the first strains of Kol Nidre were heard, was worth going many miles." The writer concluded by musing on the significance of this event. It proved that "we of the 'other half' have feelings and sympathies which can be touched," and that "if we get the best, we give up the best that is in us." He hoped that the same high standards would apply to all Alliance departments in the future.[22]

In February 1903, the same writer identified Belle Lindner as the "guardian angel." A performance of Mendelssohn's oratorio *Elijah,* the Alliance's most ambitious concert to date, had brought an overwhelming response. There were "women of a past generation bedecked with 'shaitel' and bonnet and bejeweled according to notions that have long since been relegated to the era of wonders." Their husbands, full-bearded and showing the signs of their struggle for existence, peered anxiously at the English program, which they could not understand. Their children—now Americans—acted as interpreters. Still more in evidence, the writer continued, was the "younger immigrant generation, the young workers, for whom the Alliance has been nothing less than a Godsend." He concluded:

> we of the "slum" owe much to the broad and liberal spirit that has been manifested within the past year or two in the management of these Sunday night concerts. From a series of deplorable farces, these affairs have been raised by a sane and intelligent understanding of the people for whom they were intended to the high position they now occupy. . . . Individually we are indebted to Miss Belle Lindner, the superintendent of entertainments, and the committee, of which Mr. Blumenthal is chairman, for the wonderful progress that has been made in these affairs. Since the advent of Miss Lindner upon the scene, the improvement has been marked, and it is largely to her intelligent appreciation of her audiences that these happy results have been attained.[23]

By raising the caliber of entertainments, Belle Lindner taught the benefactors of the East Side an important lesson. They need not patronize the people of the shtetl. This success encouraged her to reach further. In March 1903, she developed a plan to involve the people themselves in the selection of programs. According to the plan, those who subscribed to the next season's concert and drama series (at a cost of ten to fifty cents) would be entitled to vote for a committee of "directors"; this committee would choose the actual programs in consultation with the Alliance Entertainments committee, which would furnish the hall, light, and heat, and pay the artists. As a final note, she left herself an escape clause: if demand for such community-planned events fell off, the plan would be dropped.[24]

Belle described this project as putting Alliance entertainments on a "cooperative and self-directing" basis. These watchwords would recur frequently in her later projects for social change. The words exemplified not only her refusal to patronize but her conviction that the dole, monetary or cultural, degraded the recipient. Her Alliance entertainment project, while offering funds and coordination to those in need of help, left room for initiative and self-reliance. Her many later projects for social reform repeated this strategy, which combined idealism and practicality with sensitivity toward people in need. It would win her many allies, including Alfred E. Smith.

Belle Lindner's idealism extended also to her work in children's drama. She structured this work around moral and educational goals. Trustee Julia Richman had originated children's drama as part of the Alliance's Sunday afternoon entertainment hour. The hour at first consisted of "illustrated songs," amateur comedians, stereopticon views, and a children's orchestra playing toy instruments. Occasionally a professional company appeared in a play. Eventually some of the children mounted plays themselves, such as the *Pudd'nhead Wilson* Belle directed.

Belle kept all her children's drama work on an amateur level. For her, drama served the social work's greatest goal, "character-building." After she left the Alliance, some East Side children moved on to the stage or into films and found an escape from poverty. From a modern perspective, such an occurrence seems the quintessential "success story." But to Belle Lindner, two factors made it indefensible. First, the "Jewish child of the East Side is half play-actor anyhow. The strenuous cultivation of this instinct and its over stimulation leads nowhere but to eccentric ends." Here is one example where, instead of treating her charges as individuals, she stereotyped them in ways similar to the Alliance trustees. The second factor was moral, and showed the abiding strength of her Victorian values. Belle was not at her most tolerant or open when dealing with the career she herself had been denied. To her the theater's late hours, rouge pots and tights (the prostitute's paraphernalia), backstage intimacies, and the excitement of public appearance and approval led children down the wrong path. Evidently some parents agreed with her, for they expressed not only moral qualms about seeing their children, especially daughters, enter theatrical life but preferred that their boys stay up late studying books, not lines.[25]

Belle approached other social work activities with fewer prejudices. Her first summer at the Alliance she worked on the roof garden with Paul Abelson, a product of the University Settlement where, along with Henry Moskowitz, he had been introduced to Ethical Culture. In 1898, Paul, Henry, and other members of their University Settlement club established the Downtown Ethical Society, later called Madison House, the first settlement founded exclusively by Lower East Siders. Abelson graduated from City College in 1899, spent a year at Teachers College, then joined the Alliance staff. Having studied the educational theories of G. Stanley Hall, the Clark University psychologist, and Joseph Lee, a playground and recreation expert, Abelson planned to establish an Alliance recreation program where the "jungle law" of the streets would have no place. Instead, games of "recreational and educational value" and principles of "harmony, cooperation, and respect" for the weak would reign.[26]

Working with Abelson, Belle Lindner learned about recreation theory and group work. One of her assignments was to organize a club of Alliance girls. This group, called the "Central Alliance Guild," helped new girls who came to the roof feel at home. At the end of the season, the Alliance trustees praised her club for "raising the tone of the Roof Garden as a social center."[27]

Her success inspired her to write her first published article, "Social Work Among Young Women." Its topic was the role of clubs in girls' socialization. Belle wrote that the club leader has two goals: to help the club cooperate in a mutual project and to instill in club members a sense of self-reliance. The first is achieved not by the leader's imposing her own views on the group, turning the club into a "school room," but by showing a real interest in each member's point of view, "recognizing in each that part which makes for mutual forbearance, tolerance, and self-control." With "ready wit and tact," the leader then harmonizes individuals into a unit. The second goal is never achieved by patronizing club members, a common mistake of the overzealous social worker. Such a worker puts the new girl "on the rack," and with a dozen girls listening asks her personal questions about her friendships, family, disposition of earnings and the family history," all the while offering labored expressions of sympathy and much bad advice. At the other extreme is the worker who tells a girl she needs no help from the outside. This approach, while flattering a girl's sense of independence, reeks of "patronage." It ignores the years during which the girl tried to solve her problems and yet failed. The "real helper," Belle advised, is the leader who sees the danger of too much help and yet avoids the error of offering none at all. Social workers should encourage self-reliance but realize that it "is not a power that comes of itself." It needs to be nurtured.

Belle used these premises of social work to develop a vision of social progress in which the settlement played a central role. No one settlement can hope to save an entire neighborhood, she wrote. It can only spread its influence on the "widening of circles" reached by its club and group leaders. These leaders must not only live in the community they wish to help but be "trained" to teach others "the way of self-help." She gave several examples: a club of Russian girls learning English under the guidance of one of their countrywomen, who "knows no other aim for them than that they shall in turn . . . do such work for other girls shut out from such help now." The students of the late Professor Thomas Davidson, many of whom (including the philosopher Morris Raphael Cohen) would go on to distinguished careers, were "more advanced in aim." Then there was the Federation of East Side Boys' Clubs (founded by settlement worker Henry Moskowitz), reaching out "into the interests of the community," taking up questions of "outdoor recreation places and better economic and sanitary conditions for their neighborhood." Thus, she predicted, "an endless process of socialization is . . . being fostered and encouraged through the primary socialization of one group." In addition to group work, one woman might take a college domestic science course and return to the neighborhood to impart her skills to others. Since the home life they create forms future generations, Belle surmised that most of the results of social work depend on such women. In fact, she concluded, the more women are influenced by "right ideals, social, moral, artistic, intellectual, the higher becomes their standard of living."

In her summation, she included men:

> When several hundred young men and women whose energies have been cultivated for themselves and then for the community . . . , whose vitality is overflowing in enthusiasm to do for others what has been done for them, whose individualism has been moulded into the broadest of socialism in its real sense, . . . the social idea and those who labor in its service will have reason to congratulate themselves on the progress made toward the final uplifting of the standard of living. . . . [28]

Thus, in her mind, cultivation of the values of "association and cooperation" would lead eventually to "the final triumph of the social idea," or the improvement of the standard of living for all.

Belle was twenty-three when this optimistic vision saw print. Later in the decade she would come to recognize the limitations of the "rational altruism" of individuals and would call upon the commonwealth as a whole to undertake political action. Even so, she would hold on to her youthful, settlement-inspired ideals of "widening circles of influence," "association," and "cooperation." Much of her later strength as a social reformer lay in her ability to invoke ideals that inspired others and at the same time to devise practical plans for their realization.

Belle Lindner resigned her Alliance job a few weeks before marrying Charles Henry Israels, an architect twelve years her senior and a volunteer leader of the settlement's Robert Fulton boys' club. Though she never considered staying in social work after marriage, she had reason to take pride in the work she had done. Her salary, which started at $500, had doubled, and at the beginning of 1903 was supplemented by an apartment she shared with other women administrators on the top floor of a new Alliance annex. This compensation indicated that the Alliance trustees had approved her work. When she married, they collected contributions for a gift in recognition of her "faithful services and intelligent work." After her death, trustee Benjamin Tuska still remembered what she had done there. "Despite the lapse of nearly thirty years," he wrote Henry Moskowitz, "her association with this Institution is still a matter of note. It is not too much to say that some of the activities which today constitute so important a part of the work of The Educational Alliance owe their inception to her pioneer efforts."[29]

In the early 1900s, few Jewish women took up resident settlement work. Belle explained why in an article she published in 1907, "Jewish Women as Settlement Workers." Jewish women, she wrote, who seldom think of fulfillment outside of married life, fear that if they become settlement workers they surrender all opportunity to have their own homes. Yet, according to her, the Jewish woman was particularly suited to settlement work: "She has the emotional understanding and sympathy belonging to her race, the dramatic intensity so essential for getting into the lives of others, and where Settlement work is being carried on among Jewish people the Jewish woman can appreciate their lives and their necessities with an insight born of common fatherhood."[30] Perhaps here she was describing herself and the reasons behind her success. Whatever the case, in 1900 she was one of so few that the Alliance trustees were especially appreciative of her contribution.

On leaving the Alliance, she could not know the extent to which the experience would follow her into maturity. It brought her in contact with many individuals who would later play significant roles in her life. She met both her husbands there. Other lifelong friends from the Alliance included the City College philosopher Morris Raphael Cohen, lawyer Arthur B. Spingarn of the National Association for the Advancement of Colored People, and physician Sigismund S. Goldwater, later New York City Health Commissioner. According to Judith Bernays and Monroe Goldwater, these last two were "smitten" by Belle, but the romances had not led to proposals. Lawyers Abram Elkus and Joseph Proskauer, who brought Belle into Al Smith's circle, led Alliance clubs. She met Rabbi Stephen S. Wise, Council of Jewish Women leader Rebekah Kohut, and settlement leader Lillian Wald through the Alliance. Paul Abelson became a close friend. Other Alliance people became Smith supporters, including New York politician Stanley Isaacs, social worker Jacob Billikopf, and Judge Jonah Goldstein.[31]

For Belle, her years at the Alliance were thus a culmination and a beginning. She had found "useful work" and performed it well. She had developed expertise that would serve in later endeavors. And she had forged lifetime bonds with a group of talented, ambitious, and concerned individuals who would enjoy lives of accomplishment and renown. These friends, all Jews, became part of her network of personal and professional contacts. While her network would later take on more ethnic and religious variety, this Jewish group remained at the center, loyal to her in success and failure.

CHAPTER 2

A New Woman

It is the Woman's Age.

National Council of Jewish Women, 1900

In March 1903, the Educational Alliance honored its volunteers and staff with a dinner-dance at the Manhattan Hotel. Eighty-five club leaders and teachers and ten staff members attended. Belle Lindner and Charles Henry Israels may have met or become better acquainted on this occasion. Both had resisted marriage, Charles even longer than Belle. She was twenty-five, he thirty-seven, a confirmed bachelor who had decided that "his" girl did not exist. Meeting Belle changed his mind. According to Grace Goodale, when Charles first saw Belle he was struck by her appearance and said to himself, "If she's as good as she's beautiful, I'm going to marry her."

They were married November 11, 1903, in the Lindner home on Third Avenue. The night before, Grace sat up with Belle in the "Poster Room," Belle's top-floor aerie, and in the morning helped her dress and get breakfast. Sensing that Belle's "strong soul" was "stirred to the depths," she tried to be "flippant" to ease the stress. The two women joked as Belle continued her preparations. The ceremony over, Charles and Belle left for a hotel. The next day, after breakfasting with Grace, they embarked on a honeymoon to the West Indies.[1]

Charles's background differed from Belle's. His Sephardic ancestors had been expelled from Spain in the late fifteenth century and then settled in Holland. Sephardim, the first Jews to migrate to America, considered themselves to have higher social status than the German Jews or Ashkenazim who arrived in the mid-nineteenth century. Even after upwardly-mobile German Jews had conquered American finance and culture, Sephardim were still called the "aristocrats" of American Jewry.

To Grace Goodale, Charles Israels even looked aristocratic, an interesting mixture of "Spanish gentleman and Dutch businessman." His grandfather had owned a small bank in Groningen. Neither of his sons went into the firm. Josef, born in 1824, studied art in Paris and then set up a studio in Amsterdam. Later, to recover from an illness, he moved to a little fishing village where he lived with a ship's carpenter and began to paint village scenes. His tender, eloquent depictions of plain lives brought him fame. He received medals from the Paris Exposition in 1867, 1878, and 1889, a gold medal from the Munich International Exposition in 1883, and the Order of Leopold from the King of Belgium. By the time he died at the age of eighty-seven, he was one of Holland's most popular naturalist painters.[2]

Josef's brother Lehman, Charles's father, emigrated to America in 1851 at the age of eighteen, got a job as a proofreader, and eventually became Washington correspondent for the *New York Evening Express.* He served in New York's Fifty-Fifth Regiment during the Civil War, and wrote dispatches from the front. Wounded at Fair Oaks, he was commissioned a second lieutenant in 1862. After the war, he returned to journalism, working for several New York newspapers—the *World, Herald,* and *Tribune.* He wanted to become an editor, but in 1875 the *Tribune*'s city editor advised publisher Whitelaw Reid that since Lehman was "vain" and impulsive he opposed hiring him in a managerial role. Reid took Lehman on as an investigative reporter, but later suspended him under suspicion of falsifying information. Lehman resigned in a huff, but begged to be restored when the charge was dropped. Reid never complied. Lehman earned his living thereafter with a trade paper, the *Real Estate Record,* which he cofounded, and wrote dispatches for Scottish and English newspapers.

Lehman married Florence Zilla Lazarus. Her family had gone from England to Kingston, Jamaica; from there, her father Solomon had moved to New York to set up an import business. One day, while walking in New York City, Lehman's war wound acted up and he fainted. Florence, who lived nearby, ran out to assist him. A romance blossomed, the two married, and on December 23, 1865, Charles was born. A sister, Josephine, came later.

Florence was attractive, but fussy and a complainer, unhappy in her marriage with Lehman, possibly embittered by his inability to advance in his career. When he died in 1896, she became totally dependent on her son Charles, then almost thirty but still living at home. His mother irritated Charles. "She has no mind," he used to say. Yet he accepted responsibility for her, and so did Belle. During the first years of their marriage, the young couple shared a New York apartment with Florence, and took her with them when they moved to a house in Yonkers. This arrangement seems to have been awful. Florence may have helped with childcare and household chores, allowing her daughter-in-law freedom to pursue interests outside the home. But Grace Goodale called Florence a "dreadful" woman, an epithet strong enough to suggest storms at home.[3]

Charles Israels was educated at a private boarding school, the Irving Institute, in Tarrytown, New York. Like his uncle Josef, he showed an early gift for art. After high school, he thought of studying architecture at Columbia, but instead enrolled in the painting course at the Art Students League. Uncle Josef had run off to Paris to study art instead of going into the family firm. Charles also went to Paris, but on his return faced reality and took a job as an architect's draughtsman. He vacillated between painting and architecture the rest of his life.

His first job was with the firm that drew the plans for Manhattan's Surrogate Court, or Hall of Records. He designed its façade in an imposing, Parisian "classic-eclectic" style of carvings and statuary. Later, when he became a partner of Julius F. Harder in his own firm, his taste became less elaborate. He specialized in apartment hotels and wrote about them frequently in the technical press.[4]

Charles held high professional ideals. In his articles he urged his colleagues to maintain their artistic standards, even under pressure from speculators to make cheap designs. He recognized that no architect had complete artistic freedom. They must pay scrupulous attention to legal codes, the needs of tenants and neighborhoods, the permanency of surrounding structures, economy in the distribution of interior spaces and utilities, and the appropriateness of decoration. In this last regard, Charles hoped his colleagues would develop a more native style. In an article called "Socialism and the Architect," he theorized that American architects copied European styles for lack not of ideas but of time. The problem, as he saw it, was one of outmoded organization. American architects believed they were still living in an "age of individualism" when architects were both "master artists" and "master builders." But in the complex modern business world, such combinations had become inefficient. Charles envisioned the business world moving inexorably toward a "cooperative commonwealth" (he also called it "socialism"), in which work would be increasingly subdivided. He concluded that architects should subdivide their work too, turning business matters over to professional managers and thus freeing themselves from the mundane details that distracted them from art.[5]

In addition to his technical writings, Charles wrote for the popular press, publishing in *Good Housekeeping* a design for an inexpensive summer bungalow for a city family of middling income. For the national journal of social work, *Charities,* he wrote on "civic improvements," focusing on park and playground development, the extension of city fire codes to the suburbs, the Manhattan "skyscraper menace," control of billboards, and waterfront reclamation—all causes shared by reformers trying to preserve the quality of life in the burgeoning metropolis. He pursued these causes through active membership in "good government" groups, such as the non-partisan Citizens' Union, and in political clubs affiliated with the Republican party. He expressed his reform interests further by designing model tenements for housing competitions sponsored by the Charity Organization Society, a coordinating body for social and community services, and by leading a boys' club at the Educational Alliance. He never received, nor asked for, recognition for these deeds, but professional groups awarded him honors later in his life. The American Institute of Architects made him an associate member in 1908. He also served as secretary of the Municipal Art Society, treasurer of the New York chapter of the Architectural League, as a member of the Municipal Art Commission of Yonkers, and of the New York City Building Code Revision Commission.[6]

Thus Charles was not only accomplished but civic-minded. He was also good-looking, with warm and gentle eyes. Grace Goodale reserved special praise for his character. She found him generous, a "*giving* person," little disposed to a row. In this sense, he was quite different from his parents. On the other hand, Grace observed, if someone else's rights were threatened, "he could put up a good fight, with no flinching." Even more in his favor, he never shirked what life put before him, no matter how hard or disagreeable, including the care of his mother. In his disfavor, Grace continued, his mind was not as powerful as Belle's. He could not see "as far around a problem" as she did, and Belle "had to help him sometimes." To Grace, the limits of Charles's vision made some aspects of his social activism quixotic. But other strengths compensated. Raised in a family with "little of real home happiness," he had managed to refrain from harsh judgments and respond to beauty and nobility in others. Fighting "the forces of evil in the city, shoulder to shoulder with righteous men," he was always glad "to make the best of everybody."[7]

Charles Israels and Belle Lindner came from different ethnic and class backgrounds—Belle from shopkeeping Ashkenazim; Charles from merchant, banker, and professional Sephardim. But their shared social and civic commitments and intellectual interests bridged these gaps, which neither considered important in the first place.

They were in fundamental agreement on religion. To them, Judaism was more a shared history and perspective than a system of rituals. For a time they attended New York's Sephardic synagogue, Shearith Israel, but it followed the orthodox tradition of seating women in a balcony away from the men. Belle and Charles did not like that. Both admired the Ethical Culture Society and attended its meetings. Felix Adler, a rabbi's son who had rejected Judaism for philosophy, had founded the Society in 1876. Faithful Jews considered Adler an apostate and resented his attraction to the educated and assimilated of their community. In a letter Charles Israels wrote in 1908 to the *American Hebrew,* the organ of the German-Jewish community, he defended Ethical Culture's "strong sense of civic duty," daring the editors to find one Ethical Culturist on the rolls of organizations "responsible for much of our city's degradation," rolls that "swarm with men who are pillars of the church and the synagogue." He called himself one of the "churchless" who look upon "strong attachment to a religious organization without other visible proof of ethical standards as synonymous with hypocrisy." Belle agreed. She participated in the Women's Conference at Ethical Culture and sat on its executive committee. She also served as president of a chapter of the Society for the Study of Child Nature, which had been founded within the Ethical Culture context.[8] On the other hand, when she and Charles later moved to suburban Yonkers, they found a synagogue they liked and sent the children to its Sunday school, in which Belle taught. Thus, despite the secularism of the Israels family, in the end they raised their children as Jews.

In addition to agreeing on religion, Belle and Charles agreed on domestic arrangements: he would be the breadwinner, she would keep house. Few women of her social class worked after they got married. Settlement work was especially incompatible with marriage, as it usually required residence in the neighborhood house. But Charles's income, adequate at the time they were married, declined the following year when his business slumped. As Belle prepared to deliver their first child, she discovered she could no longer afford the kind of domestic help she wanted. She proposed supplementing the family income by writing professionally at home. Charles agreed. She soon earned enough to hire the baby nurse she wanted, a woman she kept with her until her last child, born in 1909, was two and a half years old.[9]

Charles's acquiescence to Belle's desire to earn money made her feel lucky, a "new woman" living in a "woman's age" of growing opportunities for useful work and service. Other women complained about their spouses. "The New Woman!" exclaimed a correspondent to the *American Jewess,* a monthly newspaper associated with the Council of Jewish Women, a moderately feminist and middle-class organization that Belle joined. The "cry" was being heard on every side, the writer went on, but "where, oh where, lingers the New Man? . . . will the feminine phoenix still have to content herself with the Old Specimen?"[10] Belle did not have to. Charles was a "new man." As she moved from earning a small income by writing, to ever-expanding projects in social welfare and reform, he shared her interests and supported her efforts. When she traveled—to meetings, conferences, or to speak to legislators—he supervised the home and sent her loving, newsy letters.

Belle Israels began earning money by writing publicity for the United Hebrew Charities and doing editorial work for *Charities,* the social work journal. She was not paid highly, but she earned her expenses. Her work also kept her in touch with the world of social welfare. By 1904, her chief organizational loyalty was to the New York Section of the Council of Jewish Women. Her joining this group may have reflected a strengthening of Jewish identity, an effect perhaps of contemporary events. The previous year, the number of "pogroms" or massacres of Jews in Eastern Europe and Russia had risen, stepping up the pace of Jewish immigration. As were other Jewish women's groups, the New York Council was committed to helping these immigrants.

The national Council had been formed in 1893 by women meeting under the auspices of the Parliament of Religions at the Chicago World's Fair. Its chief organizer was Hannah Greenebaum Solomon, a widely respected Chicago clubwoman. As a member of the Women's Committee of the Parliament, she had asked the Men's Committee to schedule a Jewish Women's Congress. She told them that American Jewish women needed to meet in order to reaffirm their religious identity, which was threatened by assimilation, and to coordinate their philanthropic work.

The Men's Committee suggested that the women serve the Parliament only as hostesses. At this, Solomon resigned from the planning group and, with the help of a small committee of Jewish clubwomen, organized the congress herself. The women's meeting was a great success. Out of it grew a design for a permanent organization, of which Solomon became president.

The Council of Jewish Women was thus founded, like other women's organizations of the era, out of the energy released when men refuse women independent channels of action. Throughout the nineteenth century, the numbers of such groups had multiplied, many of them inspired by the "Declaration of Sentiments" pronounced in 1848 at Seneca Falls, New York. This document listed women's reasons for protesting their lack of rights, and gave rise to the first women's suffrage associations. After the Civil War, when the federal government refused to grant women the same rights as blacks, more suffrage groups and other women's organizations formed, some seeking social and moral changes in American society, such as temperance, prison reform, educational progress, and industrial justice. Hundreds of literary or "culture" clubs were also established, most for purposes of self-improvement but by the 1890s starting to federate and promote social reforms. Clubs gave women experience as organizers, speakers, and reformers, increasing their confidence and acclimating them to public roles. The Council of Jewish Women served these functions for Jewish women, who themselves felt excluded from the mainstream of American women's organizations.[11]

At its founding, the Council announced a double mission—the study of Judaism and the rendering of social service—and made no distinction between the two. As Sadie American, the Council's corresponding and later executive secretary, observed, "Judaism has always taught that religion lies not merely in the field of speculation and sentiment, but in that of active expression in deeds." This teaching was especially apt for Jewish women. Women were not essential to Jewish ritual. Only men made up the minyan of ten worshipers necessary for a religious ceremony. Jewish women therefore tended to focus on "good works" to express their faith. During the first ten years of the Council's existence, however, the two sides of its mission came into conflict. Social service won out, with consequences for Belle Israels's development as a reformer.

Conflict developed first because the early religious study programs were dull. Minnie D. Louis, who chaired the committee on religion, decreed that the first year's topic would be "the history of the Jews as a nation" from the Babylonian exile to the Christian era. There was considerable cultural arrogance in her expressed hope that "every woman in Israel may be fitted to comprehend and appreciate her inheritance of Judaism in all its past and present grandeur and superiority, and be convinced of its unqualified adaptation to the highest development of man, materially and spiritually in every age." Study circles of ten women would appoint a leader and meet for two hours or more fortnightly, discussing readings and hearing lectures, and enforcing "the most stringent avoidance of hospitable entertainment."

In contrast, Hannah Solomon and Sadie American, who wrote the social service program, proposed a foray into modern philanthropy. Council women, they wrote, should not dole out charity but serve the needy. This meant using the new "charity organization" approach, coordinating projects, studying social issues, and bringing in professional assistance when necessary. Thus Council women would join the vanguard of "scientific" welfare workers. At the Council's first triennial convention in 1896, Carrie Benjamin, chair of the philanthropy committee, included among her welfare goals the following: factory inspection, tenement and prison reform, public works for the unemployed, efficiency in public institutions, and the establishment of industrial schools, night schools, kindergartens, free baths, district nursing, employment bureaus, libraries, and lecture programs for immigrants. Benjamin dared suggest that Council women take up legislative reform, an area traditionally reserved exclusively for men. To the Council's growing numbers of "new women," Minnie Louis's religious study circles paled beside the prospect of Carrie Benjamin's crusades. As early as 1896, some sections reported that interest in study had declined.

A controversy over Sabbath observance also caused a decline in religious study. For many assimilated Jews, Saturday was no longer a convenient day for worship. At the 1896 triennial, after attacking Hannah Solomon for failing to consecrate the Sabbath, the delegate from Montreal nominated Minnie Louis as the next president. Solomon's reply—"I do consecrate the Sabbath. I consecrate every day in the week"—became legendary, and won her reelection to the presidency by acclamation. But the lines had been drawn between Reform and Orthodox worshipers. At the second triennial in 1900, despite pleas from Solomon and American, a resolution passed that no Council section should participate in "any public action involving a violation of the Sabbath."[12] Under such a threat of religious schism, Council members thus drifted away from religious study.

After the pogroms of 1903, Council philanthropy became more necessary than ever, especially in New York, where most immigrants landed. Thanks to the influence of Sadie American, the New York section of the Council was ready for the challenge. When Belle Israels became a member, religious study had almost vanished from its program of work.

Minnie Louis had founded the New York section in 1894. In two years it had grown from 70 to 590 members. Because Louis and Rebekah Kohut, the section's first president, both believed in the primacy of religious study, the section concentrated on the Bible, Sabbath observance, and religious instruction to children. Its philanthropy committee organized lectures by settlement workers, a boys' club, a cooking school, two home libraries, a convalescent home, and a directory of Jewish charities. But when Kohut defined the section's mission as unfulfilled "until every Jewess shall have learned to read her Bible and to know the history of her people," her priorities were clear.

The section began to shift its focus when it decided to help "wayward" or delinquent Jewish girls. Rachel Sulzberger, the section's delegate to the triennial of 1900, reported that over one-third of the girls in the State Reformatory for Women at Bedford were Jewish. The section had determined to take action, but when they approached the male leaders of the Jewish community for help, the response was negative. The men either denied the seriousness of the problem, or refused on moral grounds to commit funds to help criminals, especially prostitutes.

New York Council members decided to do what they could on their own and began organizing Sunday schools at the Bedford Reformatory and the House of Refuge on Randall's Island. They hoped, naively, that training in Jewish ritual would help stiffen the girls' resolve to reform. After running the schools for a while, they had to admit that religious instruction neither rescued the delinquent nor kept others out of trouble. They started more systematic rescue programs, placing released girls with "good" families as domestics or wet nurses, or finding them jobs as sales clerks. If the girl was pregnant or had a young infant, the section helped her keep the baby and arranged for extra help from United Hebrew Charities. The section also began preventive work. It opened up "Recreation Rooms" in immigrant neighborhoods where young working women could gather in their free time for "wholesome" activities under the supervision of mature, sympathetic matrons. Rescue work could be frustrating, even disheartening. Many girls simply disappeared or ended up recommitted. Their babies, improperly cared for, often died. Further, the work caused strife among the section's board members, some of whom, perhaps following the lead of their husbands, disapproved of helping "immoral" girls and demanded that such work cease. When they failed to get their way, they resigned, leaving behind disunity and resentment.[13]

While the New York section coped with these problems, Sadie American was moving from Chicago to New York. She arrived at a moment when, because of her background in reform and position of power in the National Council of Jewish Women, she could help the section reach stability. Her success in this role made her an important model for Belle Israels.

Sadie was an unusual woman. Her father was responsible for the family name, changing it from "Abraham" to that of his adopted country. He provided his daughter with a good religious and secular education through high school but refused her permission to go on. She therefore began philanthropic work, at first organizing a group of young women who visited hospitals, and then turning to volunteer settlement and relief work. After being invited to join the prestigious Chicago Woman's Club, she took up educational reform, starting the city's first Vacation Schools for the children of working mothers. She also persuaded the municipal government to fund its first playground, and was active in designing playgrounds for the blind. Hence by the time she moved to New York in 1900 her work as a reformer and her role as secretary of the National Council of Jewish Women had shaped her into an experienced and accomplished leader.

Not everyone liked her. Some Council members thought her aggressive and abrasive. Her eyes "glaring," she would hold forth for hours telling vivid anecdotes and making scriptural analogies to inspire Council members to action. Other members feared that her emphasis on social welfare distracted the Council from its original purposes and cost too much money. American, who described herself as a person with "rhinoceros hide," usually managed to defend herself against her critics. According to a self-righteous autobiography she wrote many years later, only "the inefficient and the inert" ever dared oppose her. In short, Sadie American was a charismatic, powerful, and controversial leader. Small wonder that when she moved to New York, where the local section was "on the brink of extinction," some of its members begged her to take hold. Other members objected, arguing lamely that a National Council leader should not also run a local section. But this argument dwindled to a whisper and died away.

As president of the New York section, American persuaded the board to dedicate itself to social service. Not only should it continue rescue work among wayward girls, she said, it should make that work more "efficient" through systematic preventive programs. Saving girls already in trouble was important but preventing them from getting into trouble seemed equally urgent. By the time Belle Israels joined the New York section in 1903, Sadie American had already won this point and had been in control more than a year. Board members who opposed her had resigned. Israels approved of American's policies and was ready to serve in her crusades. This conjuncture of leader and acolyte set the stage for Belle Israels's emergence as a social reformer.[14]

American, recognizing Israels as a worthy lieutenant, got her elected to the section's board, and appointed her to chair the philanthropy committee. In this role, she supervised the work among sick and destitute children hospitalized on Randall's Island, the religious instruction and visiting of Jewish girls in reformatories, follow-up aid to maternity cases, and police court probation work, including visits to institutions and private homes. The work with children touched her, but she was moved even more by the plight of wayward girls, especially the pregnant ones.

Like other welfare workers of her day, Belle Israels granted that female delinquency had environmental causes. She had heard the stories of innocent girls drugged, kidnapped, or held against their will in brothels, debauched by their own families or bogus "fiancés." Jewish girls had told Council agents such stories themselves. Some of the girls had been trying to escape intolerable conditions at home, either in the sthtetls of Eastern Europe or in American ghettoes, only to find themselves trapped in circumstances from which they could not, or believed they could not, escape. Others had turned consciously to prostitution, perhaps only as a temporary solution. When they tried to leave it, financial or other dependencies made departure impossible. Their stories eventually gave rise to a movement a few years later to halt an international "white slave" traffic.[15] Belle Israels and her contemporaries in the Council of Jewish Women were thus alert to the causes and conditions of female waywardness. Despite this awareness, however, Council women often blamed the girls for their own troubles, or considered them permanently soiled.

In Israels's committee reports, therefore, girls who had become prostitutes were "hopeless" or "lost." Applied to the habitual, unrepentant prostitute, this terminology might have been understandable for the times. But the terms often spilled over into descriptions of women who were not prostitutes but simply sexually active or pregnant out of wedlock. Victorian morality, which condemned all premarital and non-monogamous sex, still dominated the thinking of Israels's generation and class. As she studied the cases of Jewish girls, she hoped for their redemption, but did not always believe it possible. The best chances belonged to the girls contacted early in their "careers." It is "the seed sown while the girl is in the institution that brings the harvest," she wrote in her committee report. She felt special optimism for pregnant girls. They "come to us determined to rid themselves of their shame," ready to become "the best and most devoted mothers."[16]

Belle Israels thus approached wayward girls with a mixture of empathy and condescension typical of her time. At least she wanted to help them, agreeing with Sadie American that the community not reject them. In 1905 she launched her first major project, the establishment of a home where pregnant girls could live temporarily and learn how to earn a living. The Lakeview Home for Girls, situated at Brighton Heights on Staten Island, opened in December. During its first fifteen months it sheltered forty girls and twenty-two infants. Israels arranged for the cooperation of several hospitals and persuaded Dr. Fannie Donovan Conklin to volunteer her services. She also hired a woman superintendent and an assistant. Israels's committee then acted as a clearinghouse for material and monetary donations, the largest portion coming anonymously from Theresa Loeb Schiff, wife of financier Jacob H. Schiff. Israels found the work inspiring. "We have in our hearts," she wrote, "the knowledge that many little children otherwise abandoned to their fate are loved and cared for by mothers—erring, but saved by that grace of all women, mother love."[17]

After three years of operating Lakeview on an experimental basis, Israels went before the Council section to explain, "earnestly and emphatically," the need for a permanent arrangement. A long waiting list of applicants proved the need for more space. More urgent, the owner of the present building had refused to renew the lease. Both Belle Israels and Sadie American worked hard to find a solution, prevailing upon Theresa Schiff to make another large donation toward a permanent home. This finally opened at Arrochar on Staten Island in April 1911.

By then, Israels was no longer heading the committee, and indeed had moved on to other projects. She did not attend the Lakeview building dedication, and no speaker at the ceremony acknowledged her role in its construction. Sadie American made no reference to her in her review of Lakeview's history in her 1911 president's report. It is possible that the two women had fallen out. They had once openly disagreed over the running of the philanthropy committee. In her president's report for 1908, the year Israels began to expand her horizons beyond the Council, American had written: "I have arranged that [next] year the work in philanthropy shall proceed under various committees, according to my original plan, which Mrs. Israel [sic], when Chairman, preferred to change. The change has not worked well, since it put all of the responsibility upon one person for the entire philanthropic work, and I have been able to find no one who was willing to undertake so much, and I have been compelled to do it myself this winter. I hope that next year all the committees . . . will be running." Her complaint reveals a disagreement if not breach between these two strong women.

A clash was perhaps inevitable. Neither bent easily to the will of others. Israels wanted total control over her area of responsibility. As for American, because of opposition to her, the Cleveland section threatened to resign from the National in 1905. In 1906 the section was expelled for refusing to pay new dues American had imposed. In 1911, the protests of the Baltimore section led other sections to withdraw. They were reconciled to the National only after American's forced resignation from Council offices in 1914.[18]

Despite her break with American and move into a wider sphere, Belle Israels never forgot her loyalties to the Council of Jewish Women. Her years with the group made possible many of her later accomplishments. She learned how to analyze a social problem, develop it, and follow it through to a solution. Serving as a one-person committee to monitor proceedings in Albany on concerns such as child labor and probation, she also learned how to approach legislators and testify before committees. Traditionally, the probation system was a source of political patronage. Its officers were usually ill-prepared and indifferent, rotating from district to district along with the magistrates, making perfunctory examinations of offenders, and then letting them go. In 1907, Israels represented both the Council of Jewish Women and United

Hebrew Charities at hearings on the "Davis bill," the result of recommendations made by a State Probation Commission. The aim of the bill was to improve the quality of probation officers and reduce incarceration costs to cities. The bill passed. In 1909, she organized a conference to deal with questions of duplication in probation work with Jewish girls. And in 1910, she was still working for an improvement in the quality of probation officers. Speaking to a group of reformers, she expressed the hope for yet a new law that would create a "better class" of officers, not necessarily college-trained but experienced. "I think experience in humane work is the best knowledge they can have," she said, "but I do feel that the ignorance and stupidity displayed by seven out of nine of the present probation officers can be avoided with a new law." She also demanded, in view of the number of Jewish girls appearing in city courts, that a salaried probation officer "be a Jewess."[19]

Council work also provided Israels with administrative experience. In May 1904, she was elected to the section's board of directors and helped organize the local Travelers' Aid Society, on whose executive committee she served. In 1905, in addition to her Lakeview work, she went as a delegate to the National triennial. When the section decided to incorporate in order to solicit funds more easily, she organized the process, enlisting her brother Walter Lindner to donate legal services. She won reelection to the board in April 1909 but resigned—as she had her other committee assignments—by October.[20] Health, domestic concerns, her family's move to suburban Yonkers, and new social reform priorities all contributed to her decision.

No matter how far Belle Israels strayed from the Council, she remained indebted to it. Its all-female, all-Jewish environment gave her unmatched opportunities to develop leadership skills. In the early 1900s, charitable institutions seldom elected women to their boards, and unless specifically Jewish or dealing with Jewish problems, rarely gave Jews top positions. As in other single-gender, parochial institutions, such as women's clubs or church sisterhoods, no one in the Council presumed anyone incompetent because of sex or race, and the chief basis of judgment was ability. Further, the Council provided Israels with inspirational role models. Sadie American and Hannah Solomon were only two Jewish women among many whom she watched in action, chairing meetings, making speeches, and heading campaigns. Most were married, with several children—Sadie American, who was single, was a rare exception. Other Council leaders may not have been as dynamic as either Solomon and American, but their examples were instructive.

Finally, and perhaps most important, by adopting the care of women and children as its special concern, the Council helped Israels and other active women resolve inner conflicts about playing public roles. Feminists, male and female alike, supported women's activism, but when critics cried "unsexed," women shuddered. According to Israels, Jewish women were especially worried by this cry. The Jewish woman, she wrote, hesitates to become a professional because she believes that "her only place [is] in the home, and that, going outside of it, she unsexes herself and gives up all opportunity as well as all thought of founding her own home in the real sense of the word."[21] By focusing on women's welfare issues, Council women rationalized that they were merely doing what any feeling mother would do for her child. Other women's organizations served a similar function. The motto of the Women's Christian Temperance Union was "Home Protection." The suffrage, settlement, and club movements all appealed to the rhetoric of motherhood to justify their crusades. In the "woman's age," there were few Amazons.

The Council of Jewish Women provided the main context in which Belle Israels learned how to be a reformer. But this organization itself functioned within a larger generational context of charity organizers, settlement workers, and other crusaders actively pursuing social reform. All used similar methods for breaking through public apathy and inertia; they took surveys, collected statistics, and presented graphic conclusions to potential supporters. Next they mounted publicity and lobbying campaigns to get laws passed. Even then their job was not finished; the enforcement and social consequences of the laws had to be closely monitored.[22] Two of the many institutions that exemplified this spirit of "scientific" reform were the New York State Conference of Charities and Corrections and the social work journal *Charities.* During the same years that Israels learned the rudiments of reform in the Council of Jewish Women, she received from the conference and the journal numerous opportunities for applying her skills. More important, when the Council of Jewish Women proved too narrow a base for her expanding ambitions, these broader-based institutions provided bridges into the larger reform community.

The New York State Conference was modeled on national conferences that, since the early 1870s, had been providing an annual forum for discussing the problems of charitable and penal institutions. The New York group started meeting in 1900. During its first few years, from three hundred to six hundred delegates attended, hearing papers on such topics as the causes of destitution, the care of mental defectives, preventive social work, probation, industrial safety, and delinquency among girls. From 1905 to 1909, Israels attended every conference, emerging early on as a leader. For the November 1906 meeting in Rochester, she served as an assistant secretary. In this role, using the skills she had learned at the Educational Alliance, she organized the conference's first exhibition.

Graphic presentations are essential to reform campaigns. Statistics alone cannot communicate as effectively as photographs, graphs, charts, dioramas, or models. Likewise, conferees hearing formal papers need a place to observe concrete results. For both these reasons, the New York

State Conference approved the idea of an annual exhibition. The one Belle Israels mounted in 1906 was a great success. Despite limited time, she solicited and received displays from forty agencies and firms, including that of her husband, which submitted a design for a charitable institution. Models, charts, photographs, and posters depicted ways to prevent tuberculosis, promote the physical welfare of tenement children, restrict child labor, reform housing codes, extend parks and playgrounds, encourage "providence and thrift," provide handiwork for the institutionalized, promote "productive" recreation, and aid travelers, especially women and children. The most moving exhibit was on child labor. A hundred photographs, probably taken by Lewis Hine, whom Israels had met at *Charities,* aroused more "horror" and "consternation" than any list of facts and figures possibly could. Photos of armless sleeves and weary faces old before their time were enough to convert even the most hardened observer. A "charming" miniature park with sandpiles, see-saws, games, and tiny dolls representing happy children, set up a vivid contrast. Most real parks were designed for promenades, not play, and had "Keep off the Grass" signs everywhere. In the miniature park, the conference report explained, "Though an officer of the law was not lacking and trees and green grass were abundant, no signs of restriction were evident, and the bright little spot suggested an oasis in the desert." The park drew "much admiration."[23]

In spite of qualms about preparing exhibitions two years in a row, Israels organized one for the Albany meeting the following year. She worried that, while her first one had "attained certain standards and established a guide for future exhibitions," the next would have to surpass the first and establish future policy. It "fully justified itself," her report concluded. Determined not to hear visitors say, "We saw all that last year," she insisted that exhibitors show only new things or new developments in old projects. The rule worked. No delegate complained of repetition. More important, the hundreds of Albany residents who attended the exhibition were educated by it. They corrected their impressions of institutional life, noticing especially the concern even the largest agencies showed for individual cases. Israels closed her report by enjoining future exhibitors to adhere strictly to the rule of displaying only year-to-year progress: "Just as the sessions of the Conference illustrate the intellectual progress made in dealing with the complicated social situations confronting our institutions, so the exhibition can show . . . the material progress made, and . . . stimulate effort to . . . do the right things and the best things."[24] Belle Israels's call for the "right" and the "best" became a leitmotif in her later appeals for reform. In her view, all programs for change should establish the highest standards for which society should strive, even if the chances of reaching them were slim.

Israels was justly proud of the Albany exhibition, but probably understated its impact. Held in the lobby of the State Senate Chamber, it had attracted throngs of visitors, including legislators and agency chiefs. One surprised senator called it "an eye opener."[25] After 1907, a standing committee organized successive exhibitions. In 1908, the Elmira conference put together an exhibition so large that it covered three floors and a basement of the Federation Building. Attracting over 29,000 visitors, it had to be kept open an extra two days. Belle Israels had established a tradition that many valued.

The New York State Conference was a male-dominated organization. Its first women officers were two assistant secretaries appointed in 1901 and 1902. Israels achieved that rank in 1906, and then was named third vice-president in 1908, the first woman to be so honored. After that year, women served as first vice-presidents, but none as president until 1924.[26]

Over a four-year period, Belle Israels worked on and off for the journal *Charities.* She joined its staff part time in December 1906 at a salary of $20 to $25 a month and eventually became an editorial assistant, publishing over a dozen signed articles and reviews. The journal started out as the official organ of the Charity Organization Society, founded in 1882; by 1906, after merging with three other publications, the *Charities Review, The Commons,* and *Jewish Charity,* it had become an independent publication and America's leading forum for the discussion of social work and reform. From 1908 to 1909, its staff, supported by the Russell Sage Foundation, conducted a landmark study of the conditions of industry and working-class life in Pittsburgh (the "Pittsburgh Survey"). The purpose of the study was to provide a firm statistical basis for industrial reform. After finishing the study, the journal changed its name to *The Survey,* the name it kept until it ceased publication in the 1950s. The viewpoint of this magazine had a profound influence on Belle Israels as well as on many of her contemporaries.

In a foreword to a 1907 issue of *Charities,* editor Edward T. Devine summarized the journal's philosophy by contrasting old and new views of charity. In the old view, he wrote, children were to learn habits of industry; in the new, children should not be exploited for commercial gain but be taught and protected. The old view of illness was that kinfolk should care for their sick; the new view held that professionals must together develop a "general social scheme" to eliminate preventable disease. In the old view, crime resulted from natural depravity; the new view stressed causes found in the environment and in the judicial and penal systems. The laws of supply and demand were thought to determine wages; now, Devine continued, society believed employers and employees, working cooperatively, might influence an entire complex of factors that determined wages. Broader concepts of workmen's compensation and industrial safety had become common. Education, once viewed as the province of a select class, had become "one of the permanent interests of human life," open to all; and educators, instead of using the rod to deal with truancy, diversified their curriculum to make the school a "center of interest and attraction." As for philanthropy, the old view said that anyone who gave charity was admirable. Now charity was to be judged by its results, and evaluators would look for efficiency, common sense, and expertise in the practice of social work. Citing Simon Patten's *The New Basis of Civilization,* a book on which the journal relied heavily, Devine declared categorically that it was now possible to obliterate the differences between the rich and poor. No one, he concluded, was naturally depraved; better conditions would release everyone's latent gifts.[27]

The essence of the social worker's creed at the turn of the century was thus an unwavering faith in the ideas of progress and social harmony, ideas that presupposed the innate equality of all human beings and blamed most social injustices on the environment. Social workers further believed that once they had scientifically analyzed the environment they could devise prescriptions for environmental change that would benefit society as a whole. Belle Israels expressed this optimism in her writings for *Charities.*

Her articles were primarily book or theater reviews, social criticism, or straightforward reports on philanthropic topics. In most of them she gave free rein to her social outlook. She reviewed a book about the importance of "play" to children, a topic increasingly of interest to her as she developed a concern about the lack of recreational facilities for urban working girls. In one moving essay, she described the plight of the widowed mother, using the second-person singular to put the reader inside the widow's mind:

> You live in three rooms in Essex Street. It being summer, you spent the night on the fire escape, but not to avoid the heat. There is a boarder who helps out with the rent . . . you lie awake early in the mornings endeavoring to solve the problems of shelter, food and clothing on uncertain earnings of four dollars a week. . . . Rent must be paid and the Charities has no money to help you; there are so many other poor people. Willie has a running nose and they tell you at the day nursery that if it is not better to-day you will have to keep him home. . . . That means that Nellie will have to stay away from school and take care of him. You are only thirty-six years old, but you look forty-nine, and you feel so tired and your arm hurts . . . and you are never without that dull ache in your head.[28]

With economy of style she had drawn a vivid portrait. The struggle of existence among poor women—the fire escape to avoid the boarder; the need to take in a boarder at all and suffer the resulting loss of privacy; problems of childcare and the effect on older siblings, especially girls; premature aging—all were familiar to social workers.

Israels never stood aloof from any of her subjects but expressed her personal views even in articles that merely reported events. Writing about *Landsmanschaften*—fraternal societies established by ethnic groups in resettlement areas—she emphasized how these societies kept families together. By preserving ethnic roots, the societies bridged the gap between immigrants and their increasingly Americanized children, a task that settlement houses tried to perform, not always successfully. Israels's experiences with wayward girls had convinced her of the value of close family ties. In her view, the *Landsmanschaften* served a vital function in contemporary urban society.

In another article, she took a subject of public outcry and drew a positive lesson from it. In September 1908, New York Police Commissioner Theodore Bingham published an article in the *North American Review* in which he claimed that Jews made up one-half of the city's criminal population. The Jewish community reacted with excessive indignation. Israels was more restrained. Although her reply did not appear until three days after Bingham had issued a retraction, the timing did not hurt her point. To her mind, the Bingham incident exposed the inadequacy of police court records. Bingham, she charged, had based his claim on the numbers of arrests, not convictions. No rational conclusions could emerge out of poor statistics, she admonished, citing a Board of City Magistrates report that called court records so unsystematic as to be utterly useless. "An agitation urging the introduction of a new system of court records . . . would have far-reaching value," she concluded.[29] Israels's approach to the Bingham charge was typical of the progressive social worker. She neither denied nor asserted but demanded better research.

Israels's social outlook affected her perceptions of the theater. Early in 1909 she reviewed two plays, *Salvation Nell* and *The Battle,* both works that presumed to depict the realities of working-class life. The first contrasted two girls from East Side sweatshops; one bore an illegitimate child then found redemption in the Salvation Army, the other ran "the gamut of vice," including prostitution and theft. Israels's criticism of the play centered not on its hackneyed characters but on its glamorization of crime and comic exploitation of degradation. The author, she wrote, wished to show "an unpleasant, sordid side of life in New York," but then offered no "solution of the difficulties" put before the audience. Instead he played to "the laughter of the unthinking many and the tears of the knowing few." She did not criticize the author's clichéd characters, the wayward girl made good through Christianity, the other lost forever to sin. Her chief concern was that human misery was no laughing matter. The second play, *The Battle,* was about a rich man who lived incognito in a ghetto in order to win the love of his illegitimate son. Israels called the situation "sham," not a real battle between opposing forces. Worse, the author seemed to grow "weary" of the plot, and rang down the curtain before offering any solutions. But at least the author knew there was "something wrong at the basis of things," and was not stooping to tasteless humor in order to provoke his audience.[30] For Israels, the moral effect of theater was of the highest importance. It must reflect social reality and suggest ways for improvement.

Israels had no plans to become a professional writer or editor, but the skills she learned at *Charities* served her well in publicizing reform campaigns and in working for Governor Smith. The exhibitions she mounted for the New York State Conference taught her how to communicate complex ideas to lay audiences. Even more important, the many progressive reformers she met through these organizations provided her with a wide network of colleagues and friends.[31] She drew upon them for advice and support when she was ready to head her own reform campaigns.

The Council of Jewish Women, the New York State Conference, and *Charities* all expressed the reform spirit of the early twentieth century. For reasons of gender and religious identity, Israels was most at home in the Council, but to her credit she reached beyond it, beyond what even she considered acceptable for her gender. Had she not done so, the next phase of her career might never have been possible. Whatever gave her the courage to attempt so much, within a few short years she would need all that courage, and more, to survive. As one of the "widowed mothers" she had written about in *Charities,* she would have lost her "new woman" status and been forced to focus all her energies on preserving self and family. Unlike the "widowed mother" of her article, however, she could rely on expertise gained from years of experience in a large cosmopolitan community, experience that replaced the formal education she lacked. She drew upon that experience in the coming years, transforming the skills she had learned as a volunteer into those of a professional.

CHAPTER 3

The Motherhood of the Commonwealth

O Mommer, wasn't Mame a looty toot
Last night when at the Rainbow Social Club
She did the bunny hug with every scrub
From Hogan's Alley to the Dutchman's Boot!

Wallace Irwin, *Love Sonnets of A Hoodlum*
No. XIV, San Francisco, 1902

Belle Israels would have found Mame's "bunny hug" appalling. This dance was one of many invented in working-class bars around the turn of the century, and by 1912 the rage among all social classes across the country. The dance itself was bad enough. Worse, a girl willing to perform it, like Mame, was also sexually promiscuous. Willie, the narrator of Irwin's sonnet cycle, pursues Mame in vain, for she is in love with Murphy, a flashy soda jerk. Yet at the "Rainbow Social Club" she dances intimately, indiscriminately, with any sailor she meets. From Israels's point of view, Mame was already "lost." What concerned her more was how to save the innocent girls who might fall under Mame's influence.

There was no doubt in Israels's mind that they needed saving. Social workers were worried about the absence of social chaperonage in the modern industrialized city. In an article in *Charities,* Chicago settlement worker Jane Addams charged that the modern city was engaged in a "stupid experiment" of "organizing work" without "organizing play." It gathered a labor supply consisting of a multitude of youth, including three million young women, and then released them from the protection of their homes to walk unattended upon city streets and work under alien roofs. To Addams, young women had no resources to cope with such independence. They "stream along the street," self-conscious, giggling, decked in "preposterous clothing," and

> through the huge hat with its wilderness of feathers, the girl announces to the world that she is here . . . ready to live, to take her immemorial place in the world. We are quite accustomed to this bragging announcement on the part of the boy. When he restlessly looks upon the world as a theater for his self-assertive exploits, the city makes haste to provide him with an athletic field. . . . But we are much less successful in making city provisions for the girl's needs. . . . [1]

If the municipality provided recreation for girls, Addams argued, they would be prevented from "the overwhelming temptation of illicit and soul-destroying pleasures" to which they turned when given no alternatives.

Cities usually controlled amusements through liquor laws and police authorities. But as immigration from rural America and abroad swelled the cities, commercial amusements multiplied, becoming increasingly visible in residential areas and decreasingly genteel in the quality of entertainment offered. Urban cultural elites took alarm. In their view, commercial "resorts" such as saloons and music halls not only wasted workers' time and money but encouraged drunkenness, vice, and crime. By the early twentieth century, the situation had worsened. New and even less genteel amusements had sprung up: dance halls and dance palaces, amusement parks at the ends of trolley lines, and nickelodeons, some featuring bawdy scenes. By the end of the first decade, these amusements had spread to middle- and upper-class districts and drew in neighborhood youths. Because of their attraction to young girls, dance halls, reputed to be hangouts for gangs and prostitutes, were especially suspect.[2] Belle Israels and a small committee from the Council of Jewish Women, disturbed by such contact between vice and innocence, launched a dance hall reform movement that won widespread national support. It also established Israels as one of New York's most effective social reformers.

Israels drew part of her inspiration for dance hall reform from temperance and anti-prostitution advocates, who argued that alcohol weakened a girl's ability to resist sexual temptation and that promiscuity often led to prostitution. Such arguments persuaded Israels that dance halls, with their combination of suggestive music and dancing and lack of chaperonage, should not be allowed to serve alcoholic beverages. This became the key demand of her campaign. Influenced by recreation reformers, she argued that dancing, performed properly and under wholesome conditions, was a legitimate form of play.[3] Israels therefore called for reform, regulation, and even government support of dance halls. She never demanded that the halls be shut down.

Her call for reform proved controversial. Both public recreation programs and the regulation of commercial amusements cost taxpayers money. Government regulation aroused resentment among business interests. Further, regulatory issues such as whether to allow liquor in dance halls or how far to control dancing styles often provoked intense cultural and generational clashes. While Israels's dance hall reform brought about some lasting benefits, these controversies made it as ineffective in changing American mores as the temperance and anti-prostitution movements.

Israels's experiences in social work had sensitized her to the problems faced by working girls in the modern city. Many had migrated to the city from small, close-knit, tradition-bound communities. Now they were living and working in fast-paced environments in which values and customs seemed in continual flux. At work, employers supervised them. After hours, the only places they could meet boys were commercial establishments. Lonely, or perhaps estranged from their families by newly acquired American tastes, working girls fled their tenements, which lacked privacy and comfort. Looking for fun, hoping to meet "Mr. Right," they went to dance halls and amusement parks where they became easy prey for exploiters.

Working girls loved the dance halls. Around the turn of the century, many opened in urban residential areas, making access possible without carfare. Admission was also cheap, and girls without escorts often got in free. For the same price as a motion picture, which in those days lasted only half an hour, dance halls provided an entire evening of fun, including the opportunity to meet boys. In addition, dance halls exercised little control over behavior. A bouncer might guard the door, but otherwise partners chose one another informally and then danced as they wished. Two girls could dance together until two boys broke in. The couples might then pair off for the evening. Officially, minors could not drink alcohol, but few were checked. Moreover, dance hall music, consisting mostly of syncopated "rags" to which couples danced one-steps in close embrace, was more exciting than the traditional folk dances, waltzes, and polkas allowed at settlement houses or family and neighborhood affairs. In short, dance halls answered the quest of many city youth for cheap, accessible, and exciting recreation.[4]

As delinquency rates rose among working girls, many of them Jewish and of concern to Belle Israels and her colleagues in the Council of Jewish Women, she blamed the commercial establishments, as did the girls themselves. "If only I hadn't met so and so at the dance hall," they moaned; or "If only I hadn't gone to Coney Island that day." The Council had tried to make up for the lack of amusements for girls by establishing recreation rooms, but these hardly made a dent in the problem.

In June 1908, Israels's questioning of the usefulness of recreation rooms led her to organize a committee of five Council women to study working girls' leisure. She called the group the "Committee on Amusements and Vacation Resources of Working Girls" and became its leader. A preliminary survey of public amusements in New York City and environs disclosed that nine out of every ten girls considered dance halls their favorite resort, and that over 150,000 city youths patronized them in any given week. Considering the reputation of the halls, the extent of their popularity shocked not only committee members but citizens outside the Council, who learned of the survey through the press.[5]

Israels and her committee believed that the road to promiscuity was paved on the dance hall floor, and that promiscuity led almost inevitably to prostitution. In their view, young girls, once seduced, lost not only their virginity, but also their shame, and had fewer misgivings about becoming prostitutes in order to make a better living. At this time, prostitution was a hotly debated public issue. The numbers and visibility of brothels in American cities had increased to such a point that organized attacks on them began to receive wide support. Welfare workers, women's leaders, clerics, physicians, urban social reformers, and legislators all spoke out against the "social evil." Commissions on vice subjected brothels, from New York's Tenderloin to San Francisco's Barbary Coast, to meticulous scrutiny. Their reports proved that prostitution had become a major business, enmeshed in crime and graft. Muckrakers wrote exposés of "white slavery." Social hygienists spoke openly about the spread of venereal disease, which they believed only sex education and continence could control. Legislators passed laws restricting the movement of female minors across state lines. No one living in an American city and aware of current events after 1900 could fail to believe that prostitution was a major social problem.[6]

Israels certainly believed it. Not all the unwed mothers at Lakeview earned their living as prostitutes, but Israels thought their sexual self-indulgence would serve them ill later on. Whether prostitutes or merely promiscuous, their fate was the same: illegitimate children, disease, addiction to liquor and drugs, impoverishment, social ostracism, even death. Their children faced similar destinies. What disturbed her most was that adolescent girls were learning about sex in the dance halls, which they patronized in an innocent quest for pleasure and because they had nowhere else to go.

The initial survey of the Committee on Amusements received enough publicity to permit Israels to raise money for a more thorough study. She hired Julia Schoenfeld, a former Pittsburgh settlement worker, to do the job. During the summer of 1908, Schoenfeld, accompanied by an escort, visited 73 New York dance halls. Her report substantiated in detail Israels's findings. Of the 2,205 unescorted girls Schoenfeld counted in the halls, 218—office clerks, stenographers, salesgirls, and factory workers—all told her they loved dancing more than any other pastime. Worse, 49 of the dance halls sold liquor, and 22 of these were attached to "Raines Law" hotels, commonly known to be brothels.[7]

Schoenfeld's report distinguished several different kinds of dance halls. First, there were "inside" or saloon dance halls that merely cleared away space for dancing to piano music. To encourage drinking, saloon managers kept the dances short and intermissions long. "Spielers" who worked for the management brought customers in and kept wallflowers happy. The popular dances of the day—the "buck dance," "pivot," and "twist"—were as intoxicating as the drinks. Dancing in such places night after night, Schoenfeld explained, respectable girls became dissatisfied with themselves. They wanted to dance all evening, be popular, and wear new dresses. In the end, they accepted treats from rogues, in return for which they agreed to sit in the dark alcoves and balconies that ringed the dance floors. The price they paid was left to the imagination.

The second type of dance hall was the "outside" hall, a casino that private clubs or groups rented during the summer. Charging from one to five dollars, casinos offered unlimited beer and dancing during the day. At night, Schoenfeld charged, when the price dropped to twenty-five cents, "immoral persons" arrived to use the adjacent parks for assignations. Schoenfeld saw unescorted girls leaving the Staten Island Ferry being whisked off to casinos by strangers.

Yet another menace was the "dancing academy." Some had good reputations; others were fly-by-night operations that kept little order, allowed "tough" dancing, improper dress, smoking, and even gambling and prostitution in back rooms. Schoenfeld charged that girls, some as young as fourteen, paid high fees for lessons, often learning more than they bargained for.[8] The picture was bleak. A young city girl seeking fun could avoid liquor and bad company only with difficulty. The result was too often seduction and ruin.

Israels and her committee planned a two-pronged attack: first, to regulate existing dance halls through licensing; then, using both private and public funds, to substitute decent halls for the sleazy resorts, and proper for "tough" dancing styles. While public regulation of commercial amusements was still controversial, the use of public monies to fund amusements was even more so. Israels told her critics that a larger social need, the protection of youth, was at stake. Industrial and urban conditions strained the ability of youth to resist temptation, she argued. Rising rates of illegitimacy, prostitution, and other crimes among girls proved her point. Since families were no longer providing social chaperonage, girls now needed "the general motherhood of the commonwealth."[9]

Most Progressive Era reforms had three phases: investigation of a wrong; publicity of the results; and mobilization of public opinion to get lawmakers to act. Dance hall reform followed this pattern. The investigatory phase was over by 1909, and Israels began to hold press conferences, write articles, make speeches, and mount public demonstrations of improper and proper dancing styles. These tactics, as well as her complementary themes of regulation and substitution, drew prominent adherents to her cause. She added them to the Committee on Amusements, enlarging it eventually to seventeen members and, in a step important to her later career in politics, removing it from the narrow confines of the Council of Jewish Women.

The new members represented a cross-section of New York's intellectual, social, financial, and political worlds. Coming from a variety of backgrounds—education, welfare work, journalism, civic reform, women's clubs, and business—they all had connections, either directly or through their spouses and relations, to centers of power. Some of them were rich: a Miss M. A. Parsons contributed most of the Committee's capital stock of $3,000; other members solicited subscriptions from wealthy friends. Others such as Marjorie McAneny, wife of the president of the City Club; Alma Wertheim, daughter of Henry Morgenthau, Sr.; and Emmeline Heckscher Winthrop, whose husband Egerton was a prominent lawyer and president of the New York City Board of Education, and whose sister was Mayor George McClellan's wife, lent their names to the cause and attended the Committee's demonstrations. Amy Einstein Spingarn, wife of Columbia professor Joel Spingarn, wrote an article on amusements for *The Survey*. Gertrude Robinson-Smith, then at the National Civic Federation, formed a committee to cooperate with Israels on the dance hall question. Louise Caldwell Jones, an active clubwoman whose husband Gilbert was an editor at the *New York Times,* arranged for favorable publicity from that organ. Three men from the worlds of vice reform and welfare work were also active on the committee: Frederick H. Whitin, general secretary of the Committee of Fourteen, a private anti-prostitution group; Michael M. Davis, Jr., secretary of the People's Institute, a Lower East Side educational organization then working out a system of motion picture censorship acceptable to the new industry; and Henry Moskowitz, head resident of the Downtown Ethical Society and also experienced on movie review boards.[10]

For Belle Israels, then in her early thirties, to have put together such a group of people was quite an accomplishment. She had a gift for attracting support to a cause. She also understood that she should not identify her campaign as a strictly Jewish or female issue. Her expansion of the committee over a wider base and her inclusion of men gave Israels credibility in circles outside the Jewish community and beyond women's groups.

In addition to enlarging her committee, Israels wrote articles justifying dance hall reform. "The Way of the Girl" appeared in *The Survey* in 1909, illustrated by Lewis Hine, the journal's photographer of working-class life.[11] Because it contained the basic tenets of the campaign, the article was widely quoted in contemporary reform literature.

It began by comparing good and bad amusements, showing how most places failed to meet working girls' special needs. These needs become acute in the summer when the girls faced layoffs and had no funds to travel. Boys had athletic fields; girls of the "less driven class" repaired to country clubs. But working girls, for whom amusement was incomplete without attention from boys, could go only to places where the unscrupulous prowled. Coney Island was one such place. Girls went there in pairs, hoping to be picked up. The day's highlight came when a boy treated and then offered to escort the girl home. The wise girl soon found her original companion; the weak girl discovered that the boy wanted a reward. Too many "careers," Israels lamented, started at the beach and ended at a Raines Law hotel.[12]

Amusement parks were just as bad. The park at the end of the trolley line in Fort George permitted indecent dancing, men accosting girls at will, and general intoxication. In contrast, Palisades Park was a model of probity. Its owner, a strict Presbyterian, had not only preserved the park's natural beauty but replaced vaudeville with musical comedy and forbade alcoholic drinks on the premises. "I am not writing a temperance argument," Israels insisted, but her experience with hundreds of girls who had learned the taste of liquor from the dance hall or amusement park made her realize its "stupefying effect on the moral sensibilities." Excursion boats offered yet another horror. Speculators bought up staterooms for two or three dollars a day and then rented them to couples for short periods. "Kind and gentle" Negro matrons had "neither authority nor moral effect, and spend most of their time in the little sitting room allotted to them."

The article ended with suggestions to settlement houses on how to make their vacation homes less dull. The average working girl resented the rules, regulations, and chores required at the cheap country places run by institutions, and also missed contacts with the "fellows" at evening entertainments. Israels suggested that low-cost campsites with water nearby and with abundant physical recreation available, would not only restore city girls' health but delight them.[13] She advised municipalities to provide amusement parks "of the right sort" and "people's playgrounds" at the seashore. As for dance halls, even the most worldly-wise, wary girl could not resist the sensuality of the modern dances, which "appeal[ed] to the worst" in her, breaking down her ability to "keep up the good fight." Even though all girls who "fell" did not become prostitutes, by trodding the downward path into promiscuity, they "lost the bloom of their youth."

Most girls were "innocent," Israels argued. They only wanted to have fun. The skating rink was passé, the moving picture show "on the wane" [!]. Since towns had become "dance mad," "let us provide it plentifully, safely, and inexpensively," she pleaded. To the argument that the state should not interfere with public amusements, she countered with charges that an ex-governor controlled an excursion boat and a state senator held an interest in Coney Island. These charges proved her point that only vested interests opposed reform. The city had already provided playgrounds and paid for "the socializing force of contact with good supervising men and women." Denial of these privileges "peoples the underworld," she concluded; providing them is "model preventive work."

"The Way of the Girl" was grounded in a post-Victorian but premodern concept of female sexuality. Like other recreation reformers, Israels accepted the reality of the female sex drive but urged self-restraint.[14] To her, male-female relations consisted essentially of "pursuit and capture." "The man is ever on the hunt, and the girl is ever needing to flee," she wrote. If the girl drinks, she loses her ability to wage war. The answer was not the repression of the female sex drive nor of the dance hall, now identified as the prime battlefield. Rather, society had to make sure the battle was conducted on more equal terms. Its best weapon, Israels concluded, was a rationally conceived program of municipalized recreation.

Over the next few years, Israels published many similar articles and addressed numerous national groups, such as the Playground and Recreation Association, the American Society of Sanitary and Moral Prophylaxis, the Academy of Political Science, and, in 1912, the National Conference of Charities and Corrections, at which recreation reformer Joseph Lee introduced her as the person who "knows more about dance halls than anybody else in the United States."[15]

The first step toward dance hall reform was to obtain a licensing act. During the 1908 state legislative session, New York Assemblyman Moritz Graubard proposed a licensing law. His colleagues rejected it in the belief that it was unenforceable. A number of settlement workers from his district, including Henry Moskowitz, advised him to consult with Israels's Committee on Amusements. By then the Committee had formulated a model bill. It required dance halls to conform to light, air, and fire codes. Further, the halls would have to supervise those who entered, keeping out "undesirables," and guarantee introductions under "proper" circumstances instead of through "breaking in." Finally, dance halls could not serve liquor, even in an adjoining room, nor could they offer a "return check" so that dancers could purchase drinks outside.

As a result of Assemblyman Graubard's conference with the Committee, a bill was proposed to regulate only dancing academies. It required periodic inspection by the police, fire, and building departments, but also forbade academies from selling liquor and admitting unaccompanied girls under sixteen. These last provisions outraged one senator, who warned, "This bill would prohibit the teaching of dancing classes in Delmonico's. It is urged by those who want to keep young men and women apart from each other. It is another indication of the untiring energy of those who desire to interfere with the pleasures of the citizens of New York."

Amending the bill so that it affected only academies that taught dancing and advertised as such, the senator severely limited its coverage. A dancing master, Oscar Duryea, then formed an association to fight the bill. Arguing that it discriminated against dancing academies, his group successfully challenged it in court the following year.[16]

Undeterred, the Committee on Amusements tried again in 1910. This time the bill covered dance halls as well as academies, prohibited the sale of liquor on the dance floor and in adjoining rooms, and mandated strict inspection.[17] Few critics of the proposed law objected to safety and sanitation codes, but many fought the proposed liquor ban. Brewers, who supplied the halls, and saloon dance hall owners, who made more money from drinks than dancing, pressured legislators to reject the bill. Immigrant groups, for whom a dance without wine or beer was unthinkable, joined in. A compromise bill stipulated that saloons could not allow dancing unless they were licensed also as dance halls. Further, no liquor could be sold in places that taught dancing and advertised as such. The bill passed, but left reformers dissatisfied. In order to give dance halls time to make the required changes, the bill delayed enforcement until March 1, 1911, and exempted all hotels with more than fifty bedrooms. In addition, since the license fee bore no relation to the size of the dance hall, the law forced small establishments out of business. There were no renewal procedures for halls that changed hands. Finally, since budgetary allocations for dance hall inspectors were inadequate, members of the Committee on Amusements ended up doing inspection work themselves.[18] This was not the effect the bill was supposed to have.

While awaiting the result of legislation, Belle Israels worked steadily on the companion side to licensing—substitution. She wanted to open model dance halls that would set standards and show businessmen that decent places could be profitable. In February 1909, she convened her committee's first conference of social welfare workers to discuss how to bring the plan into being. About fourteen welfare organizations sent representatives, and many individuals prominent in philanthropy attended. Belle Israels presided and reiterated that she wished not to drive dance halls out of business but to provide better facilities for this "necessary form of recreation." She wished to establish halls without liquor and free of other objectionable features. Her committee would keep secret its connection with the halls, "as young people would not patronize a dance hall known to be run for their benefit by an organization." Asking supporters for $1000 for each dance hall, she promised them a 33⅓ percent profit in six months.[19]

The Committee on Amusements opened its first model hall the following year. Hundreds of youngsters, unaware that they were experimental subjects, flocked to the place, located east of Fifth Avenue and advertised by a plain sign, "General Dancing, Fifteen Cents." Because of recent dance hall shootings, a bouncer checked for weapons at the door. Soft drinks sold briskly for five cents. The dance floor was large, the ceiling high, the room well lit and ventilated. In a space reserved for beginners, a stout dancing master—"mixture of artist and pugilist," a *New York Times* reporter observed—called out the "Lancers," a traditional dance with four set figures. A young man audaciously resting his head on his partner's shoulder found himself expelled. College students dancing too wildly were told to stop because they looked too conspicuous. Taking the place of "spielers," male relatives of the Committee on Amusements took care of wallflowers. "A funnier sight than a respectable middle-aged New York businessman, accustomed to other surroundings, picking out his partners, cannot be conceived," the reporter continued. In spite of his amusement, the reporter conceded that everyone looked happy, that men outnumbered girls, and that no women of "bad character" were seen.[20]

Over the next two years, the Committee on Amusements sponsored model dance halls in Manhattan, Brooklyn, and Newark. Newark's "Palace Ballroom" was its showplace. Located in a large rectangular building with an enclosure for a fourteen-piece orchestra and a balcony for spectators, the hall featured close supervision at the entrance, on the floor, and even in the balconies where an "introducer" helped young people meet under "proper circumstances." There was "barn dancing" once a week, complete with picturesque scenery and two or three farmers forking hay; German dancing or waltz tournaments took place on other nights. But the hall's most dazzling feature was its streamers of rainbow-colored lights hung from a central chandelier. Operated by a switchboard, the lights blinked on and off in time to the music and changed hues to suit musical moods. The hall thus mixed old-fashioned dancing and decorum with novelties of the electrical age. Youngsters gasped with delight as they entered.[21]

In addition to regulating existing dance halls and creating model ones, the Committee on Amusements also addressed dancing styles. In the late 1890s, the waltz, polka, two-step, and "set and figure" dances such as the German, Cotillion, and Lancers were standard fare at social gatherings. But by the turn of the century, ragtime and minstrel-show cakewalks had moved into northern cities. Cakewalks, performed in elaborately mounted plantation scenes, included prances, "buck-and-wing" steps, and even toe dancing. These styles began to appear on the music hall stage, merged with ragtime, and then moved into the dance halls. There they exploded into a variety of "animal" dances named after turkeys, bunnies, monkeys, and bears. The Turkey Trot required couples to move in fast one-step circles occasionally flapping their arms like crazed turkeys. The Grizzly Bear, Bunny Hug, and Kangaroo Dip encouraged close body contact. Soon, enterprising songwriters such as Irving Berlin put suggestive words to the new tunes, telling dancers to "hug up close to your baby," and "everybody's doin' it." To complicate matters, South American dances appeared, the "Argentine Tango" and French "Apache" version being among the most spectacular. The tango's medley of dips, glides, and sways suggested sexual conquest and submission.[22]

Denouncing the new dances as ugly and indecent, clergymen, dancing masters, and citizens united to keep them out of middle- and upper-class ballrooms. At summer resorts, where trends were set for the season, dancing masters developed tamer versions of the new forms, such as the "Long Boston." Society leaders banned "animal" and "dip" dances from their cotillions. Settlement house leaders and managers of large hotel ballrooms followed suit. But such bans affected only a tiny portion of urban youth, and were for the most part ignored. The "Long Boston," merely a modified waltz with a few glides and rapid turns, could not compete with the tango. At society balls, debutantes slipped out into the corridors to learn the forbidden steps. By 1912, letters to editors and news articles dwelled incessantly on whether to "trot" or not. On one occasion at the Waldorf, patrons at a club dance who were accused of turkey trotting issued a denial, adding that a much daintier "Chicken Trot" had been performed. The debate reached such lengths that Finley Peter Dunne's "Mr. Dooley" was prompted to ask Mr. Hennessey early in 1913 if he had "larned anny iv th' new dances?" "De ye know th' Turkey Throt, th' Tango, th' Scissors, th' Knee Holt, th' Neck-an-Neck, th' Envelope Flop, or th' Curvathure-iv-th' Spine? If ye don't ye needn't come around here anny more. Me socyal standing will be injured by me knowin' ye." When Hennessey asked what he was talking about, Dooley replied, "It's almost the on'y subjeck iv seeryous convarsation among th' white races."[23]

In a circular published in January 1912, the Committee on Amusements charged that the new dances were "not dancing at all, but a series of indecent antics" of "disreputable origins" done to the accompaniment of music. Starting out in "houses of prostitution," the dances have spread to "all but the most select" events. Even when performed in modified form, the pelvic motions encourage the hands to lower from shoulders to hips and the bodies to touch. Hence, the Committee argued, even the "milder forms" of the dances, when taught to the "unsuspecting" and combined with liquor, led to obscene behavior.[24] To get rid of the dances, the Committee took a remedial rather than repressive approach. Belle Israels had observed that strict dance codes had a chilling effect on social gatherings: they merely drove young people to "rowdy" dance halls. She had also seen the police ignore rules against "tough" dancing. From the start of her campaign she had argued that repression without substitution would fail. The Committee on Amusements thus organized a conference of citizens to alert the public to the "real nature" of the new dances.

The conference met in Delmonico's ballroom on January 26, 1912. Six hundred people came. Its purpose was to establish standards of decency in dancing: "What is good and what is bad?" When girls protested that "Everybody's doin' it," what should their elders say? To find out, society women in silks and furs mixed with settlement workers, city officials, clergymen, writers, and professional dancers. Opening the conference, Belle Israels introduced Oscar Duryea, who had decided that "tough" dancing was worse for his business than licensing. Aided by his wife, Duryea demonstrated how the dances start out innocently, only to become passionate, gliding embraces as the dancers tire and the music weaves a spell. The audience gasped in horror. When the Duryeas had finished, Oscar pleaded for the suppression of the dances and a return to traditional forms. Next, Israels presented Al Jolson and Florence Cable, his dancing partner from the Winter Garden. Jolson told how he had learned to dance on San Francisco's infamous Barbary Coast. Drunken tars who could barely skate across the floor would come alive when the orchestra "ragged it up," hitting seductive minor keys. To make his point, Jolson grabbed his partner, snapped his fingers, and drew her daringly close. The audience gasped again, but its attention never flagged. When Jolson and Cable finished, thunderous applause greeted them. The effect of the demonstration was ambiguous. Did it thrill audiences or arouse them to indignation?

One speaker, Reverend Percy Stickney Grant, championed the Committee on Amusements because, up to a year before, there were not three dance halls fit for a girl to enter. Still, in his view it was dangerous for older people to "delve into the psychology of the young." He also thought that working-class youth were "simpler in their attitude toward sex than we of the sophisticated society," and that society should not act as censors. Henry Moskowitz also spoke. He insisted that the Committee was not an anti-vice group but was dedicated to finding places where the spirits of those who "work to exist and live in their play" may find expression.[25]

The Committee on Amusements focused subsequent demonstrations on the "graceful" dances that illustrated the "best standards." It taught society girls new forms of folk dancing, such as the Ostend (a combination of schottische and glide), Dublin Jig, Danish Polka, Spanish Boston, and Scotch Lilt, but had more success with the Aviation Glide, Tangle Twostep, Dorothea Boston, Fourstep, Fascination, and Danube Waltzes. These promised to be "just as much of a romp . . . as the others," Israels declared optimistically. The Committee also won the cooperation of the "better class" of dance halls and ballrooms. In return for being on an approved list, their owners and managers agreed to post signs forbidding "immoral dancing," and to insert clauses in musicians' contracts against the playing of "rags" and "shivers." Israels reported that one proprietor was "relieved" to see the dances go.[26]

In time, cultivated elites absorbed the popular dance forms and made them their own. Professional dancers such as Irene and Vernon Castle, Maurice Mouvet, Wilma Winn, and Everett Evans "sanitized" the dances, turning them into the elegant "Castle Walk," "Hesitation," "Maxixe," and "Pousse Café." Popular magazines and newspapers published "how-to" articles, complete with photographs and diagrams. The dances became so acceptable that clothing styles changed to accommodate them, skirts becoming looser and flowing, or slit at the side to allow dips. Even the infamous Turkey Trot won admirers as physicians assured young and old that its effects benefited the circulation. "Honi soit qui mal y danse," quipped Mary Master Needham in *Collier's* in 1913. Freudian psychiatrist A. A. Brill attested that the dances, far from arousing sexuality, tended to substitute for it. The dances are "beneficial to our social system . . . as soothing to the populace as rocking is to the infant, . . . offer good exercise and enjoyment to thousands of people, and serve besides as an excellent sublimation," he said. A judge in Yonkers opined that, while he found his daughter's Turkey Trot "mildly ungraceful," it was not in his opinion "indecent." Even Belle Israels admitted in 1914 that the tango was beautiful, if danced well. "I think it is a mistake to condemn all the modern dances," she now said, but added that it was still important to "do them correctly," and offered to have her committee send teachers to special events to instruct dancers in the proper forms.[27]

Between 1910 and 1913 dance hall reform spread across the country. Belle Israels and her associates spoke at meetings of social scientists and welfare workers and published many articles, offering sample licensing laws, encouraging vigilance against lax enforcement, and campaigning for public funding of wholesome amusement resources. By 1914, many cities—Chicago, Cleveland, Detroit, Buffalo, Milwaukee, Cincinnati, Newark, Los Angeles, Minneapolis, Portland (Oregon), Kansas City, Oakland, and Duluth—had passed dance hall ordinances or planned to. Before Belle Israels's reform, several other cities had already included dance halls in their amusements laws: Philadelphia in 1885; San Francisco in 1903; and Boston and Pittsburgh in 1908. As a result of the recent agitation, these cities were now considering amendments.[28]

The new ordinances shared certain features: license fees, compliance with city codes, and controls or bans on liquor. But on other issues of social chaperonage there was less agreement. Closing hours ranged from 11:30 p.m. in Buffalo to 3:00 a.m. in Chicago: ages at which minors should be excluded varied from under sixteen (in Kansas City and Buffalo) to under twenty-one (in Minneapolis, although the mayor vetoed this version of the bill). Eight cities expressly forbade "moonlight" dancing, insisting on bright illumination at all times. Cincinnati and Duluth restricted the location of dance halls; San Francisco prohibited "marathon" dance contests; Kansas City, Portland, and Duluth specified correct dancing positions; and Portland prohibited free admission for females. Kansas City took social chaperonage most seriously. Its Board of Public Welfare appointed dance hall inspectors and refused permits to anyone who would not pay the required fifty cents per inspection. The board gave its inspectors wide latitude to judge whether dances were within the law, but suggested the following guidelines: no "shadow" or "moonlight" dances; no "undue familiarity between partners" ("The lady should place her right hand on her partner's arm and not on his shoulder, and partners should keep their bodies free from each other."); no intoxicated people; no "return checks" that allow dancers to buy drinks outside; and, if in doubt about a girl's age, inspectors may submit the name to the board for verification.[29]

As dance hall reform spread, moving further from its more cosmopolitan origins, stricter interpretations of chaperonage were perhaps inevitable. Belle Israels had argued repeatedly that excluding minors and repressing dancing styles would fail. But while social workers like herself had raised the alarm about dance hall dangers, they could not retain control over the aftermath. This was in the hands of local legislators answering to more varied constituencies.

In the end, dance hall reform proved disappointing. After the initial excitement, enthusiasm for enforcement waned. Official inspection forces were never properly staffed or funded. Public-spirited citizens, such as the members of Israels's committee, could help, but working alone they were inefficient and easily worn out.

There were some highly publicized successes. New York's notorious "Haymarket" made repeated attempts to get licensed and ultimately failed. Located at Sixth Avenue and Thirtieth Street, it had "as wide a national reputation as the 'Moulin Rouge' of Paris had an international." College boys from Maine to California knew it as a sure spot to find a prostitute. The place was so popular that, when the Committee of Fourteen asked Anheuser-Busch of St. Louis to stop supplying it with beer, the company refused on economic grounds. The Haymarket's lawyer, a friend of Mayor William J. Gaynor, put pressure on the mayor to be lenient. Gaynor told Israels that the Haymarket was no better and no worse than other places of its kind, and that by denying prostitutes a place to assemble the city was driving them into the streets, where they would be more offensive. Further, since the women were already plying their trade in the Haymarket, they had a "right" to dance there. Israels, replying at length, reminded Gaynor that he had promised to "keep New York City clean." By saying that the city should tolerate solicitation in the Haymarket but not outside, the mayor was condoning "modified segregation," the concept of legitimizing prostitution within a specific, controlled location. To give in on this point, she declared, "would be a blemish on the present clean conditions of the city and a distinct retrogression from the present point of decency so hardly won." By making the mayor responsible for the city's moral condition, Israels made it impossible for him to grant the Haymarket a license. The place never reopened as a dance hall.[30]

Gaynor, a grudging supporter of dance hall reform, considered Israels a prude, warning her "not to be too strict. When I was young I went to dances also and I am quite sure that I am none the worse off. We must not be too straight-laced in this world."[31] No matter how unpopular with the mayor, Israels and her Committee on Amusements persisted in badgering him. Whenever she received letters from citizens complaining about violations of the licensing law, she verified the facts then pressured the mayor, license bureau, and the police commissioner until she got action. One of her investigators, a Madison House worker named George Landy, reported on a place that neighbors said corrupted young girls. He found out that the place had no fire exits, and received beer from a dumb-waiter which it then "gave away" to minors for the price of admission. As for ruining girls, Landy heard of no actual case of rape or sexual intercourse in the hall itself, but was "sure that many of the plans are laid there and many of the victims are picked up at these dances."[32] This kind of report kept Israels occupied with dance hall reform long after the licensing law had passed.

She also pursued the mayor on issues other than violations. She wanted more reforms. All-night restaurants were inimical to the "moral tone of the city," she charged, asking the mayor to close them down. In 1913, because a proposed law would remove the mayor's discretion in granting licenses, she opposed efforts to license resorts solely on their conformity to fire and building codes. Although she would not "rush to the panacea of the woman policeman," she wanted women police officers posted in public places so that they could protect girls who cannot distinguish "what is or what is not good." She pressed for better lighting in parks, and for an increased budget for the Bureau of Licenses. Yet another request was to exempt dance hall inspectors from the civil service law. Her argument was that, since they lost their effectiveness as soon as they became known, they needed to be changed often. Gaynor's responses were always polite, but once he allowed himself to suggest that certain people "are always of the opinion that things are better somewhere else than at home." He continued, "They are always persons who are in search of better bread than wheaten. They are always persons who go out for wool and come home shorn. Conditions which you mention are just as good in this city as in any other place on the face of the earth. The trouble is you do not want to believe it."[33]

Israels's efforts to get dance hall licensing in her own town of Yonkers, where she had lived since 1909, almost failed. In 1912, she became a Progressive party district leader in her ward, helping to elect the city's only Progressive alderman, Benjamin F. R. Adams. After the election, she persuaded him to introduce an ordinance. His bill forbade entry of persons under sixteen without parent or guardian, "indecent" dancing, the selling of liquor, and dancing after 12:30 a.m. without special permission from the mayor. Despite vigorous lobbying by Belle Israels, women's organizations, religious institutions, and settlement workers, the city's Democrats succeeded in eliminating a liquor ban from the final version. A Democratic judge, Joseph H. Beall, warned that too strict a bill would become a dead letter. No government could make dance hall managers responsible for the actions of third parties, he said, or the mayor a moral censor over dancing styles. The bill squeaked through, six to four, with Adams voting against it because it lacked a liquor ban.[34]

There were some successes on the substitution side of dance hall reform. Several cities funded supervised dance platforms in parks or on docks. Boston, Philadelphia, San Francisco, Cleveland, Milwaukee, Cincinnati, and Denver experimented in this vein, but not without opposition. San Francisco clergy opposed the use of public monies to support sinful amusements, and in Denver and Milwaukee, after the defeat of progressive city governments in 1914, the experiments foundered.[35]

In general, because municipalities could or would not enforce dance hall laws, many provisions remained dead letters. Chicago was a case in point. Dance hall reform had started there in 1910, led by Louise de Koven Bowen of the Juvenile Protection Association, a group associated with Hull House. It inspired a dance hall law of 1911 that controlled liquor consumption more strictly and trained dance hall inspectors to report violations to the police. But by 1917 Bowen saw "no real permanent improvement." When the association took violators to court, juries acquitted them. Police regularly ignored violations of fire, sanitation, and ventilation codes. They might interfere in fights, but since they were usually outnumbered they did so rarely. In one dance hall, when a lone officer tried to stop liquor sales after hours, the crowd attacked him, a shot rang out, and he was killed. Association investigators frequently reported "indecent" dances such as "Walkin' the Dog," "Shaking the Shimmy," and the "Stationary Wiggle," yet authorities did nothing. The lack of liquor controls worried Bowen more than the dances. Liquor, she said, was "like setting a match to a flame." In sum, after six years of dance hall reform, Chicago's halls were "a disgrace to our city and . . . feeders for the underworld."[36]

Belle Israels's dance hall reform was thus only partially successful, its demands memorialized in law codes but its efforts at social control largely ignored. Nonetheless, agitation for dance hall reform continued throughout the 1910s and 1920s, resulting in hundreds of ordinances and state laws. When the United States Children's Bureau surveyed these laws in the mid-1920s, it found that a combination of municipal and state laws covered 348, or 70 percent of all American cities over 15,000 in population. As in the Progressive Era, the laws shared features in common—licensing for halls, permits for dances, and prohibitions on certain kinds of dancing and conduct. No matter how similar or different, however, all the cities surveyed reported difficulties with enforcement. Parents were often uncooperative. Children resented controls. Substitution programs sometimes worked but failed to reach the mass of young people who, as one dance hall inspector put it, "want to go where they can do as they please."[37]

Some features of dance hall reform received wide acceptance, including the principle of licensing commercial amusements, limits or bans on alcoholic drinks, and publicly funded "substitution" programs, many of which municipalities still enjoy. Looking back, the Committee on Amusement's attempts to censor dancing and musical styles appear naive, but differ little from the anti-rock 'n' roll campaigns of a generation or so ago, or continuing efforts by religious groups and government to discourage adolescent sexuality. In the early 1900s, the special problems of adolescent girls were not unreal. The dissemination of birth control information was illegal; there were neither reliable means of birth control nor effective treatments for venereal disease; unmarried mothers had few recourses; and there certainly was a lack of clean, safe, and vice-free recreational facilities for working girls. We may interpret Israels's dance hall reform as the last, futile gasp of an outmoded Victorian sensibility, but we cannot ignore the validity of its rationale.

From the perspective of recreation history, Israels played several important roles. She insisted that citizens, social workers, and officials pay attention to the needs of girls. Recreation theorists had much to say about boys, and in the early 1900s physical education experts had begun to develop limited sports and recreation programs for schoolgirls. But only dance hall reformers addressed the needs of working girls who no longer benefited from school programs. Israels also insisted that the community recognize dancing as a legitimate form of recreation, and must provide it in order to protect its youth. In making this point, however, she was not—what some Progressive Era reformers are sometimes accused of—trying to replace family controls with those of the state. She only wanted the state to act "maternally," nurturing and protecting those who, for reasons beyond their control, needed its care.[38] This is the major thrust of her phrase, "the motherhood of the commonwealth."

We will never know how many girls her reform actually "saved." Israels never invited working girls to serve on her committee. In that sense, she shared with other progressive women reformers a tendency to "matronize" those she wished to help. This tendency was epitomized in a statement she made to a conference of social workers in 1912. "These people work to exist and they must play if they are to live," she explained, borrowing Henry Moskowitz's phrase. "And they must play in the only way they know how to play, and that is by means of the commercialized amusements that we offer them. But we must protect them at every possible point, . . . [teaching] them what things are good and what things are bad. . . . There is no regulation that can be applied . . . like the formation of standards of public taste, public opinion, the great social chaperonage that every man owes to his brother and every woman to her sister."[39] Israels's protective motives were decent, but in 1912 she could not imagine that working girls could formulate their own standards of taste. In this sense, dance hall reform, like anti-prostitution, was probably more important for her peace of mind than for that of working girls.

To the extent that other responsibilities allowed, Israels remained active in dance hall reform throughout the 1910s and 1920s. Whenever the issue came up in the press, reporters sought her out for advice or an informed opinion. They recognized her as the fount of the reform and its acknowledged authority. Other preoccupations, some of them growing out of dance hall reform—service on this or that committee, calls to testify before legislative bodies, requests to investigate conditions—also kept her busy. When personal tragedy struck, such involvements probably prevented her from despair. They also helped her mature as a public figure, preparing her for even greater challenges that lay ahead.

CHAPTER 4

Beyond the Committee Stage

The old theory that in order to be lovely we must be lazy has given place to the new doctrine that we cannot be respected unless we are worthy.

Hannah Solomon
Council of Jewish Women, 1905

Cleaning up dance halls was Belle Israels's way of expressing the reform impulse of her age. Like other progressives, she wanted the state to act as a beneficent parent, sheltering the weak and leading the strong to serve society as a whole. While working for dance hall reform, Israels worked for this larger goal as well. Her method of achieving change eschewed social disruption and favored the voluntary cooperation of diverse social elements. Such a method demanded a great commitment of time and energy. Reformers like herself had to overcome public inertia, follow through on the politics of change, and remain alert to laxity or corruption in enforcing the changes achieved. Israels seemed to have unlimited resources for this kind of activity.

Her energy persisted despite a heavy domestic load. In 1904, she nearly died delivering her first child. The pregnancy had been hard. After some months, her doctor discovered an ovarian cyst and decided to operate, even at the risk of losing the child. The pregnancy continued, but the delivery was awful. "He's a bad lazy baby and won't push," Belle complained to Grace, who stayed with her, helping to keep Belle's dreadful mother-in-law out of the room. Fearing death, Belle recited the Twenty-third Psalm over and over. Her son finally came on November 21, with the aid of forceps, the marks of which he bore all his life. In honor of his mixed Sephardic and Ashkenazic heritage, his parents named him Carlos Lindner. Charles painted an illuminated copy of the psalm that had helped his wife survive the ordeal. It showed a tree of life on the rim of the "valley of the shadow."

The following New Year's Day, Grace wrote a poem for Belle that referred to Carlos as Belle's "far dream" that, out of a "black abyss," an "agony of empty years," at last came true. As involved as she was in social work, until she was married and had her own children, Grace thought, Belle's life had been "empty." While motherhood alone never completely fulfilled her, she derived enough satisfaction from it to bear another child almost every two years thereafter: Miriam on February 3, 1907; Josef on April 25, 1909; and Judith, who died at birth in February 1911.[1]

In the summer of 1909, the Israels family moved from Manhattan to Yonkers so that the children could grow up in the country. Yonkers was a good choice. Situated on the Hudson north of the Bronx in Westchester County, it offered large homes at reasonable prices and frequent train service into Manhattan. Belle and Charles bought a home in a new development called Park Hill. Located in the south-central part of town on terraced and wooded land, Park Hill presented spectacular views of the Hudson and Palisades. Its large homes on spacious lots convenient to schools and markets at the bottom of the hill made it especially attractive. The 155th Street station in Manhattan was only a twenty-minute ride on the Putnam Division of the New York Central Railroad, which stopped hourly at the Park Hill station, accessible by a funicular that the children loved. Park Hill residents often complained of poor service on the Putnam, but there were other ways to get to the city, including a trolley to the subway at Van Cortlandt Park. In later years, when Belle was working full time, she bought an automobile and hired a chauffeur for her daily trips.

The home, not as pretentious as some on the hill, had three stories and nine rooms, and stood on a small promontory of granite overlooking the Yonkers valley. The house was subject to a $5,500 mortgage. To protect Belle, Charles put the house in her name. He loved the house because its builder had been influenced by Frank Lloyd Wright, whom Charles admired. He designed colonial-style furnishings for it and had them made to order. Outside the house a huge oak sheltered the front lawn, which one reached from two flights of cement steps that zigzagged up from a narrow, quiet street. A wide veranda, where of an evening Charles and Belle would smoke cigarettes in the dark (so that neighbors might not see her), stretched across the front of the house and gave a view of the valley below. The children could play for hours on the large boulders and in the woods nearby.[2]

Belle and Charles were among the first Jews to move to Park Hill. A few years later, Belle applied to the local country club so that her children would have a place to swim but was turned down. She was convinced the grounds of the snub were religious or ethnic. The family joined a synagogue, perhaps in response to this experience of prejudice. Belle lit candles every Friday night, reciting in English a simple prayer: "Grant Thine almighty protection unto me and mine, guard us against all accident and evil, . . . cause the light of joy to burn in our hearts, and the light of love and peace to shine in our homes. Amen." Years later, her daughter found the original of this prayer and discovered that Belle had cut out the words, "aid Thou us to conquer every temptation and allurement of sin," an omission that testifies to Belle's lack of orthodoxy. Still, Belle insisted that her children stay home from school on Jewish holidays, arguing that this asserted the right of the more orthodox to do so.

Belle's social service work kept her away from home several days a week during the season, and occasionally for days on end when she attended conventions in other cities. She took little time out for her confinements. For Miriam's birth, she stopped her activities only two months before the birth. When Josef was on the way, friends joked that the doctor would have to send out to a committee meeting for Belle to come and have the baby. When the babies were born, Belle nursed them as long as she could, then turned their care over to others—a nurse, housekeeper, or grandparents. The Lindners had sold their Harlem shop and lived in a boardinghouse nearby. For good or ill, Florence Israels lived in the Park Hill house. Still, no matter how much domestic help she had, or how involved she became in matters outside the family, Belle thought of herself primarily as a wife and mother. "Husband, then children, then house, and then career," she said in an interview years later. It was she who organized the family routine, marketed, planned the meals, supervised the children's education, cuddled them, and took them for long wildflower walks along the railroad tracks with their spaniel Spot. In the years before she lost Charles, this is the image her children had of her.[3]

An early family tragedy took its toll on her own health. The crisis involved Charles's sister, Josephine, or "Josie." She had married a man named William Poey de Luna and had a son by him, Adrian Carlos. Sometime between 1908 and 1909, de Luna infected Josie with syphilis. De Luna ran off, never to be heard from again.[4] When Josie became an invalid, insane and dying, her mother tried to persuade Charles and Belle to take Josie into the Park Hill house. Distress over this experience, added to the usual complications of pregnancy and lactation, led to Belle's first excessive weight gain and other health problems. Josie's death was also a determining factor in her decision to fight prostitution and ignorance about venereal disease, campaigns she pursued through a long association with a citizen anti-prostitution group—the Committee of Fourteen.

Possibly because of her weight, Belle's fourth pregnancy ended badly. In February 1911, the baby's placenta separated prematurely from the uterine lining and began to bleed. Such an occurrence is often related to the mother's high blood pressure or to eclampsia. The attending physician probably did not dare perform the Caesarian required to save the baby, as the operation performed at home would have killed Belle. Baby "Judith" thus bled to death. "God ought to be spanked," cried four-year-old Miriam when told she had lost her sister. Three months later, Belle donated a new baby's outfit to the Council of Jewish Women's Lakeview Home for Girls.[5]

The family associated happy times with the presence of Charles, a soft-hearted, loving figure. Charles's favorite hobby was playing Gilbert and Sullivan tunes on the piano, the children around him, struggling to keep up with the words. This activity became a family tradition. Since he had a large studio on the third floor of the house, Charles often worked at home. In his last years he was showing renewed interest in sculpture and painting. The children remember when he called them up to his studio to see their first rainbow. On April Fool's Day, he made them roar with a tale of a cow on the porch roof. They remember "great romantic harmony" between Charles and Belle, and "excitement and joy" whenever he was at home.

Only one of Charles's letters to Belle survives, but it is enough to show his devotion to her and their children. Writing of Miriam's cold, he told how he kept her in but walked "our big boy" down to the schoolyard. "I am attending to your mail faithfully," he continued, "and looking after the kiddies lovingly, though no one can take the place of Mama." Not confining his letter to domestic matters, he wrote also about politics and business. He was pleased she had gone to a convention of insurgent Republicans in Chicago, "for when you tell me all about it I will know better how to insurge in harmony with the Republicans of the west." He reported on his talk with the chairman of the West Side Republican Club's Building Committee. The club needed new quarters, and since he had done the first building for them "sans fee," he hoped to get the new one "without competition." He also conveyed anxiety over money. Hoping that her "mission" (a money-raising quest of some kind) was succeeding, he ventured that it probably would "because we are always successful in getting money for *other people*."[6]

Only one of Belle's letters to Charles survives. It was written in 1907, four years after their marriage. They had endured painful and frightening childbirths, illnesses, family strains, and tragedies. Yet her love for him had grown stronger. "Dear," she wrote, "the days are very precious with you and sometimes it seems as if I'd like to hold each happy hour and keep it back from slipping past too soon." She continued:

> We have nearly come to four years since we ventured together and sometimes it is like four days—and again, four*teen* years. It is curious to think how each of us has become a vital part of the other and how the four years are just a steady growth of love and confidence. You are an inexpressible something to me that is just as real a necessity as the air I breathe and you always seem to be about me surrounding me with a cloak of love and devotion and "work-for-you" that is too beautiful to be told about. I only hope I am worth it all and that I do in some measure bring you a compensating help. You often say I do and if we can but both remember how something *might* take one from the other, and how *very* precious our life together is, we should stay so happy in one another.

She closed with the comment that the feelings she had just expressed "couldn't have been half so real or half so true, four years ago."[7]

The grim prophecy of this otherwise happy letter came true. Four years later Charles was "taken" from her. The letter suggests they knew their time together might be short. Such knowledge may have intensified their relationship, perhaps even strained it in ways the letter did not acknowledge. Still, their marriage seems to have been good, with domestic and outside interests energetically shared and mutually supported.

Charles's death followed a short illness. In November 1911, he caught cold during the building of the Yonkers Temple Emanu-El, which he had designed. The cold turned into bronchial pneumonia, and then pericarditis. On November 13, two days after their eighth wedding anniversary, his heart gave way. Grace Goodale was there, as she was in all family crises. Until then she had never seen Belle break "her steady calm in the face of trouble." According to her recollection, Belle "fought every inch of the way at the doctor's side and at the end held Charley in her arms saying over the Hebrew prayers with him." But

> when he was gone she stumbled into Walter [Lindner's] arms and slid into a heap on the floor with a moan. Then she said "I want my baby," and I ran and snatched Joe from [the nurse's] arms, and rolled him in a bath towel and brought him to his mother and she hugged him tight and rocked back and forth weeping. . . .

"Steady the Buffs, old man," Grace cautioned, quoting Kipling. "Bite on the bullet and don't let 'em see you're afraid." Half an hour later Belle had Grace telephoning about getting work for her to keep the family going. "That was my James," Grace concluded, calling Belle by the nickname of their youth.

The funeral, held at home, was private and simple. Stephen Wise, the controversial rabbi of the Free Synagogue, and Reverend Madison P. Peters, both friends, conducted the service. Julius Harder, Charles's partner, and a representative from the Institute of Architecture, spoke. Representatives of the Yonkers City Council and Municipal Art Commission, on which Charles had served the previous spring, attended. Wise quoted what Horace had said of Virgil, that he "never knew a whiter soul." At the cemetery in Tarrytown, Belle, "always one to see things through," remained until the last shovelful covered the plain coffin and flowers were spread on the grave. A workman picked up a loose bunch of violets that Grace had brought, tucked it in, and gave it a little pat. This touched Belle, who felt then and in the hard months that followed that Charles was with her, not in the way of "manifestations," but in her "heart and mind."[8]

Charles left his family with only moderate means. Belle and Charles's mother shared insurance policies totaling $13,000 and Charles's real and personal property. Belle received two-thirds. In 1907 Charles had served on New York's Building Code Commission but never collected his $5,000 fee; Belle's brother Walter, executor of the estate, began proceedings to get it but in the end had to settle in 1914 for only half the debt. There may have been a settlement from the architectural firm, and there may have been savings; nevertheless, Belle felt she had to earn money, and right away. She must have communicated this urgency to her oldest son, Carlos, then seven years old. When told his father had died, he went to his mother and said, "I'm only a little boy now, and you'll have to get the pennies for a while, Mama, but I think in ten years from now I can do it." As the oldest child, Carlos bore the heaviest burden. Trying to take the lost breadwinner's place, his young mind grew up too fast.[9]

For Belle, the future looked hard but not bleak. She was an accomplished, mature woman of thirty-four. Her gentle husband was gone, but his mother would go too, and that had to be some consolation. Years later, Oliver H. P. Garrett wrote a profile of Belle for the *New Yorker*. In it, he took special note of her reputation for self-control and calm even in provocative circumstances. "Only one person," he wrote, "so far as is known, ever cut through her reserve to the quick of her soul." That happened when

> Charles H. Israels . . . found her . . . and married her. Even then her beauty gave her unusual poise. But in this period of her life she encountered in an older woman one who could hurt her and throw her off her stride. Few know what passed between them but it was bitter and frequent: the short, heavy-set girl, with her Madonna-like face, standing flushed and deeply angry before the other, hot words, like lava, pouring through her lips from burning, undiscovered places within.[10]

This portrayal may have been exaggerated, but it can refer to only one person, Florence Israels. After Charles's death, Florence moved into a New York boarding house where she lived until her death in 1919.

Belle's parents took Florence's place. Esther kept house; Isidor—once again the patriarch—repaired watches brought in by neighbors. Social reformer Frances Perkins remembered dining at Belle's one Sabbath evening:

> Her father and mother were charming, lovely German-Jewish people who lived in the old style. . . . After her husband died, [they] made common household with her, I suppose to help her with expenses, and so forth. They had a house in Yonkers. I was invited there to supper on a Friday night once. All the Jewish customs prevailed. The father, the patriarch, was at the head of the table, with the children around him. The youngest child did something. . . . The candles were lighted. The old man put his cap on and his shawl and he said a special prayer. It was perfectly beautiful.[11]

Perkins was impressed with how well Belle managed her affairs after Charles died. Her success showed "what a good and orderly mind can do. She was a good housewife. . . . She just had that quality. That was all. She was self-made. She developed magnificent judgment."

Belle knew she could make her way in the field of social welfare. But even with help at home the strain of being her family's sole support oppressed her. Her children felt her absences more keenly now. First their father had left them, and now their mother was away from home "getting the pennies" every day and soon long into the night. Her son

Carlos once told a friend his keenest childhood memory of his mother: he was sick in bed, miserable with a fever; she came into his room, felt his head, clucked her tongue sympathetically, but then left.[12]

Belle Israels did not flounder too long after her bereavement but set herself early upon a career track. Consulting with friends, she planned a career in the field she knew best, social welfare. But at the outset of widowhood, Israels turned to writing for income. Appealing to friends, such as her old schoolmate Fannie Sax Long, to share their housekeeping tricks, she composed "how-to" articles for ladies' magazines. She even traveled to Fannie's home in Wilkes-Barre, Pennsylvania, to examine her efficient kitchen.[13]

Another contact, Lee K. Frankel, formerly of United Hebrew Charities and in 1912 a vice-president of Metropolitan Life Insurance, commissioned her to write a pamphlet on child care for policy holders. "The Child," thirty-one pages long, covered preparation of the home, pre-natal care, care of the baby up to one year, and care of the "runabout" baby. Israels consulted three doctors, two professionals from the Russell Sage Foundation and two from the insurance company, and seven books on pediatrics, baby care, and hygiene.

The pamphlet was aimed at the working-class mother who could not afford post-natal nursing care. It was written in simple language with explicit instructions on everything from preparing the birthing bed to home pasteurization of milk. Even so, it insisted unrealistically on constant attention from the medical profession. Israels believed that reliance on patent cures or a neighbor's advice often led to disasters that a doctor could prevent. Of a piece with this perspective was her firm stance against midwives. "Everything depends on the care that mother and baby get at this time," Israels wrote. "If anything goes wrong, it may mean death or a long serious illness that will make the mother too weak ever to work again. A DOCTOR OR THE HOSPITAL WHEN THE BABY COMES, BUT NOT A MIDWIFE."

This caveat was less an attack on the traditional world of female self-help than an expression of Israels's bias toward modern scientific professionalism. Midwife associations retaliated, writing to Metropolitan Life early in 1913 that their members would cancel policies unless the warning was excised. By mid-year, Lee Frankel was calling Israels's statement "too drastic," but justified on the ground that "eminent specialists" had approved it. By fall he conceded defeat and agreed to remove the statement from future editions.[14]

The chief focus of "The Child" was not doctors but the mother's responsibility for her child's welfare. Before baby arrived, the mother was to concern herself with two things: her own health and a meticulous cleaning of the baby's future environment. She should eat wholesomely, not starving but not overeating either. She should keep clean, rest, and wear loose clothing. On this last subject Israels commented: "Motherhood is so beautiful that it need never be hidden." In another note unrealistic for working-class mothers, Israels advised them to quit work, especially factory work, two months before delivery. "Having a baby is natural and normal," she wrote, but should be well prepared.

"Dirt means Danger," the pamphlet cautioned. Enjoined to use plenty of hot water and soap, boric acid, and boiling water to remove dirt and germs, the mother learned that failure jeopardized her family. Israels also made it clear that a mother cannot clean as she should unless she strictly regulates the baby's life. Feedings, begun at two-hour intervals and gradually increased to four, should occur at definite times; even the "runabout" baby should expect food regularly. The mother should enforce regular sleeping hours, never keeping the baby awake for the sake of parental enjoyment of late excursions. Parents must give up excursions until the baby is old enough to enjoy them. With "a little patience on the part of the mother," Israels advised, "the baby's life can be made to run as evenly as a little machine." The mother can then "count on the hours she will have for other work."

Israels completed her picture of efficient motherhood with other suggestions, some of them quaint now. Toilet training should begin at three months by holding baby over a chamber at regular times. While admitting that daytime training was seldom complete before eighteen or twenty-four months, she never questioned the early start. Pacifiers were "unclean" and made the infant "nervous"; they were strictly forbidden, as were thumb-sucking and the use of "soothing syrups" or paregoric to ease teething. The perennial remedy for digestive complaints was castor oil. Since colds came not only from infection but also drafts, babies should be kept off floors in winter. On the other hand, stressing an idea only recently returned to repute, nursing the baby was the mother's "most important duty." "She does her child a great wrong if she does not make every effort to nurse it," Israels continued, implying in her next statements that the mother who failed or refused to nurse would be responsible for a child's illness from cow's milk.

Writing for an insurance company, Israels necessarily concentrated on preventive care. Nevertheless, there was no discussion of the mother's or child's psychological and mental development. At Teachers College, she had studied psychology and as a young matron had attended and organized chapter meetings of the Society for the Study of Child Nature. This organization had grown out of women's concerns to understand their children not as "little machines" but as dynamic human beings. These influences do not show here. In commissioning Israels to write the pamphlet, Frankel may have had only narrow purposes in mind. The end result was a stark manual, softened by neither beauty nor joy. Worse, the pamphlet put a heavy burden on mothers. In summarizing her rules, Israels wrote: "The baby that has been well fed, has regular habits and plenty of fresh air gets well quickly and does not often get sick. The baby whose mother has not looked after these things, may be sick often, will take a long time to get well, and may die."[15] Maintaining an absolutely clean, regulated home environment was the mother's chore, and hers alone. The price of failure was illness, perhaps death, certainly remorse.

Directed at working-class mothers, Israels's imperatives seem heartless. Metropolitan Life had not hired her to write a polemic, but given her background as a social reformer it seems odd that she made no reference to the environmental causes of childhood mortality. Perhaps the domestic values she expressed in the pamphlet were so deeply ingrained that she believed all mothers would find them essential, regardless of the external conditions of their lives. From her own perspective, only a regular, controlled, and non-indulgent home environment could give her the peace of mind to express herself outside of it.[16]

Early in 1912, Belle Israels landed a regular job. At a salary of $3,500 a year, she became "Commercial Recreation Secretary" for the Playground and Recreation Association of America. This was the first occasion on which she united her volunteer work with a profession. Founded in 1908 by educators concerned about the lack of urban play space for children, the organization lobbied for playgrounds and supervised recreation funded by the state. In 1909 it hired Howard Braucher as executive secretary, a post he held for forty years. By 1910 he had expanded the association's focus to include the development of recreational resources for all ages. In doing so, Braucher attracted new adherents, including Belle Israels, then building her reputation in dance hall reform. Perhaps influenced by her, Braucher took a brief interest in commercial recreation. He formed a committee to investigate the issue and appointed Israels its secretary, thus giving her a forum through the association's congresses, regional meetings, and national publication, *Playground,* from which to argue the case for municipal regulation. But as soon as Israels turned to a new field in 1913 the Playground Association lost interest in the issue of commercial recreation.[17]

While working for the Playground Association, Israels kept up her contacts with other social reformers. Some of them had joined the anti-prostitution work of the Committee of Fourteen and invited her to join in February 1912. Israels responded eagerly, for she was convinced of the direct link between disorderly dance halls and prostitution. Unlike the playground movement, however, anti-prostitution was controversial. Its most avid supporters came from the middle and upper classes. No one openly opposed them: it was not possible to advocate prostitution publicly. But some argued that the trade was an eternal part of civilization that should be tolerated, if not regulated. Tolerationists believed that using prostitutes to satisfy male lust saved "decent" women from ravishment or excessive pregnancies. Regulationists argued that, since prostitution could never be abolished, society should confine it to districts, registering and inspecting prostitutes at regular intervals. Anti-prostitutionists countered by pointing out the double moral standard: why should women be registered and inspected and not men? Further, inspection could never prevent the spread of venereal disease. Even so, regulationism still had followers in the early 1900s.

There was yet another side to the debate, represented by those with financial interests in brothels—Tammany Hall ward leaders, police, and other officials using the excise system to exact protection fees. The profits of brewery, surety, and real estate companies also grew along with the saloons, gambling houses, and brothels. All of these naturally opposed crackdowns on vice. Others, believing that poverty caused prostitution, argued that only vast social and economic changes would rid society of the evil. Still others found the thought of such topics so disgusting that they shunned all talk of anti-prostitution and wondered at the motives of those who joined the movement.[18] Anti-prostitution thus evoked strong responses.

By joining the debate on prostitution, Belle Israels identified herself with a reform of the most controversial sort. Yet her decision to work through the Committee of Fourteen, which represented a moderate wing of anti-prostitution, indicated she was seeking a balanced approach. In 1912, the Committee included just the kind of people she had on her Committee on Amusements: lawyers, clerics, social reformers, settlement leaders, a publisher, and several manufacturers. Another factor that may have attracted her was the number of women members. After she joined, there would be six out of a total of twenty-two, a rather high ratio for the period. The original Committee had only one woman, Mary Simkhovitch, director of the Greenwich House Settlement. After 1912, Ruth Standish Baldwin, a clubwoman interested in probation work among blacks, Mrs. John M. Glenn, a Charity Organization Society leader, Mrs. Barclay Hazard, a municipal reformer, and probation worker Maud E. Miner had joined. John D. Rockefeller, Jr., a powerful figure in the war against sexual vice, had opposed putting women on committees concerned with the issue, lest they be exposed to unseemly topics. In 1910, while setting up a mayoral commission on white slavery, an evil which primarily victimized women, he had suggested to philanthropist Felix Warburg that women should serve only in an "advisory" capacity. In 1912, writing to real estate investor Allan Robinson, head of a citizen's committee to investigate the murder of a gambler, he had asked whether the committee ought not to be composed entirely of men.[19] Contrary to sharing Rockefeller's concerns, the Committee of Fourteen felt that women's presence in their midst was not only right but essential to the success of their mission.

All the Committee members had been involved in reform, acting through neighborhood, religious, and settlement associations. As lawyers, probation officers, and welfare workers, many of them had worked with prostitutes, whom they pitied rather than abhorred, saving their outrage for those who exploited them. They saw prostitution as a "business conducted for profit," "an elaborate system systematically fostered by business interests rather than a consequence of emotional demand."[20] Such "commercialized vice" corrupted the young that lived within its vicinity, sapped the foundations of the public trust by encouraging official collusion, and brought the terrors of a dread disease to the innocent, something Belle Israels knew all too well from her sister-in-law's tragedy. Part of the Progressive Era mentality was to discuss openly matters usually avoided by polite society.[21] The members of the Committee of Fourteen all agreed that the consequences of venereal disease should be exposed, and that society had to begin to think creatively about this social problem that few had been willing or able to tackle.

Thus the Committee of Fourteen should not be seen as prudes launching a moral crusade. They stressed traditional moral values about human sexuality, but not revulsion against modern values. They were responding to conditions that they thought made urban life intolerable and ruined the lives of innocents. They had some blind spots. They used the ambiguous term "disorderly" to condemn places of public amusement that failed to meet their standards of decency. This kind of blanket condemnation aroused charges of fanaticism, hurting their relations with those businessmen who were willing to cooperate. Further, no Committee of Fourteen member could imagine that a prostitute might refuse to leave her "life of shame." They ignored evidence that, in spite of its risks, prostitution offered an independent life style envied by many working girls.[22] But there were few middle-class Americans without similar blind spots. For over twenty-five years, the Committee of Fourteen sustained itself on the voluntary donations of hundreds of contributors, large and small. Its vision of a society without commercialized vice, of wholesome amusements for young people protected from the "moral contamination" of liquor and illicit sex, and of a government that would "hound to their undoing" the greedy businessmen whose cupidity was a "baser sin than the lust on which it preys," formed a cornerstone of the Progressive movement.[23] Belle Israels found this vision completely compatible with her own.

When first founded in 1905, the committee sought only to wipe out Raines Law hotels. The 1896 Raines Law allowed Sunday drinking in hotels of at least ten rooms. Taking advantage of the loophole, saloons bought or rented adjacent apartment houses and called them "hotels." Fake ones, the Committee of Fourteen charged, and soon transformed into brothels. An earlier citizen's group, the Committee of Fifteen, had called for the elimination of these hotels in its pioneer study *The Social Evil* (1902), but since this group disbanded after its study appeared, another formed to take action.[24]

The Committee of Fourteen's approach to the Raines Law hotels showed its compatibility with a reform program like Belle Israels's

Committee on Amusements. Beginning in 1906, the Committee of Fourteen secured an amendment to the excise law requiring hotel license applicants to prove that their buildings complied with the hotel code. This immediately eliminated about half the Raines Law hotels. Next, the Committee established liaisons with state and local authorities, offering advice on interpreting the law, establishing a clearinghouse for complaints against disorderly places, investigating them, and informing the proper agencies. The Committee also appealed to legitimate businesses to check carefully on the places they served. Surety companies, for example, supplied bonds to saloons applying for liquor licenses. The Committee asked them whether they really wanted to bond brothels. Similarly, the Committee told brewers that, if they failed to assist in suppressing vice, more stringent laws, higher license fees, or even total prohibition might result. Such methods, combining enforcement of existing laws, cooperation among public and private agencies, creation of communication networks, and appeals to the moral consciences of businessmen, were precisely those being used by Belle Israels in her dance hall campaign.

As it pursued its work, the Committee of Fourteen became aware of other issues related to commercialized vice. While pressing for higher bail and longer sentences for the owners of disorderly houses, it supported a Women's Night Court to expedite prostitution cases. Then the "white slave" agitation arose. Tales of an international traffic in girls, drugged, kidnapped, and forced into prostitution, brought calls for action by social reformers, welfare workers, and women's groups. In response, the Committee of Fourteen decided to attempt the total suppression of commercialized vice. Its first step in this direction was to increase its numbers, asking representatives from cognate agencies, like the Committee on Amusements, to join.[25]

Belle Israels accepted the invitation. She was already cooperating with the Committee of Fourteen. Frederick Whitin, its executive secretary, had joined the Committee on Amusements during 1908 and investigated complaints about prostitutes soliciting in dance halls. Israels later called on him to testify as an eyewitness at legislative or licensing hearings. In the fall of 1909, when Israels's committee faced a financial crisis, Whitin, seconded by Michael Davis of the People's Institute, advised her to expand the membership of the Committee on Amusements, including the names of wealthy patrons who would attract further subscribers.[26] In return, Israels provided the Committee of Fourteen with information on dance halls and prostitution, thus giving further weight to the cry for a city-wide anti-prostitution campaign.

The Committee of Fourteen and the Committee on Amusements, sharing common goals, thus nourished one another. Israels benefited from the organizational know-how of Committee of Fourteen leaders while providing them with the documentation they needed. She also helped them lobby for more money for excise authorities, a goal that concerned her because dance hall reform would fail without adequate support for inspectors. She helped get an ordinance that required hotels, saloons, and dance halls to post the names and addresses of their owners. Since maintaining vigilance in Albany and pressure on the Excise Commissioner were essential to Committee of Fourteen tactics, Israels did her part by following up on all dance hall-related prostitution questions. She supported the Committee of Fourteen idea of abolishing the Sunday drinking ban, a law that reformers argued drove liquor into back rooms where gambling and prostitution thrived. Later, as a member of the Committee's executive board, she participated in the group's attempts, ultimately futile, to develop a law aimed at a prostitute's client.[27] Finally, she served as the Committee's link to wealthy Jewish contributors. In 1913 for example, when Whitin was in need of more funds, he used Israels to approach financier Jacob Schiff. She allayed Schiff's fears that Whitin's group was duplicating work being done by Jewish groups on the Lower East Side.[28]

In the summer of 1912, the Committee of Fourteen placed Israels on an executive subcommittee responsible for its summer program. That summer, then, the first of her widowhood, she was actively pursuing enforcement of dance hall licenses, spreading the dance hall-reform message in other cities, doing freelance writing, and attending Committee of Fourteen executive meetings. In August, after the murder of a gambler named Herman Rosenthal precipitated a civic crisis, she became even busier in anti-vice campaigns and made her debut in party politics.

On July 16, 1912, Herman Rosenthal was to testify before a grand jury on police graft. He and Police Lieutenant Charles Becker, head of the vice squad, had had a falling out. Having cooperated for some time in Rosenthal's gambling operation, Becker, under pressure from Police Commissioner Rhinelander Waldo to show more arrests, had raided a Rosenthal game and arrested one of his nephews and two others, promising them quick release. But the grand jury indicted the men. Thinking Becker had betrayed him, Rosenthal went to District Attorney Charles Whitman to complain. When Whitman rebuffed him, he took his case to the press, contacting Herbert Bayard Swope, a young muckraking reporter at the *New York World.* The *World* was running a series on gambling and snapped up the story. The publicity forced Whitman to arrange Rosenthal's appearance before the grand jury. He never kept the appointment. In the middle of the night, after bragging to friends about his revenge on Becker, Rosenthal was gunned down in the street by four men who escaped in an automobile. Gangland murders were common enough in New York, but, when preliminary investigation exposed Becker's involvement not only in graft but also possibly in Rosenthal's death, the public cried out for reform.[29]

The Rosenthal murder had a long-term ripple effect. In the past, police scandals periodically ended in the election of reform administrations, but after one term the old crowd routinely got back into office.[30] After Rosenthal's murder, a similar chain of events was set in motion. It had a major impact on Belle Israels's life.

No less than five agencies launched investigations of the police in the summer of 1912. The Board of Aldermen created a committee that promised to expose, prosecute, and prevent police graft. District Attorney Whitman announced that his office, while pursuing Rosenthal's killers to the electric chair, would also investigate. Judge John W. Goff convened a special grand jury, and the Bureau of Municipal Research, founded in 1907 to study city government and propose administrative reforms, turned its attention to police-department reorganization. Finally, at a Cooper Union "mass meeting" of 5,000 concerned citizens, brought together "without regard to race, creed or party politics," a "Citizen's Committee" came into being.[31]

Among the signers of the "call" for the mass meeting were Belle Israels, Henry Moskowitz, financier Jacob Schiff, and lawyers Eugene Outerbridge and Raymond Ingersoll. At the meeting, these individuals and others protested the "treasonable connection between some members of the Police Department and organized crime." Israels was the sole woman speaker. She hoped that citizens would take this opportunity to "cure allied evils while . . . battling against gambling and graft." The dance halls, for example, were linked to police-assisted gangsters who profit from vice. Calling for a "constructive system of effort" that would give the city "the right kind of law," she ventured that discussing the social evil might offend "our old-fashioned, hypocritical, Puritanical conscience." But citizens must set aside their revulsion. "This is a time for plain talk." She continued:

> We are not men and we are not women; we are human beings. We need to care for and to look out for the weakest in the community, and this network that has simply shown itself on the surface is so infinitely more rotten underneath [than] anything that you have touched as yet, is so infinitely more vile in the degradation that it supplies to the holiest and most beautiful thing in the whole world, and I mean a young, innocent girl.

Concluding, she called for a city "so clean so straight and so wholesome" that everyone could walk the streets unafraid to enter a candy shop, cigar store, or dance hall.[32] Other speakers sounded the same theme: a city in which people die by the hands of men presumed to be protecting them was a city unsafe for anyone.

When the speeches were done, the assemblage resolved that the signers of the call form a Citizen's Committee authorized to increase its membership to thirty, solicit funds, engage counsel, and do whatever was necessary to "vindicate law and order." It planned to urge officials to pursue Rosenthal's murderers energetically, convince the Board of Estimate it should approve the $25,000 cost of the Aldermanic investigation, urge the Aldermen to be "unsparing" in their study of graft, and call upon those with knowledge of graft to step forward without fear of reprisal.[33] Of course, the Citizen's Committee could not guarantee their safety, a fact that underscored its lack of legal authority.

Even without such authority, the Citizen's Committee contributed over the ensuing months to the public debate about the constabulary's relationship to vice. It pressed officials to keep up the investigative work, and kept alive public interest on the question. The Committee numbered twenty-six members, of which Israels was one of only three women, and the only woman on the Committee's executive board. Her special assignment was to examine the "social evil."[34]

As chair of an executive board sub-committee that included Henry Moskowitz and Frederick Whitin, she brought together experts who had studied vice and solicited their views on how the police should best deal with it. She was especially anxious to consult Abraham Flexner, who had just returned from an exhaustive survey of the relations of European police administrations and prostitution. But John D. Rockefeller, who paid for Flexner's work through his support of the Bureau of Social Hygiene, argued that Flexner should be protected from all outside influences and therefore refused to let him meet with the Citizen's Committee.[35] Nevertheless, Israels was able to complete a study that turned out to be the core of the Citizen's Committee's final report.

Released on February 26, 1913, the report began by reviewing New York's chronic dissatisfaction with its police department. Corruption, always present, was now rife. The Committee recognized the folly of extensive legislation on "moralities," and therefore promised to avoid the topic of "vice repression." But in recommending police reorganization, it argued that the result would be an improvement in the city's moral standard.

First, the Committee attacked the city's Sunday drinking ban. This ban illustrated "the American habit of carving moral ideas into statutes, which represent aspirations and not practicable rules of conduct." Fundamentally law-abiding citizens ignored the ban, thus greatly enlarging the possibilities for extortion. Plainclothesmen entered bars on Sundays, bought drinks, then threatened arrest unless they were paid off. The Sunday ban had to go.

Turning next to prostitution, the Committee observed that the repeal of the drinking ban would automatically eliminate Raines Law hotels. As for the various evils associated with sex, no "anti-toxin" existed, but citizens could insist that the police enforce the law. Thus far, police attempts to control vice had only corrupted them. The Committee therefore recommended that the police maintain peace and order, leaving matters of vice to a "morals commission" appointed by the mayor and serving without pay. The members of such a "Board of Social Welfare," as the Committee euphemistically called it, should be "representative citizens, whose motives will be above suspicion." They would work closely with "bodies of citizens organized for the betterment of civic conditions." In other words, the Board, which would oversee "sex evils," gambling, excise laws, and places of commercial amusement, would be the enforcement arm of the moral reformers who had been active in the city since the turn of the century.[36] The Citizen's Committee recommendation was no less than a bid for power by the righteous.

For the Citizen's Committee, and this included Belle Israels who signed its report, vice was a "problem" to be analyzed and solved. For others, it was a question of law and order. Emory Buckner, counsel for the Aldermanic investigation, said privately that the Citizen's Committee plan for a morals police was a remedy worked out "at the lunch table," a "patent nostrum" proposed "without any adequate study of conditions here or of conditions abroad where Morals Police are employed." In one sense, he was right. Joseph P. Cotton, the Citizen's Committee counsel, had hit upon the idea as early as the previous October after "thinking about" police affairs and "'theorising' about them on the basis of certain general facts."[37]

The Citizen's Committee report was totally compatible with Belle Israels's idea about solving social problems through cooperation between public and private agencies. She assumed that all such agencies were striving for the general good of the community. With "cooperation" her theme, one that had worked in dance hall reform, the idea of trying new, albeit "academic" or even "radical" solutions did not worry her. In fact, to her mind, "radical" ideas might facilitate cooperation by freeing agencies from old ways of doing things. This approach to reform, one that rose above both experience and politics, characterized Israels's work over the next few years. She would not abandon it easily.

The city's Board of Aldermen issued their own report in June, unequivocally opposing a special morals department. The state legislature held hearings and introduced a bill that created a "board of social welfare," removing from the police all jurisdiction over gambling and prostitution. As was his right in state legislation regarding New York City, Mayor Gaynor vetoed it. The idea died.[38]

Although the city dismissed the Citizen's Committee plan, the movement that the group represented found other forms of self-expression. It became a center around which a "fusion" movement gathered that ousted the Democrats in the next city election. By the fall of 1913, the Democratic party was in disarray. Police Lieutenant Becker had been found guilty of planning Herman Rosenthal's murder. As Mayor Gaynor persisted in defending his police department, his candidacy for reelection faded and Tammany Hall, which had once supported him, nominated someone else for mayor. Gaynor hoped to run as an independent but died in September. Meanwhile, the impeachment of Democratic governor William Sulzer further disgraced the party. Reformers began to focus their hopes on a young lawyer and alderman named John Purroy Mitchel, who had been prominent in the Aldermanic investigation of the Rosenthal murder. Ultimately victorious, he brought social reformers such as Henry Moskowitz into city government for the first time.

The salary Henry received as president of the Municipal Civil Service Commission allowed him to propose marriage to Belle Israels, whom he otherwise would have been unable to support. Others involved in the aftermath of the Rosenthal murder also experienced career boosts: Samuel Seabury, Emory Buckner, and Henry H. Curran all went on to distinguished legal careers; Charles Whitman became governor in 1914 and held this post for two terms until ousted by Alfred E. Smith.

Belle Israels's career flourished, too. A few days before her Cooper Union speech exhorting citizens to clean up the police force, she had joined her first political organization, the Yonkers Progressive party. Unlike other parties, the Progressive party, newly born of insurgency against the Republican old guard, encouraged women's participation, at least to a limited extent. Israels's dance hall reform and involvement in the Cooper Union mass meeting must have enhanced her reputation in Yonkers's Eighth Ward. Ward party members elected her associate leader of the ward's second district, and a member of the party's City Committee. As that body's only woman member, a singularity she would often claim, Yonkers's newspapers gave her considerable notice.

In praising Theodore Roosevelt, the party's national leader, Israels said she supported him because of his "wonderful experience" as governor and president, and because the poor needed his help. Of all the candidates, only Roosevelt was sufficiently acquainted with the efforts of "our social workers" to know how to aid them.[39] For Israels, then, ridding the Republican party of its bosses and fighting the trusts were less compelling reasons for joining than Roosevelt's promise of social welfare programs. She also supported Roosevelt because he favored votes for women. Israels was a late convert to suffrage. She had thought that women did not need the vote in order to accomplish what they envisioned. "It took comparatively little practical experience with legislation," she recalled, "and a suffrage meeting addressed by two speakers whose point of view I respected, to make me understand that I was not the kind of person that the anti-suffragist represented . . . , that after all I was making a mistake and I belonged in the rank of those who were struggling for justice."[40]

In September, Belle Israels was among the local delegates who boarded trains to Syracuse for the State Progressive convention. John Kingsbury, a former teacher and charity worker, now Yonkers Progressive party chairman and a fellow Eighth Warder, was already there as a member of the platform committee. So was Henry Moskowitz, on the same committee and also acting as chairman of the draft resolutions subcommittee. He had performed similar services at the national convention the previous month. Accompanying Israels, among others, was Homer Folks, another Park Hill resident and secretary of the State

Charities Aid Association.[41] And there were other social workers and reformers whose paths she had already crossed or would in the near future: Paul U. Kellogg of *The Survey,* Mary E. Dreier, president of the New York Women's Trade Union League; Frances Kellor, investigator of immigrant social conditions; Maud Nathan of the New York Consumers' League; and Walter Weyl, former settlement worker and author of *The New Democracy.* College professors, including Joel Spingarn, E. R. A. Seligman, Carlos Alden, Samuel McCune Lindsay, and George Kirchwey, all of whom Israels either knew personally or through their writings, were present as well. They rubbed shoulders with more strictly political elements in the state Progressive party movement: Roosevelt followers such as Oscar E. Straus and William Prendergast; followers of another former state governor, Charles Evans Hughes, such as William H. Hotchkiss, state Progressive chairman and strong contender for that party's gubernatorial slot; and businessmen George Perkins, Frank Munsey, Tim Woodruff, and Francis Bird, who had been shut out of the Republican party's inner circle.[42] Israels had little or no contact with the political men, but instead took her cues from her social worker friends, including Henry Moskowitz, who at the convention's start was spoken of as a possible candidate for Secretary of State or State Treasurer.[43]

The national progressive movement had been brewing for almost two decades. In New York, one of its chief sources was the movement for civic reform, which took shape as a protest against machine politics. The machine was Tammany Hall, the well-organized Democratic club that controlled the party through patronage and which kept returning to power in spite of periodic efforts to oust it. By the turn of the century, however, New York's woes had exceeded any single group's ability to solve them. Dirt, noise, congestion, dangerous streets and flammable factories, decaying and filthy housing, the spread of commercialized vice, unregulated monopolies controlling public utilities at the consumers' expense, labor violence and—at the foundation of the rest—a weak governmental structure seemed incurable as long as any party machine interested chiefly in self-perpetuation held sway.

"Civic reform" in New York came to mean "anti-Tammany." Reform groups such as the Citizens' Union, the National Municipal League, the City Club, the Bureau of Municipal Research, and independent political clubs attracted adherents who differed politically but were economically and socially homogeneous. There were Republicans, of course, rebel Democrats, and independents from the ranks of the recent college graduates, professionals, social welfare workers, and the clergy. All relatively young and middle class, these men—and at first they *were* all men—sympathized with the plight of the laboring classes yet feared their potential for creating a social revolution. On the same count, even as reformers abhorred the injustices perpetrated by bankers and industrialists, they relied on these same interests for their own financial and professional success. Caught between radicals and conservatives, civic reformers suffered the contempt of both as well as of the professional politicians whose power they sought to destroy. Politicians called them "goo-goo's," a derisive term for "good government" men.

During the early 1900s, New York's civic reformers increasingly became identified with the broader causes advocated by social and welfare workers—pure food and drugs, child labor, factory safety and sanitation, workmen's compensation, tenement codes, utility rates, public recreation, and others. None of these goals could be realized without the cooperation of politicians. Civic and social reformers thus began to join forces to cleanse the political system of those who had won power through party connections instead of merit; to change administrative methods so that efficiency, economy, and expertise would be brought into government; and finally to institute programs intended to correct the social and economic imbalances of modern industrial times. These were the aims of New York "Progressives" in the election year of 1912. They fought a losing battle at the polls, but the campaign they waged awakened American voters to new political possibilities, many of which would be fulfilled in later decades.[44]

As the national election approached, the Republican party was torn by controversy. Insurgency against the party's old guard had arisen in the Midwest. Roosevelt factions squared off against Taft factions, the former accusing the latter of corruption, obstructing reform, and succumbing to the "special interests" of big business. In New York State, similar controversies raged, fueled by bribery scandals among the party's state legislators and factional fights over policies such as the direct primary, tariff, and income tax. In 1910, moderate reform forces had captured the Republican gubernatorial nomination, running Henry L. Stimson, then United States District Attorney for New York. Henry Moskowitz had worked in that campaign, along with his friend Felix Frankfurter, Stimson's assistant. But Stimson lost heavily, bringing down with him Republican control of the state legislature and twelve congressional seats. In 1912, after Theodore Roosevelt declared himself a presidential candidate on the Progressive ticket, reform forces in the New York State Republican party finally broke with the old guard, and organized a new party by joining with other reform forces in the state. The result was the gathering Belle Israels attended in September, her first political convention.[45]

She played only one documented role, seconding a motion to nominate by acclamation the party's surprise candidate, Oscar Straus. The convention had been split between Prendergast and Hotchkiss, the latter supported by social workers and professors. Straus, Roosevelt's former Secretary of Commerce and Labor, and the convention's temporary chair, had declined to be a candidate. Finally, a delegate's rousing speech in his favor ended in a seventeen-minute demonstration by a crowd overjoyed at the prospect of not having to choose between the two major contenders. In Belle Israels's little contribution to this event, she once again stressed the social worker's theme: "It seems almost too good to be true that we should have the man who for years has stood as champion of the cause of the men and the women, and above all the young girls and the little children who are entrusted to our keeping."[46] For her, the fatherly, philanthropic figure of Straus, a wealthy German-Jewish businessman, fulfilled her idea of government extending the care of the family to society at large.

There were 1,500 delegates to the New York Progressive party convention, only 10 percent of them women, despite the promises of "full equality." And this was the party that most welcomed them. The 10 percent represented social workers, business women, club officers, society and suffrage leaders, even young girls and housewives. They were "purposeful," journalists said, animated by "intense conviction" and the spirit of the Crusades. Marching in the muddy streets, they sang "Onward Christian Soldiers," taking joyous advantage of their new opportunity for political self-expression.[47]

Despite the enthusiasm of Progressive party women, Straus lost in December. Party weakness upstate and Straus's lackluster personality gave the Democrats a clean sweep. Israels's own ward went for Progressives in all but the state assembly race, but elected only an alderman, and not because of any special leadership on Belle's part: Yonkers's Eighth Ward was already known as a "hotbed of Progressive voters."[48] Henry Moskowitz had tried to organize Manhattan's Lower East Side for Straus but found its residents unresponsive to a rich German Jew from uptown. Moskowitz himself ran for the 12th Congressional seat but also lost. These experiences were only temporarily discouraging. Belle Israels and her friends had not dared hope for success; they saw the election as a chance to educate the public and to tell legislators that a substantial body of voters favored reforms.[49]

For New York Progressives, getting men into office was, of course, desirable. But a separate resolution passed at their convention showed that they set their sights more on programs than on election victories. This resolution established a "legislative committee" to draft bills or amendments that would carry out their platform pledges, and to present them before the legislative and executive branches of government. According to William L. Ransom, an upstate reform lawyer and one of the resolution's framers, Belle Israels "took the greatest interest in the committee idea of the convention, and said that it represented the most important action taken by the convention." Writing to Straus on election day, he submitted names for the committee, including that of Henry Moskowitz for Manhattan. Believing there should be a woman on the committee, he suggested Belle Israels as one of four women he thought suitable.[50]

Israels's interest in the legislative committee is characteristic of how she saw her own relation to politics. The legislative committee represented a break with professional politics in which getting a man into office to reap and dole out the spoils was the highest priority. For her, winning or losing mattered less than the success or failure of a reform program. Later in her career, as Al Smith's strategist, she would necessarily show more interest in election-day tallies.

In her first year as a widow, Belle Israels could count many achievements. She had earned a national reputation in social reform, deepened her knowledge of social issues, and entered party politics for the first time. Each task she undertook extended the previous one. She had begun with her own home experience, writing articles on household economy and the child care pamphlet. Hired by the Playground Association, she earned an income for doing what she had done without pay before. Because she believed that bad dance halls led girls into prostitution, she joined the Committee of Fourteen, agreeing with the view that fallen women were innocents exploited by business interests. Through this group she learned more about the politics of social reform and the inner workings of law enforcement. Her service in the police investigation after the Rosenthal murder provided further insights along these lines. It made sense to enter politics, the ultimate arena of social change.

By the time 1912 drew to a close, she had a much clearer idea of the direction her life would take. If, as she believed, progress occurred when diverse groups bridged the gaps among them and worked together on social problems, then society needed people like her to initiate and facilitate communication. Her specialty seemed to be the development of new perspectives on old or recurrent problems. She carried this image of her role into her next field of endeavor, labor relations. There, moving away from the peripheral concerns of working-class life, such as the uses of leisure time, she attacked the more central issues of workplace conditions. Over the next three and a half years, she would face great professional challenges. The lessons she learned led her away from reform and back into party politics.

Had Charles Israels lived, Belle would probably have continued as a volunteer on a slew of committees. She might never have addressed issues beyond those of young girls' recreational resources. When Charles died, social service as a means of self-expression gave way to social reform as a professional career. In the ensuing years, she still volunteered to aid causes such as dance hall reform, suffrage, and anti-prostitution. But professional goals now took priority. As she remarked to her daughter Miriam years later, contrasting herself to Henrietta Franklin, Miriam's future mother-in-law, she herself had long since passed "the committee stage."

CHAPTER 5

Apostle of Industrial Peace

We cannot go back to savagery in industry, whatever it costs to go forward.

Julius Henry Cohen
Law and Order in Industry (1916)

March 25, 1911, was a date few New Yorkers of the time would ever forget. On that day, a fire broke out at the Triangle Shirtwaist Factory. It started on the eighth floor and then spread to the ninth and tenth. Some workers escaped down the elevator or across the roof. Others reached emergency exits, but they were locked from the outside or led to dangerously weak fire escapes. Employees burned or jumped to their deaths. The final count of the dead exceeded 140, most of them young immigrant girls.[1]

Conditions in New York's garment industry were chaotic and sordid. Workers, primarily recent immigrants from southern and eastern Europe and from Russia, were poorly paid and often mistreated. Women suffered special indignities. Paid less than men, they also endured sexual and psychological abuse from employers and foremen. Since the trade was seasonal, workers had no job security. Worse, while larger shops boasted safe, sanitary conditions, gross violations of lighting, ventilation, and fire codes occurred elsewhere. The fire at the Triangle surprised no one who knew about sweatshop life firsthand.

Since the late 1890s, unions had been trying to improve conditions but could not win the right to collective bargaining. Lower East Side social workers, such as Henry Moskowitz and Lillian Wald, tried to mediate between German-Jewish employers and their employees. The Consumers' League, Women's Trade Union League (WTUL), and other groups also agitated for change. The former called consumers' attention to the conditions under which goods were made, and developed a system whereby approved factories sewed a special label into their goods. The WTUL promoted unionization among women workers, who were harder to organize than men, and raised money for strike pay and picketed alongside striking workers.

From the employers' point of view, conditions were not much better. Cutthroat competition drove prices down, forcing shops to skimp on essentials or to cut wages, which in turn led workers to retaliate by stopping their machines or striking. Employers then locked them out and hired others, or closed down altogether. Employers' associations formed to control competition and to counter the unions, but industrial warfare kept on. Its continuation meant loss of income for all sides.[2]

To complicate matters, the garment industry had three related yet independent branches. Until 1919, when lighter dresses became fashionable, the cloak-, suit-, and skirtmakers dominated the trade; the dress and shirtwaist industry came second; and the so-called "reefer" trade (housedresses, kimonos, corsets, and children's wear) third. What happened in one segment of the trade affected other segments, but did not rule them. Hence, a strike in the dress and waist trade meant cloakmakers might strike, too, but an agreement in dress and waist would not necessarily affect cloakmakers.[3]

In the fall of 1909 a major battle broke out in the dress and waist trade. The Triangle Shirtwaist Factory was at the center of it. Claiming a slack season, it laid off workers sympathetic to the new Local 25 of the International Ladies' Garment Workers' Union (ILGWU) and immediately advertised for new help. Local 25 declared a strike; a general strike in the trade soon followed. Some employers settled quickly, accepting workers' demands for collective bargaining. Larger firms held out and formed the Dress and Waist Manufacturers' Association (DWMA). At its first meeting on November 27, Samuel Floersheimer, owner of a large dress concern, condemned the unions and charged that a trade agreement with them was not "worth the paper it is written upon."[4] Eventually, public opinion forced the association into negotiations, but these broke down, and one by one the firms settled with their own employees. By mid-February 1910 the union declared the strike over, its victory only partial. As a result of the strike, Local 25 now had a large membership, but still lacked the right to bargain collectively.

In July 1910 the cloakmakers' union organized its own, more carefully orchestrated strike. The manufacturers' association in the cloakmaking trade was counseled by lawyer Julius Henry Cohen, a proponent of industrial peace. The association agreed to address grievances and to improve factory conditions but refused to yield other managerial rights. During the often violent months ahead, various outsiders struggled to make peace. They feared that continuing warfare would drive the garment industry out of the city, depriving immigrants of their livelihoods. Peacemakers, asserting that the industry could govern itself, given the proper spirit and agreements, sought not only an end to this particular strike but a permanent peace.

Two of the peacemakers were lawyers: Louis Marshall from New York, a German-Jewish philanthropist, and Louis Brandeis from Boston, fresh from his triumphs in progressive causes. Brandeis was brought in by department store owner A. Lincoln Filene, another Bostonian and an innovator in labor-management relations who was anxious for stability in the industry. Three young Lower East Side social workers acted as intermediaries: Paul Abelson, Meyer Bloomfield, and Henry Moskowitz. Conferring with Cohen, counsel for the cloak manufacturers, and Meyer London, a socialist lawyer representing the ILGWU, the peacemakers devised the first "protocol," an agreement that settled specific grievances and also announced general principles for future labor-management relations.[5]

Signed in September 1910, the protocol established a fifty-hour week, double pay for overtime, and a minimum wage. It raised the number of paid holidays, and required shops that charged workers for electricity to provide the power free. The protocol left the fixing of piecework prices, disputed at every change in style, to representative shop committees, and declared it illegal both to take work home and to subcontract work to shops where prices and standards were lower.

The protocol also established Brandeis's concept of a "preferential" instead of a "closed" union shop. In return for union promises to keep members under strict control, employers would "prefer" union members when distributing piecework or hiring. Finally, the protocol outlawed strikes and set up two mechanisms to prevent them: a Joint Board of Sanitary Control run by a medical doctor, who would inspect shops and supervise sanitation and safety measures; and a grievance system to settle worker and employer complaints.[6]

The dress and waist trade took more than two years to cooperate at the same level. About a year after the 1909–10 general strike, individual shop agreements expired and union membership fell off. Since general strikes yielded members, some unionists proposed that the ILGWU call one. The fire at the Triangle in March 1911 intensified their demand. But the ILGWU membership was not ready. They no longer had faith in spontaneous strikes and wanted to organize first. A general strike was not authorized until June 1912 and did not take place until January 1913.

Meanwhile, department store owner A. Lincoln Filene had convinced Samuel Floersheimer of the Dress and Waist Manufacturers' Association that accepting a protocol in his trade would be better than another general strike. Early in 1912, Floersheimer revived the association, enrolled thirty manufacturers, and made two appointments: Julius Henry Cohen as counsel and Walter H. Bartholomew, head of another merchants' association, as manager. Beginning in November, secret conferences between the ILGWU and DWMA forged a protocol similar to the cloakmakers', with one important difference. Instead of leaving piecework prices up to shop committees, the dress trade would settle them by "shop tests." Worker and employer representatives would pick "average" workers to sew pieces of new garments. The time these workers took to finish the pieces, multiplied by the minimum wage (then thirty cents an hour), would be the piece price. In case of disputes, a "wage-scale board" would render decisions, meanwhile conducting an industry-wide investigation of piece prices.[7]

A general strike in the dress and waist trade was called for January 15, 1913. But for union leaders it was merely an exercise to show how well they controlled their members. The next day the dress manufacturers issued a statement urging workers to join the very union they had reviled a few years before. Two days later, union and association signed a protocol of peace.[8]

Not everyone was happy with these events. Socialists condemned the union for cooperating with its class enemies. Some smaller firms, after organizing their own protective association, accused the union of taking a $10,000 bribe from the DWMA to call off the strike. (They later retracted the charge.)[9] But the protocol framers believed that energies that had once been channeled into class conflict would now work for the benefit of all. They knew the protocol was not a panacea. They saw it rather as a major advance over "anarchic" and "irrational" methods for resolving labor disputes. They also perceived themselves as having modernized the industry, bringing efficiency where chaos once ruled.[10] As such, the protocol framers soon clashed with conservative forces on both sides of the industry they hoped to reform.

For Belle Israels many of these events unfolded against a backdrop of personal tragedy. In 1911 she lost both her last baby and her husband. After the Triangle fire, she took part in protest meetings, the largest of which was held on April 2 at the Metropolitan Opera House. Its purpose was to raise money for the fire victims' families and to decide on a course of action. Henry Moskowitz was one of the speakers. As a representative of the public on the Joint Board of Sanitary Control in the cloak and suit industry, he had read surveys of fire conditions in the garment shops. The Triangle fire, he said, dramatized what the public already knew and had done nothing about. Another speaker was Rose Schneiderman, a garment worker who, with the aid of the WTUL, had been trying to organize women workers. She excoriated the crowd, made up mostly of social reformers, philanthropists, civic leaders, and clergy, for supporting the unions so weakly that workers' cries for safe conditions had gone unheard. "We have tried you good people of the public and we have found you wanting," she whispered. The crowd strained forward to hear. Fighting tears, she said, "The life of men and women is so cheap and property is so sacred. There are so many of us for one job it matters little if a hundred forty-three of us are burned to death."[11]

How did Belle Israels react to this chastisement? Dance hall reform must have seemed trivial next to the life-and-death issues Schneiderman raised. The Council of Jewish Women, with whom Israels identified most closely, had kept aloof from labor questions. Many of its members had garment factory owners in their families. During the great strike of 1909, women shirtwaist workers had addressed the Council, asking for support. Council minutes show no response to their plea.[12]

Belle Israels joined a delegation of citizens who went to Albany to press Governor Dix for action. The citizens had formed a "Committee of Safety," which philanthropist R. Fulton Cutting financed with a $10,000 grant. Henry Stimson became its president; Cutting, Henry Moskowitz, and Mary Dreier of the WTUL sat on its board. Frances Perkins, then secretary of the Consumers' League, and an eyewitness to the fire, became executive secretary. Arguing that factory regulation and inspection were inadequate, they asked the governor to investigate factory conditions throughout the state. During the years that this investigation took place, the Committee of Safety served as a pressure group and kept information on fires and fire prevention before the public.

Two powerful Democrats, Assemblyman Alfred E. Smith, chair of the Ways and Means Committee, and State Senator Robert F. Wagner, president pro tem, co-sponsored the bill that set up the New York State Factory Investigating Commission. Wagner became chair, Smith vice-chair. Between 1912 and 1914, the commission sponsored thirty-six bills on fire regulations, working hours for women and children, factory sanitation, organization of the Labor Department, and more. Belle Israels observed the commission closely. It sensitized her to the working conditions of the women whose moral condition had previously been her greater concern.[13]

Thus the Triangle fire led to the reeducation of two people whose paths had not yet crossed, but who would later work together for over a dozen years. Al Smith, until then a protégé of Tammany Hall without much knowledge of conditions outside his Lower East Side district, would travel the state and learn the broader picture. This experience, and the tutelage he received from social reformers such as Frances Perkins and Mary Dreier, convinced him that government must sometimes interfere to make conditions more humane. Belle Israels, already of the view that government had to abandon laissez-faire policy toward public recreation, refocused her energies from the peripheral realms of working-class life to its center. Dance hall reform and working girls' vacations would still appeal to her "mother heart." But when in January 1913 a job as grievance clerk opened up in the Dress and Waist Manufacturers' Association, she leapt at the chance to have a direct impact on the quality of industrial life.

The protocol grievance procedure was the linchpin of the entire system: without it, unresolved disputes would drive the industry back into anarchy. Modeling their staffs on those in the cloak trade, the DWMA and ILGWU created offices to handle worker and employer complaints. As "chief clerk" of its "Protocol Division," the union elected Solomon Polakoff, a Russian immigrant who had long been active in unionism. The association appointed its general manager, Walter Bartholomew, as chief clerk of its "Labor Department," and on the recommendation of Julius Henry Cohen, who may have gotten Belle Israels's name from Henry Moskowitz, hired her to assist Bartholomew at $4,000 a year. Since 84 percent of the dress work force was female, the hiring of a woman clerk seemed imperative.[14]

Dress and shirtwaist manufacturers were notorious for resisting the ideas of reformers. But Israels thought that some employers were well-intentioned, and that others could be taught that the cost of strikes was higher than that of constructive reform. As she explained in a 1915 speech to women workers, when the employers, knowing of her "affiliations with the girls" gave her "almost absolute autonomy" in policymaking, she felt she could do "much good" as their "industrial diplomat." Indeed, she could hardly resist this chance to mediate between her ethnic confrères, the German-Jewish employers, and her immigrant coreligionists.

Apart from a desire to do "good," she also believed in the protocol idea and in herself as uniquely suited to working for it. In her speech she explained that the protocol had come about because employers and workers wanted peace. In return, each side gave something up. Workers gave up the strike, taking complaints to grievance instead of the streets. Employers gave up being "masters" in their own houses, hiring and firing at will, without heed to conditions, and surrendered to joint control some cherished managerial prerogatives. It is right, she said, that this development of "industrial democracy" took place. But, she warned, democracy has a price. Groups strong enough to bargain collectively for their rights must renounce "selfish interest for the good of the individual in order to preserve the good of the whole mass."[15]

Israels argued that the "selfish interest" workers had to renounce was the right to be non-union. As she well knew, most young women hoped to leave their jobs as soon as they married, and thus resented paying dues, however small, for benefits that seemed intangible unless there was a crisis. Others, content with conditions in the "better" shops, refused to pay weekly dues of sixteen cents to help "sweated" sisters elsewhere. Yet, Israels warned, the protocol will fail unless workers agreed to union control. On the other hand, in her view unions, run exclusively by men, did not know how to attract and retain female members, and they must learn how to do so. Further, they must renounce the "old-fashioned masculine idea" that saw unions as "fighting engines" that ran down unless powered by constant "little victories in the shop." Instead of trying to "hammer things" into the employers, she advised, unions should "study out the workers' side" in the whole industry, get the facts, and make intelligent, rational demands. When unions were convincing, she declared, business would listen.

She saw no less responsibility for employers. They must apply the preferential shop clause and cease discriminating against union activists. They must abide by wage and hour rules, maintain health and safety standards, submit every dispute to mediation, support the association financially and serve on its committees, and pay back wages if found guilty of discriminatory discharge or paying under scale. They must do this, and more, no matter how bitter the pill, because industrial peace was cheaper than war. The employers were surely not "angels," she avowed, and some were more difficult than others. But she planned to "train" them to meet their new duties, and predicted more management concessions if the union kept its side of the bargain.

Belle Israels Moskowitz believed that the protocol system of mutual sacrifices for mutual gains would work, but not without enforcement, which was a job for professionals trained in fact-gathering and social analysis and possessing a knowledge of the law. They should also be free of the "masculine" need for "little victories" in the shop—*in fine,* persons like her. Although not formally trained, she felt her practical experience as a social reformer had qualified her. Being "sympathetic to the girls," she believed she could be "fair" to them while defending their bosses "just so far as they are right."[16] Despite her preparation and attainments, sustaining this mediating role was harder than she anticipated.

In the dress and waist trade, the protocol grievance process worked as follows: when an employee or employer filed a complaint, Polakoff and Bartholomew assigned deputy clerks to go to the shop. There the deputies tried to settle the dispute; if they failed, the chief clerks reinvestigated as soon as they could. If they too failed, the case went to a grievance board composed of equal members of manufacturers and workers. If this board deadlocked, a permanent arbitration board of three members agreed upon by both sides made a final ruling. As in the cloak industry, Louis Brandeis chaired the dress arbitrations until he became a member of the Supreme Court in 1916, when Judge Julian Mack of Chicago succeeded him. Henry Moskowitz served as the board's clerk and kept the chair up to date on developments in the industry. He kept this post until fall 1915, about a year after becoming Belle Israels's second husband.[17]

The grievance process worked well. During its first year it handled almost five thousand cases, 90 percent of them union complaints. Of these union cases, the deputy clerks resolved over 80 percent; the chief clerks, about 14 percent; and the wage-scale and grievance boards, about 5 percent. Of the many fewer cases initiated by the manufacturers, the deputy clerks adjusted over 95 percent. But the process took its toll. It was time-consuming, and emotionally and physically draining. The staff worked long and random hours, including weekends (most shops were open either Saturday or Sunday). Joint boards, staffed by union and association members with other jobs, could meet only after 5:00 p.m. These meetings, at first irregular, soon became weekly and sometimes continued over two evenings. One meeting in February 1914 went on until 2 a.m., and most lasted until 10:30 or 11:00 p.m. As the hour grew late, tempers flared, for the emotional content of the mediations was highly charged. Referring to his "breakdown" during the summer of 1913, chief clerk Bartholomew confided to Brandeis that his job had aged him five years.[18]

One can only imagine the toll on Israels and her family. As Bartholomew's assistant, she went to shops for on-site adjustments several times a day. In addition, she attended all grievance board, wage-scale board, and DWMA executive committee meetings, taking part in special investigations, inspections of recurrent problem areas, and countless informal sessions with associates. After each investigation, she and her union counterpart compared notes and issued joint memoranda. During all of this, while mediating between workers and employer, she was expected to defend the employer's rights under the protocol. Given the difficulty of performing conciliatory and advocacy roles at the same time, it is no surprise that she failed to please her employers whenever she pleased the union. In the summer of 1913, shortly after dispensing with Cohen's legal services, the DWMA fired her.[19]

The manufacturers fired Cohen, with whom Israels was closely allied, when he sent them a bill almost twice the size of the DWMA's annual budget of $48,400. To make matters worse, Cohen had demanded an annual retainer. The manufacturers were already angry at him for inadequately preparing them for the ramifications of the preferential shop. His large bill was the last straw.

While the exact circumstances of Israels's ouster are unknown, the most likely cause was the high proportion of grievance cases settled in the union's favor. Of all cases filed by the union, 62 percent were settled in the union's favor as against only 9 percent for the association (the balance were compromised, dropped, or withdrawn); of the association complaints, only 30 percent were in its favor, whereas 7 percent of them were settled in the union's favor with 44 percent compromised and the rest dropped or withdrawn. Since settlements involved reinstating discharged workers or paying back wages, employers found them humiliating and costly. They therefore sought a more strict interpretation of the protocol and fewer adverse rulings. To these ends, Bartholomew replaced Israels with three male, non-Jewish clerks from law school. Their legalistic approach to grievances immediately antagonized the union. As Polakoff complained, the first clerk "didn't know how to handle Jewish girls"; the second knew nothing about the garment industry; and the third was in office only eleven days before cross-examining an experienced union clerk. Harking back to the early protocol days, a woman union clerk opined that with Belle Israels "we used always to patch things up," but now "we don't get anywhere."[20]

Bartholomew and Israels held divergent views of labor mediation. From her perspective, Bartholomew's "masculine" approach viewed victory as more important than conciliation. Her approach was "feminine." She sympathized with the "girls," but in defending employers' rights tried to bring both sides together.[21]

Her vindication came soon. Relations between union and association deteriorated rapidly. In fall 1913 the DWMA accused the union of failure to prevent work stoppages, refusal to discipline members for protocol violations, and unwillingness to promote efficiency in joint functions. The ILGWU responded that it would not treat workers as "inanimate things," forcing them to obey rules that employers violated. The key issue was the preferential shop. Since the association could not force its members to prefer union workers, the union claimed inability to control workers' actions. In November the arbitration board met to resolve the impasse.[22]

In its decision, the board blamed "misunderstandings" as the cause of troubles, but reasserted the union's obligation to prevent work stoppages and the association's to apply the preferential rule. Privately, Brandeis "lectured" Bartholomew about abandoning his legalistic approach to grievances. He insisted further that he rehire both Cohen and Israels. Concerning Israels, Bartholomew grudgingly promised "to manage things so that she will be used in situations in which she will prove of great value, and tactfully eliminated from situations where she would do harm." As for Cohen, Edward Filene prevailed upon Felix Adler, who knew Cohen well, to urge him to lower the huge bill he had presented the association. After awkward financial negotiations with Cohen, Bartholomew once again retained Cohen as counsel. In January 1914 Israels returned to work at the association, no longer as assistant but chief clerk. She turned immediately to clearing an enormous case backlog.[23]

Her relations with Bartholomew did not improve. When in late February the protocol seemed again bound for shipwreck, Bartholomew asked Polakoff and Cohen to join him in a conference with Brandeis, then in Washington, D.C., on business. Cohen insisted on bringing Belle Israels along, and as a result of a "little chat" with her discovered that Bartholomew had been "playing both ends against the middle" in order to preserve his position. Cohen told Brandeis privately that Bartholomew had not only "fought hard" to keep Belle Israels "out," but had been using Paul Abelson, who was close to both Israels and Henry Moskowitz, to slander Cohen. At the same time, Bartholomew was telling Cohen that Moskowitz had joined the anti-Cohen faction. "When I compare notes with Henry," Cohen wrote, "we shall both laugh—if we do not weep—or kick ourselves for our too ready trusting a man we have long suspected." Cohen then praised Israels to Brandeis: "You got, I feel certain, a fine impression of [Israels] last night . . . She has more than a feminine intuition. She will be the one, I am sure, who will in the last act reveal Iago (the appellation is her own)."[24]

How she revealed Iago is not known, but by the end of March Bartholomew had resigned. He claimed "ill health," but his friends said that "developments in the inner circle of the association" had taken a course that Bartholomew believed violated the protocol's fundamental principles. Floersheimer, the DWMA's president, "emphatically denied" a shift in association policy.[25] George S. Lewy, the association's business manager, became its general manager, and Israels was named "Manager, Labor Department." The Israels-Cohen diarchy appeared triumphant, but there were more Shakespearean plots to come.

Belle Israels, shortly to become Belle Moskowitz, now had complete charge of DWMA labor policy. In addition to running the grievance department, she "trained" the manufacturers in their protocol duties, worked with the union on membership problems, and promoted the industry's modernization. As her responsibilities grew, her earlier conciliatory approach gradually became less flexible, which brought her into conflict with both union and association. Union leaders found her patronizing or threatening; association members, especially those who produced cheap waists and thus could not easily absorb rising labor costs, rankled under the new concessions she demanded. The antagonism she generated can best be illustrated by the two issues that strained the protocol most: the preferential shop and piece prices.

In response to continuing arguments over the preferential shop, the arbitration board met in November 1914. The immediate issue was the plummeting of union membership from twenty-five thousand to sixteen thousand, a drop caused, the union charged, by the employers' failure to apply the preferential clause. Since union membership was not bringing preference, union leaders argued, workers had little reason to pay dues. The association answered that, first, many workers suspected a union their bosses promoted and refused to join it; second, employers were loath to fire faithful help just because they were not in the union; and third, employers willing to apply the clause could not get current lists of members from the union. In sum, emotional as well as logistical problems troubled the preferential system.

Louis Brandeis offered as a solution the deduction of dues by employers. Belle Moskowitz refused on the ground that, if manufacturers become collection agents for the union, girls opposing the union would go to non-association shops. She argued further that if manufacturers could collect the union would be strengthened falsely. Union membership is "their problem, not ours," she announced after a long debate with Brandeis at the November hearings. In ways she possibly would have avoided in the early days of the protocol, she now emphasized the dichotomy of interests between union and association, asserting that it was not "her" place to push "their" grievances to a conclusion. The union people must "learn how to govern themselves," "to solve their own problems."

The union had a different view. Polakoff argued that manufacturers lacked faith in the protocol. Fearing future strikes, they did not want their workers organized. Benjamin Schlesinger, ILGWU president, conceded that employer collection of union dues was a bad idea, but charged that employers sabotaged union organizing and refused to treat the preferential system as a duty.[26] Later, both sides agreed to exchange up-to-date lists of employees and union members.

In addition to devising a bureaucratic form to expedite the exchange of lists, Moskowitz proposed a number of "practical" ideas for unionizing women. She suggested that the union explain the protocol better to workers, arousing in them a spirit of duty and self-sacrifice, and that the union send better informed people to meetings, holding these in "clean, wholesome" places, not "saloons." This infuriated Polakoff. When Moskowitz named a church, a West Side hall, and a school as three likely locations for a union meeting, he rejoined that his people would neither enter a church nor venture to the West Side, and that schools always claimed they had no room. This exchange underscored the cultural differences between union and association personnel. Another exchange over the rising cost of living emphasized class differences that the protocol could never erase. When an employer asked Polakoff what he thought of the recent rise in the price of clothing, he replied that he knew nothing about it—he wore only second-hand suits.

As industrial consultant Ordway Tead explained, workers were being caught "between the millstones of rising cost of living and the economy of employers," and thus they had been unable to learn or teach such ideas as "industrial peace," "identity of interests," or "co-operation," ideas promoted by industrial peacemakers. Referring to the rising cost of clothing, he repeated the phrase of another writer who had said, with telling irony, "Let us *all* dress and have dinner before we talk of morals."[27]

Cultural and class differences often lay at the heart of the confrontations between Polakoff and Moskowitz. Polakoff, a Socialist union leader from Russia with twenty years' experience organizing workers around the world, was contentious, aggressive, sometimes even rude at the conference table, and to Moskowitz typified overzealous males who regarded "little victories" as the essence of the workers' struggle. To Polakoff, Moskowitz's refusal to allow employers to collect dues during the critical early stages of his union, and her ideals of self-sacrifice and duty, exposed her as a well-meaning but naive social worker. Moreover, though she spoke Yiddish to the "girls" at grievance hearings, and was generally sympathetic to them, to Pokaloff she was still a middle-class German Jew who served the bosses. Their impatience and frustration were mutual.

The other major point of contention between them was piecework prices. The dress and waist protocol required that a statistician devise a "scientific basis" for the cost of making one piece of a garment. Although newspapers credited Bartholomew with the idea for this study, Cohen later attributed it to Belle Moskowitz. Whatever the case, she certainly agreed with the study's aims. Using a "test worker" to determine piece prices had failed. If a test girl could make a garment in two hours, the union told her to make it in three; the union countered that the firm picked above-average workers and "speeded them up." With every change in fashion, a frequent event in the dress and waist industry, disputes arose over who should "test" and under what conditions. These disputes often led to the dismissal of more argumentative employees, who would then go to grievance charging "discrimination." In 1915 Moskowitz told the wage-scale board that testing had failed: if the price came out too high, the firm complained; if too low, the worker. "We ought to do something constructive in the industry," she pleaded, ". . . just to go on fighting about prices in absolutely irregular and guess work fashion, seems to us an intolerable situation."[28]

Impressed by the efficiency movement emerging from Frederick W. Taylor's scientific management studies, Belle Moskowitz believed that there was a rational way to solve the piece price dilemma. She had read about the British "log system" that "divisionalized" garments into many separate operations, assigning to each a rate based on an "average" worker's skill in completing the stitches. Hoping to adapt the system to the American dress and waist trade, she also sought to take the system a step further. She wanted the "test" worker to come not from a shop seeking a piece price but from a "test shop" set up solely to determine prices. Such an arrangement, she believed, would eliminate disputes over the "test hand" and advance the cause of industrial peace.

Early in 1916 the industry revised the protocol. It adopted a "Board of Protocol Standards" funded equally by union and association to enforce wage and hour standards and conduct an experimental test shop. In March, Robert G. Valentine, an industrial consultant from Boston, became the board's director. Valentine, having studied the clash between labor and scientific managers for the United States Commission on Industrial Relations, believed that efficiency movements would fail without the consent of labor. He made "complete work analyses" of the shops and opened a "neutral test shop." In late September he issued a report that pronounced both association and non-association shops equally short of protocol standards, claiming further that conditions in the former were worse. The report proposed a major overhaul and continuing supervision of association shops to bring them up to standards. But the DWMA's executive committee rejected Valentine's report and declared his generalizations "unsafe and unsound." J. J. Goldman, then DWMA president, charged that Valentine wanted a "complete socialized control of the industry."[29] Thus, even those who favored reform thought Valentine had gone too far.

In conferences later that fall Belle Moskowitz prevented the manufacturers from abolishing the Board of Protocol Standards but barely managed to get for it a minimum operating budget.[30] The manufacturers objected to more costly modernizations and further losses in their managerial independence. Union leaders shared the manufacturers' dislike of the board. In the original debate over the test shop, Morris Hillquit had rejected Moskowitz's plea for greater rationality in piece pricing, observing that time studies led to speedups and the work of professionals in test shops lowered prices. He called her idea for a test shop "too radical," adding that her calling it "rational" did not make him "irrational" in rejecting it. In the end, he said, "haggling" is the best way to fix prices: chaotic and old-fashioned, yes, but it kept pricing power in workers' hands.[31]

Apart from the preferential shop and piece prices, other issues hampered good relations between Moskowitz and the union. Since the protocol grievance system asked clerks to conciliate and advocate at the same time, they were bound to clash. When Polakoff, for example, said he would announce a favorable board decision as a "ruling for our side," Moskowitz objected, accusing him of wanting to arouse hostility. But she herself had called union grievances "theirs," and in a 1916 hearing defended her own assistant clerks as "pretty good fighters":

> Sometimes a manufacturer is willing to sign away his rights [just for the sake of peace], and sometimes our clerk must tell him that he must not do that. That kind of adjustment, of course, irritates and annoys the Union because one inroad in that direction means some two hundred other inroads sooner or later; and we feel that we must give that kind of protection to the members of the Association.

Cohen followed her statement with an attack on Polakoff in which he said that the clerk's job "does not require fighting ability." Morris Hillquit, then union counsel, caught the contradiction but could not himself resolve it.[32]

Disputes also arose over what Polakoff called Belle Moskowitz's "inefficient" grievance system. Delays in settling grievances, he charged, were the prime causes of work stoppages. Early in 1915, after three months service as chair of a "Committee on Immediate Action" in the dress and waist trade, Leo Mannheimer concurred with Polakoff that Moskowitz's department was run in a "slovenly and inefficent manner." He was convinced that the employers' "one fixed policy" was to delay taking up employee grievances. Moskowitz answered such criticisms with lists of resolved cases (she claimed thirteen thousand by the end of her tenure), and by blaming irresponsible workers and weak union control for the continuing troubles in the shops.[33] In the end, however, it was less the union's displeasure that determined Moskowitz's future in the industry than a faction of shirtwaist makers among her own employers.

Since the early protocol days, the DWMA had suffered from conflict between the larger dress- and smaller waist-making firms. The association weathered the conflict during the early protocol years, but problems persisted. In a confidential memorandum in March 1915 to Rabbi Judah Magnes, Leo Mannheimer charged the association's administrative officers with favoring "important members" to the disadvantage of others. The issue of favoritism came to a head in 1916. In January, the ILGWU presented the association with an outline for a new protocol that demanded a 15 to 25 percent wage increase, a forty-eight hour work week, an additional legal holiday, new protocol boards, and stronger employer support for union organizing.[34] In turn, the association demanded an end to work stoppages and further encroachments upon managerial rights before it would consider any new agreement. In the midst of preliminary conferences on these points, Brandeis left to join the Supreme Court, thus depriving the protocol of its most experienced and prestigious proponent.[35]

Judge Mack took over as arbitrator, rendering decisions on February 7 that pleased neither side. Giving some immediate pay increases, he reduced the work week to only forty-nine hours, and issued baffling instructions on other wage hour disputes.[36] While awaiting clarification, on February 9 the union struck against non-association shops, hoping to win the same gains Mack had awarded under the protocol. On that day Belle Moskowitz, declaring the association "at peace with the union," announced that she had granted a union request for a "demonstration holiday" at the workers' own expense. During this holiday, she explained, workers in association shops "may show their sympathy with those on strike, and may be properly registered" at their union. She claimed further that all the association's members had agreed to this, and that the holiday would last no more than a "day or two."[37]

But the union failed to get its workers back to the shops. As the "holiday" stretched to a week, the fury of the employers mounted. They claimed losses of $60,000 a day in profits. A number of them, impatient with Moskowitz's policy of "appeasing the unions," called a secret meeting for February 28. Chaired "in an advisory capacity" by one of Bartholomew's former lawyer clerks, the association's "insurgents," mostly waist manufacturers, demanded a clean sweep of the DWMA's officers and executive committee.[38]

Meeting again on March 1, an even larger group heard remarks from Harry A. Gordon, counsel for two other garment manufacturers' associations and well known for opposing the interference of outsiders in the industry. For example, the previous year Gordon had charged that the protocol forced employers and employees into a relationship "controlled, not by economics and business principles, but by theoretical precepts which were often communicated over the long distance telephone [from Boston] . . . ," an undisguised reference to Brandeis and Filene, whom he called "alleged conciliators, legislators, and industrial Napoleons." According to a reporter, at the March 1, 1916 meeting the insurgents drank up Gordon's hyperbole like a "tonic": "You dress and waist manufacturers were sold into slavery. It was a crime against law, a crime against industry for you people to have made the conspiracy in 1913 and to have agreed to collect dues for the union. It was a crime for you to have sent the workers out on strike recently." With scathing remarks about Brandeis, Mack, and other "social uplifters" who pretend to know something about making waists, he even attacked the Joint Board of Sanitary Control, a body usually immune from criticism. Gordon called it a "fake."[39]

Two weeks of stormy meetings ensued. At one, Julius Cohen was shouted down and four members of the DWMA's executive committee resigned. Calling for the "democratization" of the association, the insurgents proposed new bylaws on March 9 that enlarged the executive

The new officers of the association reaffirmed the protocol and a labor policy of conciliation and mediation. They denied "war rumors" started by "those apostles of industrial peace and arbitration," "false prophets" who have led the workers astray.[49] But industrial warfare eventually broke out. The protocol arbitration board met for the last time in January 1917, awarding workers a small wage increase but denying them the eight-hour day they wanted. For the next two years, the Great War kept the trade quiet. When the protocol came up for renewal early in 1919, the union made new demands, the DWMA (through its new spokesman, the anti-labor Harry Gordon) refused to arbitrate, and a major strike erupted that lasted eleven weeks. When it was over, the union had won a shorter work week and some wage increases, and had also reclaimed the strike as a weapon. But the arbitration board, the symbol of the power of outsiders to interfere in the industry, was now gone. In addition, the manufacturers reclaimed control over the distribution of piecework and the right of discharge.[50] The protocol era had ended.

Belle Moskowitz's only public statement on these events was that an "organized conspiracy" had rid the association of the "constructive policy" for which she had stood. Affirming her belief in the protocol, she said that although it may not have been the best, it was still the only agreement that she had been obliged to enforce.[51]

On the evidence of other statements she made, she seems to have blamed the irresponsible, irrational, and selfish elements in industry for the protocol's demise. Leaders had pledged to abandon traditional freedoms for the sake of peace, but the rank and file resisted. When union officials told manufacturers that they ran their shops improperly, sparks flew. "You're not my boss and never will be," protested a manufacturer whom Polakoff had accused of lowering rates. The manufacturer was probably guilty, but he expressed feelings shared by many of his colleagues.[52] As loath as businessmen were to lose their freedoms, the ordinary worker was even less ready to give up the strike. Investigating a work stoppage, Belle Moskowitz once asked workers why, when in the past they had always let the clerks adjust their complaints, they had struck this time. One replied, "The Union had told the girls that we cannot strike. [The girls] said that we cannot work anymore and that the Protocol does not tell how it is, it is only a piece of paper."[53] This "piece of paper" was unreal to the girls in the shop and incomprehensible to businessmen. When professionals failed to meet their needs, for better or worse, both sides reclaimed control over their own affairs.

Paradoxically, then, the insurgent employers, who were fundamentally hostile to labor, had more in common with the workers than the employers who sympathized with labor and had favored industrial reform. The pro-reform employers had too easily believed in the potential of objective, rationally or scientifically derived standards to end the chaos in their trade. Further, they believed that unions and employers' associations could impose these standards on the industry as a whole. Their goals were laudable: the improvement of labor conditions and a decrease in cutthroat competition. But most businessmen and workers hoped their organizations would act as agents of their will, not as institutions to control them. They too wanted better conditions, but not at such a high cost. The return of the dress and waist trade to a confrontation framework harked back to the old days. Modernization received another blow.

Belle Moskowitz at first refused to give in to this resistance to reform. Soon after the DWMA dismissed her, she opened an "Industrial and Administrative Service" that offered management advice to business. "Get the Right Advice, the Right Viewpoint, the Right Suggestions and the Right Fact," her brochure commanded. Referring to "her long experience with the problems of industry and management in New York and expert training in organization problems," she advertised the following services: analysis of factory layouts for sanitary and safety improvements; the prevention and adjustment of labor disputes; the application of scientific management techniques; personnel management focusing especially on the problems of women and girls, the development of welfare services, and problems of training, transfer, and promotion; the explanation of labor laws and advice on pending legislation; the taking of surveys; and advice on legislative and financial campaigns. Promising that these services would help businesses outdistance competition, she prophesied that "Whatever threatens Peace threatens Profit," and "Disorder or Lack of System Means Business Demoralization."[54]

In this brochure, Moskowitz showed how much she had matured since the days of dance hall reform. She was now an expert in many areas of industrial life, including mediation. By her own count, she had settled 13,000 grievance cases. Since she left no record of her career as a private industrial consultant, we cannot know whether her expertise proved of value to clients, or whether she had any clients at all. In the years immediately following her dismissal from the Dress and Waist Association, illness, her own and within her family, forced her to abandon work for a time. Returning to activity when Alfred E. Smith ran for governor, after his election she was too busy in his service to take up private consulting.

There was continuity, however. While working for Smith, she applied her knowledge of industry and labor to government and politics, fields wider and with more authority than any in which she had operated before. In so doing, she drew often upon her protocol experience. The idea of resolving conflicts by creating mixed boards of representatives drawn from labor, business, and the public, and of systematic study to reduce chaotic conditions and bring about efficiency, were prominent aspects of her later advice to Smith. Whereas in private industry she had been forced to rely on voluntary controls, during the 1920s she would come to enjoy the power and prestige of being connected with the state. Although for her the protocol era had ended in her dismissal, she could look back upon it later with satisfaction. She had entered a field in which she had no previous training, contributed to it in ways that both friends and critics acknowledged, withstood waves of attack without flinching, and emerged with her self-confidence and (in contrast to the hapless Walter Bartholomew) her nerves intact. Since all of this experience served her later, she would regret none of it.

CHAPTER 6

From Social Reform to Politics

Every program which the social worker makes involves political action.

Belle Moskowitz, *ca.* 1930

"How can I be my brother's keeper?" asks the man who is harassed by the sight of the abyss. Point the way. A religious or an ethical movement which fails to point the way will not retain its place in the vanguard of social progress.

Henry Moskowitz, Address to the Society for Ethical Culture, May 1911

Between her dismissal from the Dress and Waist Manufacturers' Association in late 1916 and starting political campaign work for Alfred E. Smith in 1918, Belle Moskowitz experienced some dramatic transitions. Entering into marriage with an old and dear friend, she left the Yonkers house she had loved and in which her first husband died. Moving back to Manhattan, she developed a new pattern of family life. Equally important, she began to realize the key role of political action in the achievement of the social reforms she envisioned. Whether she came to the realization from bad experience or because of Henry Moskowitz's influence we will never know. One thing was clear: Henry, also a social reformer, was a deeply political person. He had carried campaigns to City Hall and Albany, led municipal fusion movements, run for Congress, and in 1914 won a place in the city government of a reform mayor. If any single influence carried Belle from social reform into politics, it was that of Henry Moskowitz.

Of the several suitors who visited Belle Israels at Park Hill, the children liked Henry the best. He was outgoing, funny, and soft-spoken, at thirty-four a man of pastoral warmth but free of cloying sentimentality. His physical appearance—he was relatively short of stature, with a round face and twinkling eyes—endeared him to the children, who called him "Dichtie" (for Doctor), and after he and Belle married, "Dad." The adoption was mutual. Henry proved a devoted father to Carlos, Miriam, and Josef, in 1914 aged eleven, seven, and five.

Settlement worker Lillian Wald, one of Henry's mentors, had advised him against taking on the care of a widow with three children. But the prospect left him undaunted. A long-time bachelor who had always cared about the children of others, he longed for his own family. He and Belle later tried to have more children, but without success. Two pregnancies ended in miscarriage. A third went to term, but "Sylvia" died three days after birth. Belle was ill for a long time, and never tried to conceive again.

Henry, born in Rumania on September 27, 1879, was two years younger than Belle. His mother Selma Wind had brought him to America as a young child. His father Meyer had immigrated a year before and sold suspenders and notions from a pushcart until he could send for his family. Meyer's original surname was "Josef." In a mixup common to immigrants, he received the name of the family in line in front of him and became "Moskowitz." Meyer earned a living as a store clerk and then merchant. Selma bore him six children.[1]

Many East Side boys sold newspapers for a living. Every afternoon young Henry went to Printing House Square, bought an armful, and raced to the corner of Grand Street and Bowery to hawk them. When his supply was gone, he bought more from the newspaper wagons and spent the evening, often until midnight, selling in front of Bowery theaters. Later he worked as a baster in a tailor shop, earning three dollars a week. After several stints running errands, he got a job as an office boy in a law office, studying in his spare time and at night.

Three streams fed into the development of his character: James B. Reynolds's University Settlement Society, Felix Adler's Ethical Culture movement, and Lillian Wald's Henry Street Settlement. Through the University Settlement clubs, he learned leadership skills and made a group of lifelong friends. Adler provided the ethical sanction for leadership, a "sort of religion of deed and not creed," as Henry later put it. Wald instilled in him a desire for "social service and right living." As a result of these early character-forming experiences, Henry came to define the chief ethical problem of an immigrant neighborhood as bridging the chasm of the generations. He built the first part of his career around this challenge.[2]

The University Settlement sponsored an "S.E.I." or "Social and Educational Improvement" Club for teenage boys. Writer Hamilton Holt, who lived at the Settlement from 1894–95, called it the "banner boys' club" of the house. Its members were a "bright set of lads" who could launch a spontaneous debate that excelled anything Holt had heard in preparatory school or college. Henry, a "short, thickset, sunny-faced fellow," was the club's "real leader." Other members included future labor arbitrator Paul Abelson, social worker Meyer Bloomfield, lawyer Louis Lande, artists Jacob Epstein and Bernard Gussow, and historian Jacob Salwyn Schapiro. This extraordinary group provided the context of Henry's adolescence and remained his close friends into adulthood.[3]

Other writers who met Henry at the Settlement included Ernest Poole and William English Walling, both of whom prowled the Lower East Side looking for local color or causes to champion. When Poole wanted to listen to recent immigrants discuss politics, Henry guided him through district cafés. Walling, an upper-class socialist and reformer, paid Henry's expenses in Eastern Europe during the summer of 1905 so that Henry could teach him German and some Jewish dialects. Lillian Wald remembered the young Henry as "a favorite and a real helper at the University Settlement," intelligent and understanding. Later a resident at Henry Street Settlement, he "gave light and brightness" to the group. Phrases that passed glibly from the lips of others—"brotherhood of man," "interrelationship of people one to the other"—were realities to him. While concerned for those handicapped by lack of privilege, he also knew that fun was "a spicy but important condiment of life."[4]

When he was a student at City College, Henry and some of his S.E.I. Club friends began to attend Adler's Sunday morning lectures. In December 1898, Adler and his associate, John Lovejoy Elliott, invited Henry Moskowitz, Paul Abelson, and Jacob Schapiro to discuss ethical work on the Lower East Side. Such work, Adler believed, would succeed only if taken up by the downtowners themselves. At the City College chapel, the three students announced a plan to open a settlement and asked for volunteers. Some thirty young men responded. They met at the Henry Street Settlement, appointing a committee of five—the original three, plus Louis Lande and Jacob Shufro—to found a Downtown Ethical Society. The committee rented three dingy rooms in a Madison Street tenement. Their first activities included a bi-monthly public lecture, two classes for younger boys and girls, and reading circles composed of club members. They set three goals: to build "character" through ethical teaching; to Americanize while easing tensions between foreign-born and American-bred family members; and to teach self-government. They hoped their little band would someday be the "center from which 'more light' may be radiated—a medium through which the Ethical Idea shall reach the Masses."[5]

After graduating from City College in 1899, Henry studied ethics, economics, and philosophy for a Master's degree at Columbia University. The next year he became involved in his first reform cause: the fight against the increased presence and police protection of brothels in working-class, residential neighborhoods. In the fall of 1900, a Lower East Side cleric reported to his bishop that police had scoffed at him when he complained about a fifteen-year-old girl soliciting on the steps of his church. In response, the bishop wrote an open letter of complaint to Mayor Van Wyck. This led to public attacks on Tammany Hall, widely thought to be behind the vice in the wards. Because of their concern for the welfare of children and the moral tone of the neighborhood, settlement workers took up the issue. Henry Moskowitz led their movement to organize an investigation.

On November 27, a group of civic leaders connected with Lower East Side institutions formed the Committee of Fifteen. Its chair, William H. Baldwin, knew that the Committee could never drive vice from a neighborhood, but he hoped to pressure the police to stop protecting it. His Committee, the prototype of later urban vice commissions, collected data, testified at legislative hearings, and cooperated for a time with William Travers Jerome, a colorful judge who led raids on brothels and gambling dens. Publicity about the Committee's findings fueled the movement that elected reformer Seth Low mayor in 1901. In 1902, the Committee published *The Social Evil,* a full and open discussion of the causes and effects of prostitution that rejected proposals to solve the problem by state control.[6]

In addition to his anti-prostitution work, Moskowitz played leadership roles in two local organizations. The first was a Federation of East Side Boys' Clubs, which he founded in 1901 and chaired. It coordinated the social and moral uplift programs of (at first) boys' and later all clubs in the area.[7] The second was more ambitious. This was the East Side Civic Club, founded in December 1901 by Paul Abelson. Moskowitz served as vice-president and later as president. The club mobilized public opinion on neighborhood issues, such as tenement law amendments proposed in 1903 which would have brought back airshafts, dark halls, and privy vaults. After collecting 40,000 signatures on petitions, the Civic Club presented them to Mayor Low and officials in Albany. In the end, the governor, mayor, and State Assembly Housing Committee pledged to defeat the amendments. The club also pursued such causes as cleaning up streetcar conditions and stopping the building of more elevated transit lines. In reviewing these crusades, Moskowitz commented, "Public opinion unorganized is impotent; crystallized it is impregnable."[8]

In 1904, the Ethical Culture Society awarded Moskowitz its first fellowship to prepare young scholars for ethical leadership. Adler, an admirer of Germany's academic discipline, wanted Moskowitz to study there. He may have hoped that Moskowitz would follow him as Society leader. He would be disappointed. Moskowitz earned his Ph.D. in philosophy, beginning his studies at the University of Berlin and finishing in 1906 at Erlangen. But, as he wrote Lillian Wald, "learning" failed to touch him. "On the contrary," he admitted, "I have become more convinced than ever of the hollowness and emptiness of academic philosophy. . . . [J]uggling with barren formulas has run against my grain and I am done with it for good." He promised she would find him "in harness again at the old stand." Upon his return to New York in August, the uptown branch named him associate leader. He gave his first address there at the end of the year. But he also became head resident of the downtown branch, now in larger quarters on Madison Street. His commitments to "the old stand" soon absorbed his energies.[9]

Despite Moskowitz's promise that he was through with scholarship, his early lectures reveal some ambivalence. He offered one series on literary and philosophical topics, discussing Socrates, Tolstoy, Emerson, Nietzsche, Ibsen, and Goethe—all normal fare for ethical audiences. Yet, showing his call to a wider sphere, he spoke (in Yiddish) to parents of settlement members, and gave a series of lectures on questions such as "Modern Industrialism and Its Neglect of Play," the "Struggle of Democracy Against Privilege," "Women in Economics," and "Economic Profit and Human Waste."[10]

Since he spoke effectively on these themes, colleagues and organizations called upon him frequently to address them. Once, lecturing on industrial justice, he evoked a vivid picture of factory life. "Come with me," he said, describing a factory where for eight or ten hours a day men pressed clothes with hot irons over burning gas jets. Breathing noxious fumes, the men send out for beer "to fight one poison with another." Or, he suggested, enter a factory where hats are made with felt modified by mercury, or matches with phosphorous, or a printer's shop, where lead is handled. There the workers soon complain of headaches, debility, symptoms of slow poisoning, or the beginning of Phossy Jaw, a dread disease which defaces and deforms. "You and I who enjoy the product are unfamiliar with this cost of blood and life entailed in the making," he said. Calling on workers to fight collectively to rid the workplace of such conditions, he stressed a theme central to the reform ideology of the times—the need for all social elements, workers, employers, and consumers, to cooperate if lasting change is to be won.

> We have become so inter-dependent that it is difficult to know where the responsibility of the worker, the employer or of the public begins and ends. One thing is certain, that when conditions of labor are such that they undermine health of a large mass of people and thus threaten the welfare of society, society has a right to step in.[11]

By 1912, when he ran for Congress from the 12th District as a Progressive, these themes predominated. His campaign literature, which named his two chief interests as social reform and industrial justice, said nothing of his links to Ethical Culture.

In his pursuit of his interests, Moskowitz joined countless numbers of groups and campaigns. He was proudest of his role as mediator in the garment industry. In 1910 he worked indefatigably to bring together workers and manufacturers in the ladies' cloak, suit, and skirt industry. That year he began five years as secretary of the Board of Arbitration in the industry, and many more years as one of three public representatives on the industry's Joint Board of Sanitary Control, of which he also served as secretary. Under the direction of Dr. George Price, the board supervised a scientific and medical survey of factory conditions and their effects upon workers' health. These activities led to a special label sewed into garments made in approved factories, and to union-sponsored health services.[12]

As a natural outgrowth of this work, in 1911 Moskowitz was invited to sit on the board of the Committee of Safety, established after the Triangle fire. Thereafter, he spent weeks in Albany lobbying for fire-protection bills and a fifty-four hour week limit on working hours for women. In the following years, he testified to the New York State Factory Investigating Commission, and helped settle a strike in the men's cloak industry. He joined a group that successfully pressured Congress to appoint a United States Commission on Industrial Relations. He served with increasing frequency in his later years as an impartial arbitrator in textile and related industries. In this role, he became aware of the difficulties arbitrators faced in making informed decisions about wages. A proposal he made in 1916, that industries set up national fact-finding boards to provide information on the relationship between local and national market conditions, was not fulfilled until the New Deal.[13]

While remaining active in industrial reform, Moskowitz also kept in touch with civic and social reform issues. He held memberships in both the City Club and Citizens' Union, where many reform projects were born. At the request of Mayor Gaynor he planned "safe" Fourth of July celebrations on the Lower East Side, advised the mayor on park management, and worked with a city-wide Committee on Congestion that pressured the city to pass stricter zoning laws. He served on a National Board of Censorship of Motion Pictures established by the People's Institute. In 1909, he helped found the National Association for the Advancement of Colored People. Shocked by the 1908 race riots in Springfield, Illinois, William English Walling published a plea for a new struggle for Negro equality. Early the next year, Mary White Ovington, then living in a New York tenement and studying Negro life, met with Walling, who invited Henry Moskowitz to join them. In their discussions, Walling spoke of the treatment of blacks in the south, Ovington about the north, and Moskowitz, "with his broad knowledge of conditions among New York's helpless immigrants," interpreted their facts within a broader context. After this meeting, whites and blacks concerned about the Negro question called a national conference, and the NAACP was established. Moskowitz does not seem to have done more in the organization beyond listing his name on the Advisory Committee.[14]

One local reform group that boasted his leadership, the East Side Neighborhood Association (ESNA), was less successful than some of his other projects. It was founded in December 1912 by Jonah J. Goldstein, a young lawyer associated with Assemblyman Alfred E. Smith. Goldstein had worked earlier for the "Bureau of Social Morals," a crime-fighting "vigilance committee" set up after the Herman Rosenthal murder. The Bureau was a division of the New York "Kehillah," Rabbi Judah Magnes's Jewish community organization. By his own account, Goldstein criticized Magnes's reliance on professional and financial help from outside the district, and resigned the Bureau in November to start the ESNA. Expressing a view with which Moskowitz concurred, that only reform initiated from within the district could succeed, Goldstein persuaded Moskowitz to serve as ESNA chair.[15]

In its fight against crime on the Lower East Side, the ESNA engaged in three types of activities: investigation, prevention, and lobbying. Using "confidential" information, it reported to the police evidence of prostitution, gambling, excise violations, and anti-Semitic harassment. On the preventive side, it held meetings at which Moskowitz spoke with dance hall owners and motion picture exhibitors about voluntary controls of "disorderly" conditions. Lobbying activities were directed at district problems, such as replacing unsanitary horse cars with electric cars, and securing physical and mental examinations of juvenile delinquents before they were sentenced. In performing all of this work, Goldstein claimed a high rate of success.[16]

In the summer of 1913, without Moskowitz's approval and perhaps without Goldstein's knowledge, an ESNA official appealed for money among uptown philanthropists. One recipient of the appeal was Felix Warburg, who made some inquiries about Goldstein. The Charity Organization Society responded positively, but Rabbi Magnes, whom Warburg also asked, reported that he had fired Goldstein for using the Kehillah as a personal springboard. He charged further that, in revenge, Goldstein had used his connections with Al Smith to get the Tammany-dominated legislature to deny a charter to the Kehillah. Finally, citing a detailed police report, he convinced Warburg that Goldstein had exaggerated the success of the ESNA programs. As for Moskowitz, Magnes had high regard for him but thought he had endorsed Goldstein because he "thinks that, by reason of his own clean-mindedness, he will be able to influence Mr. Goldstein and others of Mr. Goldstein's type." Moskowitz defended Goldstein, but Warburg, and in turn Jacob Schiff, to whom Warburg sent the facts of the affair, accepted Magnes's judgment.[17]

Whether Magnes's distrust of Goldstein was justified will never be known. Goldstein sued Magnes for libel but settled out of court after receiving an apology, albeit reluctant, from him. Henry Moskowitz's role in the ESNA continued, but the organization never grew to the stature he hoped for it. Like Goldstein, Moskowitz had eschewed support from uptown Jewish patricians. As he matured, he came to appreciate their leadership and sought a stronger connection with them. Indeed, his later correspondence with Warburg, Schiff, and Louis Marshall bordered on the adulatory. Yet, perhaps because of the Magnes-Goldstein affair, they never gave him a central role in their deliberations. Magnes's continuing regard for Moskowitz notwithstanding, the power brokers of Jewish philanthropy treated him as a well-meaning, but somewhat naive and impractical idealist who, in his mentor Felix Adler's words, "did not see far enough."[18]

During the early 1900s, many social workers and social reformers gravitated toward politics. Moskowitz was one of them. He hoped that the settlement houses would wrest political control of the wards from Tammany Hall. Thus, after organizing the district campaign for Charles

Evans Hughes's reelection campaign for governor in 1908, he took part the following year in a movement to fuse the city's anti-Tammany forces to support a reform candidate for mayor. Working with Ray Ingersoll, head of Maxwell House, a Brooklyn settlement, he organized a "Committee of One Hundred" to choose a candidate. They first selected Federal District Attorney Henry L. Stimson. Moskowitz and lawyer Winfred T. Dennison traveled to a hunting lodge in Quebec to offer him the nomination, but Stimson, currently pressing sugar monopoly and customs cases, had to decline. The Committee then offered the nomination to banker and philanthropist Otto T. Bannard. Moskowitz mounted an exhibit called "The Chamber of Horrors" that dramatized the consequences of Tammany misrule. Publisher William Randolph Hearst ran for mayor as well that year, splitting the anti-Tammany vote. Bannard lost to William Gaynor, but fusion forces elected other reformers, making a virtual clean sweep of the city's major offices.[19]

In 1910 Moskowitz supported Stimson for governor, chairing a "Stimson League of Independent Voters" that concentrated efforts on the Lower East Side. Felix Frankfurter, then Stimson's assistant, assigned Moskowitz the task of bringing Felix Adler, Jacob Schiff, Louis Marshall, and Rabbi Stephen Wise into Stimson's camp. Moskowitz published a character study of Stimson in *The Independent* that was distributed to newspapers throughout the state. Along with Ingersoll and Bannard, he was one of three individuals identified by Stimson as being in charge of campaign policy. Despite Moskowitz's prodigious efforts on Stimson's behalf, the electorate never warmed to Stimson's personality, which was notably uncharismatic. He lost badly.[20]

Two years later, having decided that there "must be a party which will emphasize the relation of economic and social problems to political action," Moskowitz became a delegate to the Progressive party convention in Chicago. He respected Woodrow Wilson, and would have been content with a Wilson victory. But he believed that Democrats were too heavily saddled with "outworn traditions and individualistic principles" in a time when social problems demanded a "collectivist" solution. Yet he also feared a "successful" Theodore Roosevelt. As a defeated Republican, Roosevelt had become "less of a politician and more amenable to real progressive influences." Moskowitz knew this from direct experience. With reformers John Kingsbury and Paul U. Kellogg, he visited Roosevelt at his home on Oyster Bay to present a "minimum" industrial reform program for the new party. Roosevelt accepted it, and most of this program later appeared in the party's platform.

Moskowitz and other reformers, notably Jane Addams and Joel Spingarn, took issue with Roosevelt on questions concerning blacks. The reformers wanted to push for black representation in delegations from the South and a plank in the platform affirming Negro rights. They failed, and the party remained "lily white." When the New York State Progressive party met in Syracuse in the fall, Moskowitz chaired the

Committee on Draft Resolutions, a post that gave him major responsibility for the platform. He directed the campaign on the East Side for the party's gubernatorial candidate, Oscar Straus, and ran for Congress himself. The S.E.I. Club turned out to help—Louis Lande chaired his campaign; Paul Abelson was a mainstay. The boys and girls of his settlement joined, too, holding forth on street corners in English and Yiddish and handing out flyers and photographs. Although defeated along with Straus and Roosevelt, he was pleased by the devotion of his young followers. He felt he had contributed to their civic training and potential for self-sacrifice.[21]

In 1913, Moskowitz became one of four vice-chairs of a new fusion movement, the Citizens' Municipal Committee. Its chair was Norman Hapgood, a writer with whom Moskowitz would have close ties in the 1920s. This fusion drive succeeded, electing thirty-four-year-old lawyer and former alderman, John Purroy Mitchel. Moskowitz had not sought personal reward from his political activities, but this time he received one. In the belief that he and Moskowitz shared the same ideals of civil service, Mitchel appointed him president of the Municipal Civil Service Commission. Moskowitz would carry into this job a social worker's enthusiasm for reform, extending the merit system, improving the examinations, and developing methods to rate employee performance. These reforms led to contact with Robert Moses, a young social scientist from the Bureau of Municipal Research whom Mitchel brought in to develop a rating system. The reforms also led to an imbroglio involving Moskowitz and opponents who wished city jobs to remain the plums of political patronage. In 1919, Belle Moskowitz brought Moses into Governor Smith's inner circle, with major consequences for the history of public works in New York State.[22]

Moskowitz's days as a low-paid settlement worker and perennial volunteer in causes were now over and he could look ahead to four years with a decent income and high local prestige. He could marry. But the future would not be smooth. The State Civil Service Commission initiated an investigation of his office which, while resulting in his exoneration and the resignation of his accusers, nonetheless absorbed energies that might have been directed more usefully elsewhere. His wife's dismissal at the end of 1916 from the Dress and Waist Manufacturers' Association, and her three failed pregnancies, meant a period of financial and personal stress at home. Near the end of Mitchel's administration, doomed to one four-year term, Moskowitz was forced to resign from the Civil Service Commission to take up a less controversial post as New York's first Markets Commissioner. This shift in duties represented Mitchel's eleventh-hour attempt to stave off the mounting chorus of complaints against Moskowitz's anti-patronage campaigns. Thus, neither Henry nor Belle Moskowitz's transition from social reform to politics would be free of painful associations.

While these events were going on, Moskowitz's ties to Ethical Culture were loosening. Throughout its first decade, the Downtown Ethical Society had expanded steadily, adding financially secure summer programs that attracted hundreds of youth. In 1910, the settlement moved into its own building at 216 Madison Street. At the same time, strains developed between the downtown and uptown societies. Henry Moskowitz's political activities disturbed Felix Adler, who had no faith in politics. At election time, Moskowitz would cancel his uptown office hours. Perhaps he neglected other duties. On April 13, 1910, the Society's Council debated the extent to which the institution should engage in reform. Henry was one of the speakers. "One of the chief reasons why the Society originally attracted many members," he said, "was because . . . it was a religious movement which discussed practical issues." As a church, the Society had energized individuals, but it lacked a "secular platform for the consideration of practical, political and ethical questions." His circle of young men complained "that they feel no vitality in the Ethical Society's interest in these practical issues; that while it may have an academic interest, it has not made the working man feel at home here, has not shown hospitality to the new developments in the great world-ethical movement." In sum, the Society had grown "sluggish." Henry asked it to throw open its doors to discussion of all sides of new ideas, and "stop being afraid of the long-haired radical." At the end of this meeting, the council moved that the Society would not "take a position on economic and political movements," but would hold meetings on important issues and "stir the enthusiasm of its members to the working point."[23]

In 1912, the Downtown Society changed its name to Madison House, reflecting its growing independence from the uptown branch. Two years later, on the eve of joining Mitchel's cabinet, Moskowitz resigned as associate leader of the Ethical Society. The step had been long in coming.

The relationship between Belle Lindner and Charles Israels had included romance and also shared values. When Belle Israels and Henry Moskowitz discovered a new aspect of a relationship that for years had been strictly professional and political, they too must have felt romantic excitement. Yet it was their intellectual companionship that forged the strongest bond.

Belle's life pattern had run parallel to Henry's. As young settlement workers, they had observed each other at work. In December 1902, for example, Belle Lindner recited a Christmas story at the Downtown Ethical Society. Early in her marriage, she probably heard one of his addresses at the Ethical Society. In 1908, she directed a production of *The Story of Joseph and His Brethren* for the tenth anniversary celebration of the Downtown Ethical Society.[24] Shortly thereafter, Henry joined her dance hall reform movement. Both were involved in the aftermath of the Triangle fire and Rosenthal murder, and crossed paths in the New

York State Progressive party. Their work under the Protocol of Peace threw them together. Finally, they both believed in three tenets of the current reform creed: no reform would succeed without the organization of public opinion; reform must be initiated by those most affected by its results but would last only with cooperative effort across all layers of society; and, when social welfare was at stake, the commonwealth must interfere in the private sphere. Few marriages could boast a stronger foundation of shared experiences and dreams.

Belle and Henry announced their engagement on October 16, 1914, and were married in the Park Hill house at 3:00 P.M. on November 22. Belle, who had slimmed down for the occasion, wore a traveling costume of chiffon velvet. Solomon Lowenstein, superintendent of the Hebrew Orphan Asylum, performed the rite, which according to the local press was "of the utmost simplicity." Aside from immediate families, only Lillian Wald and John Lovejoy Elliott attended. Of the three children, only Carlos, the eldest, was allowed to observe. Henry and Belle honeymooned for a week in Atlantic City. The senior Lindners moved to a boardinghouse on Manhattan's West Side. Belle kept the Park Hill house, but sold it a few years later to Eli Bernays, Edward and Judith Bernays's father, who bought it as an investment and to relieve Belle of the burden. The Moskowitzes took up residence in a narrow, four-story house at 147 East 38th Street.

They enrolled the children in the Friends' School on East Fifteenth Street. Miriam recalled roller skating there, making a "horrendous noise" down Third Avenue under the "El." When the family moved uptown to a brownstone at 147 West 94th Street, the younger children transferred to the Ethical Culture School, and Carlos went to school first in Riverdale and then at Townsend Harris High. They attended Sunday school at Temple Israel, where Miriam was confirmed and Carlos bar-mitzvahed. Joe played hookey. When caught, he complained that the Sunday school was "awful." Henry visited and agreed, and Joe was released.[25]

"Now, what are you going to do about your name, change it again?" asked Belle's feminist friends when she married. "I've put two names across," she replied, "I guess I can manage a third." Thus she became Mrs. Henry Moskowitz.

Years later, Belle complained to her daughter of tiresome interviewers who asked how she was able to combine career and home. She developed a ready answer: "A career is a splendid thing for a mother. It gives her a wider contact with life . . . through which she can better develop the budding personalities of her children." Furthermore, a mother who is "a figure of some consequence in the world" is more precious to her children. They sentimentalize her less, and are more likely to come to her confident she will understand them." But, Belle warned, "it isn't easy running a home and a career. It requires above all a tremendous physical energy and a complete indifference to nerves."[26] To her daughter Miriam, however, she confided a slightly modified opinion. In her view, women should do whatever they want, but as mothers they should not also be burdened with the financial responsibility for their families.

Her income combined with Henry's made it possible to engage live-in help—a governess, cook, and maid. Later she hired a couple, the woman to cook and do housework, the man as butler and chauffeur. This was a normal staff for a family of their size and status. Even with this help, Belle maintained control. She did her own marketing and made her own preserves. The rest was a matter of routine: family breakfast at eight o'clock; discuss the plans of the day; consult with cook; market. She tried to be home at five for the "children's hour," to read and talk to them while they ate supper. In later years, when they dined with the adults, new arrangements were made. Arguments at table became unbearable: Carlos a "know-it-all," Miriam too "critical," Joe a troublemaker. Belle and Henry called a meeting and announced that henceforth the children would finish eating and leave the dining room to the adults. Peace reigned.

As Miriam grew older, Belle leaned on her for help at home. "Schatzie" (as the family called her) took over the marketing, put away the laundry, mended. But she gave Belle more than household help. She adored her mother, even idealized her, and longed to be her emotional support. Every night, instead of writing in a diary, the adolescent Miriam wrote a note to her mother. She reviewed events of the day and poured out feelings about her family and others. She wrote of her "little talks" with her mother, in which Belle had taught her

> To control my temper,
> serve others with good-will,
> to kill the demon that within me walks.

The "demon," Miriam's tendency to criticize, was (according to her brothers) the cause of sibling wars, but it was certainly no worse a fault than those of her brothers.

Instead of squelching a daughter's love, Belle's stern lesson increased it. "Mother, dear, I love you with all my heart and soul," Miriam wrote; ". . . you are my ideal, my greatest ambition is to win in order to place my victory at your feet." In another letter, "Mother, they say that time destroys ideals—but my ideal, and that is mother, I seem never to have discovered the whole of, or the depth of beauty in." She especially appreciated her mother's capacity for sympathetic listening, writing in praise: "You always understand; and you never laugh." She wrote warmly of Henry as well. "Every day I love Daddy more and he comes nearer to me. He is so big and fine and strong. Oh Mother I love him so much and when he puts his arms around me I feel so safe and protected. I think I love him as much as I do the memory of father. Which is saying a great deal for me." But she worried about Henry and Josef's relationship, which seemed at the time to be based solely on discipline. "Daddy must come half way," she suggested, sounding a warning of breakers ahead.

At once worshipful and supportive, Miriam's letters have a poignant side. Aware of her mother's periods of stress, she wanted to help. At about age fourteen or fifteen, after reporting on the mending she had done, she asked, "Wouldn't you feel happier if you told me when you were very tired or worried, or when there was something left undone that was troubling you? Won't you try it, or am I too little to know things yet?" Or, "Mother, dear, are you resting enough? Are you taking care of yourself? You still look as though you needed 'a week in Atlantic City.'" Belle must have been buoyed by her concern. Yet, along with the concern, there was also longing. Miriam yearned for more time with her mother. "I wish I could see more of you," she once wrote. "But at present I can see that that is impossible because you always have to go somewhere." Or she would refer to "that rare thing, an evening with you." As an adult, Miriam denied that her mother was unavailable to her. Perhaps the quality of their time together compensated for the lack of frequency.[27]

Despite the bittersweet tone, this correspondence shows many positive features of Moskowitz family life. No matter how preoccupied or busy, Belle sustained satisfying relations with her children. There may have been times when they felt neglected; they never felt unloved. Moreover, Henry had integrated himself fairly well. He had trouble with Josef, but so did everyone. At age fifteen, Joe dropped out of school, took a radio course, and signed on to a submarine. The family's "black sheep," he had a passion for city low life and literary erotica that, given his mother's past as a dance hall reformer, disturbed and embarrassed her. After consulting a psychiatrist, Belle realized she had no choice but to tolerate Joe's foibles. Indeed, she doted on him, which aroused jealousy in his older, more achievement-oriented brother Carlos. Joe became a writer, publishing two novels with Doubleday, Doran and Company. In the 1930s, he worked with Belle in her publicity business. Carlos graduated from Amherst College and Columbia Law School, joined a prestigious New York law office, and ultimately established his own firm. Miriam studied art, and at the age of nineteen—too young to know what she was doing—married a wealthy Englishman.[28]

Miriam gave testimony to the mutual supportiveness of the relationship between Henry and Belle. The couple led independent but parallel lives. They knew the same people, shared the same networks, and always had much to talk about at the end of the day. Each also brought the other new contacts. Introducing Henry to her spheres of active women, Belle inspired in him new respect for women's abilities in public life. Henry put Belle more in touch with the Lower East Side as well as with the talented bureaucrats in the Mitchel administration.

Yet Henry, like Charles Israels, was not a domineering male figure.

Perhaps for that reason, some Moskowitz friends believed that Belle "wore the pants" in the family. Actress Aline MacMahon, who liked Henry and had great affection and esteem for him, called him a "shnook," a lovable weakling. Frances Perkins saw the relationship of Belle and Henry otherwise: "He had a trained mind and she had a brilliant mind that wasn't as well-trained. He was stocked with learning. She had quantities of practical experience, a brilliant and penetrating mind, but not trained and not really conversant with the learning of the world; . . . whenever there was a difference of opinion with regard to political matters, or matters of legislation, programs . . . it was always Henry who won the debate. Belle listened to him. She at once recognized a superior mentality and a disinterested person." In the late 1920s, Englishman Ivor Montagu, a cousin of Cyril M. E. Franklin (Miriam's husband), was a guest in the Moskowitz home. In his view, Belle was the "obvious head of the family, . . . a twinkling, humorous person, small and round . . . who gave the impression of immense latent power. A dynamo concealed in a plain deal box. . . ." Henry was "devoted to her, of equal intelligence but deliberately self-effacing. . . . His great moment of glory came at mealtimes when with his own hands he mixed the salad like a true paterfamilias. At this he was a master."[29]

Belle "had strings that led everywhere," Montagu continued. Through her he gained entry wherever he wanted, from a literary lunch at the Algonquin to a transvestite dance hall in Harlem to Dr. Irving Langmuir's research laboratory at General Electric where he saw his first experimental television set. Belle may have respected Henry's mind and may have learned from his wisdom, but all that outsiders saw was her wielding awesome power and Henry mixing the salad. The combination of their minds and temperaments was perhaps as unusual for the times as the partnership of Belle Moskowitz and Al Smith would be.

As Henry moved from social reform to politics, Belle accomplished a transition of her own. A few months before she married him, she made her first and last foray into electoral politics as a candidate. In the spring of 1914, a group of women Progressive party members decided to select women as delegates to the upcoming state constitutional convention. At a dinner on May 19, twenty-five women formed a "Women's Committee of the State of New York for Representation in the Constitutional Convention," chaired by Lillian Wald. By August, 175 women had become members. The committee picked Wald, Dr. Katharine Bement Davis (Commissioner of Correction in Mitchel's administration), Frances Kellor (an expert on immigration and a member of several state commissions), and Josephine Goldmark (an industrial investigator) as their candidates for delegates at large. Belle Israels was one of five women chosen to represent senatorial districts in the primaries.[30]

The press approved the women's goals, but neither the Republican nor the Democratic state convention endorsed a woman candidate or passed a resolution favoring them. Early in September, Belle Israels decided to withdraw, because "she could not undertake the work or manage with the help she could get." But she stayed with the cause, serving on the Campaign and Executive committees. At a meeting on September 14, she announced that she still might receive the endorsement of Westchester County Progressives. This endorsement came through at the end of the month, and Israels ran after all, finishing ninth in a field of fifteen, her total of 1,010 votes falling far short of what she needed for election. New York State Progressives in general failed to achieve election in 1914. All across the state, the numbers of party enrollees had dropped to an insignificant remnant. The constitutional convention was dominated by Republicans, who swept two-thirds of the seats. Of the four women who ran as delegates at large, only one—Dr. Davis—had been endorsed by Wald's committee and the state Progressive party; the other three women were Prohibitionists. None of these were elected.[31]

This experience must have strengthened Belle Israels Moskowitz's commitment to women's suffrage; after it her name began to appear on lists of suffrage activists. She served in 1915 on a committee of the Women's Political Union that organized a "Votes for Women Ball"; in 1916 on the Advisory Board of the New York City Congressional Union for Woman Suffrage; and in 1917 on the National Advisory Council of the National Woman's Party, a group with which she would later break over the Equal Rights Amendment. In 1918 she chaired an election night "Suffrage Ball" held by the New York State Woman Suffrage party. During this period, Moskowitz was also reelected to the board of the New York Council of Jewish Women and became a charter member of the Women's City Club. In the 1920s, this last group became the focal point of her nonpartisan political activities.[32]

The original purpose of women's clubs, many of which were founded after the Civil War, was to give women secular opportunities for companionship, self-improvement, and philanthropy. By the turn of the century, women's clubs had become more interested in politics and reform, increasingly so as women's suffrage neared. Men already had "city" clubs in which they studied, debated, and took action on controversial public issues in their local communities. Civic-minded women began to want clubs of their own.

The New York Women's City Club was conceived at a meeting in July 1915 of about one hundred New York suffragists, including Vera Boarman Whitehouse, Alice Duer Miller, Katharine Davis, Alice Carpenter, and Helen Rogers Reid. Confident of winning the vote at the fall election, they wanted to create an environment in which women could study and debate political issues. That fall, New York men rejected women's suffrage, but the club opened anyway in a suite of rooms formerly occupied by Alfred G. Vanderbilt's family on the top floor of the Vanderbilt Hotel. On January 31, 1916, they elected their first officers. Alice Miller became president, and Mary E. Dreier first vice-president. Other officers and board members included such well-known names as Virginia Gildersleeve, Frances A. Hand, Bertha Rembaugh, Mary Simkhovitch, Mrs. Ray V. Ingersoll, Mrs. Herbert Croly, Mrs. John Dewey, and Ida M. Tarbell—all prominent in their own right or married to leaders in New York's political and intellectual life. The sixty-nine founders of the Women's City Club put $120 each into the treasury. About 440 women then joined as charter members, each paying $30 and an initiation fee. By joining this second group, Belle Moskowitz entered a network of affluent and powerful women. Her connections with them would prove useful to Smith and state Democratic party fortunes over the next decade.[33]

Before making her final shift into politics, Belle Moskowitz drew on many of these women for help in work undertaken to support the World War. She organized and managed Mayor Mitchel's Committee of Women on National Defense. When the war began, federal, state, and municipal Committees on Defense had been formed, with all-male memberships. Women protested, leading to the formation of women's committees to coordinate their war efforts. In April 1917, concerned that the war-related activities of women's groups in New York City were becoming impossibly tangled, Moskowitz urged Mitchel to appoint a women's committee. She supplied the names of potential members, set up the first mailings to over five hundred organizations, and coordinated meetings between their representatives and public officials. Those attending such a meeting on October 10 found it so useful that they decided to constitute a permanent "Mayor's Council of Women." Moskowitz described this development as "a notable step toward complete effectiveness of the volunteer effort of women."[34]

In addition to acting as a clearinghouse for women's "patriotic duties," the women's committee initiated its own programs. These included food conservation and gardening, placing women in agricultural work, Americanization of aliens, and protection of young girls around training bases. Moskowitz was in charge of Americanization. Using terms reminiscent of her Educational Alliance days, she defined Americanization as bringing to foreigners "what is best in American culture and civilization, at the same time retaining the finest and best that foreigners have to contribute to this country." Yet, as she outlined her program, the latter aspect of this definition was barely visible. Her report emphasized intensive training in English, especially to those listed on the census as being unable to speak English. "They will be followed up and brought into classes where they will receive instruction," she wrote. Further, the committee planned to follow the foreign-language newspapers and observe street gatherings in order to learn "what the foreign population is doing and saying and thinking, so that educational propaganda may meet direct needs." Finally, using such institutions as a community chorus, the committee would spread American cultural influences.

Moskowitz closed her report with the hope that her committee's program would make of New York City "one people with one language, instead of a group of foreign towns."[35] Although Moskowitz described Americanization in the language of a sympathetic settlement worker, she was not immune to the nationalistic pressures of her times.

The Mayor's Committee of Women was but one of a long string of committees and commissions Belle Moskowitz would create. In the fall of 1917, in the same election that gave New York State women the right to vote, John Purroy Mitchel lost the mayoralty to a Tammany candidate, Judge John F. Hylan. Alfred E. Smith, then sheriff of New York County and running for president of the Board of Aldermen, had helped Hylan win. As if this were not bad enough, when a group of suffragists approached Smith to ask for his support for their cause, he had shot a stream of tobacco juice at a spittoon and told them that as far as he was concerned a woman's place was at home.[36]

Less than a year later, Belle and Henry Moskowitz, inveterate suffragists and anti-Tammany "goo-goo's," announced they would support the sheriff for governor. "Will we be on opposite sides this election?" Belle wrote Frederick Whitin, executive secretary of the Committee of Fourteen. "I plan to support Mr. Smith." Some of their reform friends must have been surprised. But in their view Tammany was changing, and Al Smith, who had led the fight for labor legislation after the Triangle Fire, seemed to symbolize that change. They were not disappointed.

CHAPTER 7

Building the Partnership

If a reconstruction period means anything at all, it means the utilization of such a time to put the State's house in order, to do away with what proved under test of great emergency to be unfitted for the strain and to prepare the citizenship and the administration of its government to meet the future with better results.

A History of the Reconstruction Commission
December 18, 1920

For almost two decades, the political machine that reformers despised had nurtured Alfred Emanuel Smith. A Catholic of predominantly Irish descent, he was born and bred on the Lower East Side. In 1888, after a parochial school education that ended after the eighth grade, he went to work to support his fatherless family. Starting out as a messenger in the trucking business and graduating to shipping clerk, he spent two years as a bookkeeper in the Fulton Fish Market, a place he later called his "university." He also performed in amateur theatricals, developing a reputation in the district as a smooth talker with a dapper air and good sense of humor. He spent much of his spare time in a saloon owned by Tom Foley, the Tammany leader of the Fourth Ward.[1]

In spite of what the goo-goo's said, Tammany had its points. Operating through local branches, the club kept Democrats in closer touch with New York's working classes than any other political party. At its best, wrote former Progressive Frederick Davenport in 1918, Tammany was "the family applied to politics." Through a system called "contracts," it handed out food, fuel, jobs, and other favors to the needy. If someone got in trouble with the law, it buffered the weak from the mighty. On holidays, it put on neighborhood picnics, boat excursions, and parades. To increase neighborhood conviviality as well as opportunities for graft, Tammany also sponsored saloons, dance halls, gambling houses, and brothels, all affronts to middle-class uptowners. Some downtowners found Tammany offensive, too, but enough were grateful to reward its candidates with frequent and often overwhelming majorities. Uptowners fumed but could not shake Tammany's hold on the new immigrant citizen.[2]

The young Al Smith benefited from Tammany. Tom Foley liked him and sent him on "contracts." He enlisted Smith's oratorical skills in his own struggles against Richard Croker, Tammany's boss. As a reward, Foley got Smith his first white-collar job as a process server for the Commissioner of Jurors. Smith's ambitions grew. In 1903, by which time Foley's friend Charles F. Murphy was in control at Tammany Hall,

Smith was nominated for the Assembly from the Second District. Victory was easy, and sweet: he won by a margin of almost five to one.

As State Assemblyman, a post he held a dozen years, Smith began a new phase in his education. Frustrated by parliamentary maneuvers and jargon, Smith appealed to a better-educated colleague to teach him the ropes. He studied hard, earning the respect of his fellow lawmakers, who rewarded him each term with better committee assignments. Newspaper reporters, attracted by his funny, homey anecdotes, liked him too. More important, his constituents kept returning him to office. He remained loyal to Tammany throughout, opposing direct primary nominations that threatened machine control over appointments, and fighting anti-saloon measures. In 1911, as majority leader and chair of the powerful Ways and Means Committee, he pressed hard for a workmen's compensation system that would remove from the worker the burden of proving negligence. His efforts here, and in the next few years as vice-chair of the Factory Investigating Commission, endeared him to labor. After traveling across the state to visit factories and hear testimony, he began to issue statements that reflected interests larger than those of his immediate constituency. He became the champion of workers, especially women and minors, fighting for better fire protection, health and safety rules, limited working hours, and a reorganized state labor department.[3] He fostered some three dozen pro-labor bills that narrowed the gap between himself and the social reformers who hated Tammany.

Smith's political fortunes swung with those of his party. In 1913 he was Assembly Speaker during the impeachment of the Democratic governor, William Sulzer. After winning the 1912 election, Sulzer had become locked in a power struggle with Charles Murphy, boss of Tammany Hall. The issues of disagreement between them included an investigation of graft that Sulzer started, Sulzer's alleged inattention to Murphy's requests for patronage, and Tammany's version of a direct primary bill. A battle ensued and Sulzer lost. A bi-partisan legislative inquiry exposed misdeeds in his use of campaign funds and rode him out of office.[4] This sad affair, along with the scandal of the Herman Rosenthal murder, brought hard times to Democrats. In 1913 fusion candidate John P. Mitchel became mayor of New York, and in 1914 Republican Charles Whitman became governor, bringing in a Republican majority with him.

During the following years, Smith continued to play leading legislative roles in the Assembly. He sided with progressives on two fronts. In 1915 he helped save the Widows' Pension bill, and as a delegate to the state constitutional convention favored a streamlined administration and an executive budget, and fought Republican opposition to further social and labor legislation.[5] After the convention, civic reformers and prominent Republicans, such as Elihu Root and Henry Stimson, acknowledged his expertise on constitutional issues and his skill as a debater.

Smith began to think about more lucrative political posts. Married since 1900 and father of five, he needed more income. At the end of the five-month convention, Smith won the Democratic nomination for sheriff of New York County and then won the race. Smith earned substantial fees in this post, but the work bored him. In 1917 he put himself forward as a candidate for the mayoralty of New York. When Tammany chose "Red" Hylan, he settled for the presidency of the Board of Aldermen. Democrats swept city posts that year. Thus far, Smith had yet to lose an election. He promised constituents he would expose "the polish and the shine on the gold brick," stand on his record, and never "pass the buck." He kept his word. His open and frank personality made him seem honest, reliable, and candid. These qualities, added to his rough-hewn affability, assured his popularity. Crowds loved him. Neither a reader nor a deep thinker, he could recall facts on the spot, and at the podium spoke extemporaneously from notes jotted on the backs of envelopes. He often acted out conversations with opponents, setting listeners ahowl, and turned memorable phrases which the press loved.

In the fall of 1918 he received the state Democratic party nomination for governor, supported even by anti-Tammany forces. In mid-October, Abram I. Elkus, Ambassador to Turkey and former counsel to the Factory Investigating Commission, formed an "Independent Citizens' Committee for Alfred E. Smith." The committee revived the 1911–14 coalition that had pressed for industrial reform. Its members were well-known professionals, lawyers, business people, social workers, and reformers. Some were scions of the German-Jewish community, such as businessman Jessie I. Straus, lawyer Samuel Untermyer, and Elkus's partner Joseph M. Proskauer. About one-third were women, an important new element in the upcoming election since, as a result of the 1917 election, they were to vote for the first time in New York State.[6] By bringing them all together, Elkus delivered to Smith the money and talents of some of the state's leading citizens. This group remained the nucleus of his power base from that time to his presidential race ten years later.

Naming twenty-two women to the committee was a prescient move. Political analysts believed not only that a specifically female vote could be courted but that winning it might swing the election. Among the women Elkus chose were Harriot Stanton Blatch, suffragist; Ella O'Gorman Stanton of the National Democratic League; Mary L. Chamberlain, Factory Commission investigator and a social worker; Mary Simkhovitch, a settlement leader; Elisabeth Marbury, a literary agent and reformer; Maud Swartz, an officer of the Women's Trade Union League; and Belle Moskowitz.[7]

As the campaign progressed, Moskowitz began to worry that new women voters, many of whom despised Tammany, would not vote for Smith. She herself believed that Tammany was changing. As a result of the Progressive movement, it was decreasing its involvement with vice and becoming more cooperative in social causes. As for Smith, Belle agreed with her husband Henry that he was a "brilliant" product of the East Side, a man with a "shrewd political sense," tolerant of others, and sensitive to workers' needs. But most of her civic-minded women friends saw him otherwise. Taking her concerns to social reformer Frances Perkins, a Smith adherent, she suggested that the Citizens' Committee set up a Women's Division to focus on "the peculiar social interests of women." Elkus and his partner Proskauer adopted the idea and placed Moskowitz in charge. This Women's Division eventually became the prototype of similar divisions in the state and national Democratic parties during the 1920s and 1930s.[8]

The 1918 fall campaign was already well advanced. With barely two weeks to election day, working out of a small room in the Biltmore Hotel, Moskowitz prepared several public meetings at which Smith would speak to women. Her publicity never mentioned his connections with Tammany Hall. Instead it highlighted his work on the Factory Commission for the "emancipation of industrial workers" and the improvement of labor conditions.

The state faced tough issues, including Prohibition. Smith, who opposed the ban on alcohol, wondered what he should say to women, who in general supported it. On October 26 Smith and his advisers—Tom Foley, John Godfrey Saxe, John Gilchrist, and others—called Moskowitz into their "inner sanctum" and asked her for advice. This was her first "real personal meeting" with Smith, and she was "scared to death." Nonetheless, she told Smith to handle the wet-dry issue "without gloves," directly and honestly. Smith said nothing at the time. At a November 2 meeting for women, he "sailed into the Anti-Saloon League" as few other candidates had dared. Since many women in his audience harbored suspicions of the League and feared the repressive nature of its programs, Smith's speech that day won him many new supporters.[9]

Moskowitz organized three meetings for women. The first, held in Cooper Union and sponsored by the Business Men's League, was billed as given "by and for women." Apart from Smith, all the speakers, even the police officers, were female. Smith used the occasion to attack Governor Whitman on specific appropriations issues. The following day the Women's Division sponsored an afternoon meeting in the Lyceum Theater. The hall was packed. While awaiting the arrival of the candidate, a number of speakers gave short warm-up talks, including Elisabeth Marbury, Women's Division vice-chair, Joe Proskauer, and Moskowitz herself. Mary K. Simkhovitch spoke candidly about Smith being "wrong on the woman subject," but she thought "he will change." Finally, a band struck up and Moskowitz escorted Smith to the rostrum. Smith then attacked Whitman on the expenses of his Food Commission and on liquor, repeating a Republican State Senator's charge that "more rum was consumed in the Executive Mansion in the last four years than in the twenty years preceding."[10]

When this meeting ended, Moskowitz approached Smith with trepidation. In a "moment of enthusiasm," she confessed, she had agreed without consulting him to take him to another speaking engagement at the Women's University Club. There, women from that organization as well as from the Colony, Cosmopolitan, and Women's City clubs were waiting for him. Smith paled at the idea of addressing "highbrow" women. He said he would go, but without much grace, promising only to shake a few hands and leave the speechmaking to Bob Wagner. Moskowitz tried to persuade him otherwise. She told him he could talk to the university women in the same way he talked to professional men, discussing campaign issues frankly and openly. Frances Perkins, riding in the car with them on the way to the club, took her side. When Smith asked Moskowitz what he ought to discuss, she produced a brief list she had already written down on a scrap of paper—his legislative record on health, women's working conditions, fire prevention, and women's suffrage. Smith did address these issues, impressing his audience with his candor and grasp of detail. In closing he said, "You see, I know what it is to run a great state. You can check up on me, for if I do wrong it will not be a case of ignorance but of wilful intent."[11] Moskowitz thought this was Smith's best speech of the campaign.

Thus began the political relationship between Belle Moskowitz and Al Smith. Over time, it would blossom into a partnership. He probably did not owe his victory to her. But she was knowledgeable about issues concerning labor and women and had wide contacts among New York's professionals whom she was prepared to enlist in Smith's service. After the 1918 state campaign ended, she hoped Smith would ask for her help. With Hylan in office, city hall was closed to her. Henry was now an "industrial consultant" with the Submarine Boat Corporation, a New Jersey firm. He had undertaken "gripping work"—developing welfare programs for thousands of workers, investigating absenteeism, settling grievances, studying the causes of labor turnover and the relation of wage rates to worker efficiency, interpreting labor decisions for the company, and representing his employers in Washington, D.C. He was even planning a commissary that could feed 18,000 men in an hour. "When the central kitchen is established," he wrote his old friend Lillian Wald, "and the 12 mess halls are finished, our feeding arrangements will be a show place."[12] In contrast, except for her brief work for Smith, nothing had "gripped" Belle for some time. She was at loose ends.

Her opportunity came after the November 11 armistice. Suddenly the word "reconstruction" was everywhere. Although the United States had entered the war late, the conflict had been socially and economically disruptive. Now that it was over, the country faced further disruption. Soldiers would come home and want jobs, but industries that had turned to war production now watched their government contracts being canceled.

In New York, where so much industry was concentrated, the task of readjustment seemed overwhelming. Moskowitz knew that New Yorkers would look to governor-elect Smith for leadership on such questions as demobilization and the return of industry to a peacetime footing. But she hoped they would also see the time as ripe for attacking long-term problems: housing, education, health care, minimum wages, and safety in the workplace. Because of the war, reformers had focused their energies on the war effort. Now it was time for them to return to the problems that had engaged them before, but with a fresh perspective. Moskowitz thought Smith was in a unique position to lead. New York was the country's most populous and industrialized state. With his political power, legislative acumen, and personal magnetism, he could make New York a beacon for the whole nation. She had only to persuade him to play this historic role.[13]

Again, Moskowitz shared her thoughts with Frances Perkins. "We must do something," she said. "This administration of Smith's has got to mean something positive to the people of New York. We must pick up a program. Everything has drifted along previously. The war is over and the people expect something to be done, something new. . . . The end of the war . . . is a time for new beginnings." The non-partisan term "reconstruction" struck Moskowitz as appropriate. In New York, "we have people who are capable of planning," she told Perkins. "We've elected one of them to office. We can bring others in, . . . Republicans as well as Democrats, men and women, young and old can all contribute."[14]

Perkins and Moskowitz drew up a plan for a "Reconstruction Commission" consisting of experts in a variety of fields who would tackle the state's short- and long-term problems. Bernard Shientag, Elkus's assistant on the Factory Investigating Commission and now a Smith intimate, joined their discussions. Perkins went to sound out Smith, who liked the idea but wanted to discuss it with a few others. Organizing a dinner for about ten people in a restaurant near City Hall, he invited Belle Moskowitz, Frances Perkins, and Ida Blair (Field Secretary for the Women's City Club), along with men who had already expressed interest in serving on the commission and some of his Tammany friends. This unusual blend of reformers and politicians proved an omen of how Smith would combine contrasting elements in his inner councils. But the Smith of 1918 was not yet the Smith of the mid-1920s. Because "ladies" would be present, he brought his wife and mother along, too.

After dinner the group adjourned to Smith's office for discussion. Moskowitz explained her idea, justifying it not only as good reform but as good politics. She asked the group to remember who had elected Smith: a coalition of machine Democrats, independents and, equally important, disgruntled Republicans of the "federal crowd"—followers of Charles Evans Hughes and Theodore Roosevelt, including Stimson,

Root, and George W. Wickersham, all federal officeholders who had worked for administrative and social reform since the turn of the century. If Smith wanted to keep hold of the "federal crowd," Moskowitz explained, he would need a program that appealed to their public rather than political loyalties.[15]

Smith was impressed. Moskowitz had given him a ready-made slogan for his administration: Whitman had been the "war governor"; Smith would be the "reconstruction governor." Further, the idea made political sense. The commission would work first on a plan to streamline the state administrative structure. This would please the "federal crowd." Moskowitz suggested using the word "retrenchment" in publicity about reorganization. New Yorkers worried about rising taxes would support reorganization if they thought it would save them money. Smith saw the whole idea as political gold.

When Moskowitz next conferred with Smith, she gave him a list of people willing to serve. Smith accepted it. Abram Elkus would be chair. Bernard M. Baruch, who had recently chaired the Federal War Industries Board, had also accepted. Charles H. Sabin, president of Guaranty Trust, was joined by others in banking and finance, including Gerrit Y. Lansing of Albany, Thomas J. Quinn of the Bronx, George Foster Peabody of Saratoga Springs, Addison B. Colvin of Glens Falls, and Mortimer L. Schiff of New York. Lawyers agreed to serve, among them John Alan Hamilton of Buffalo and John G. Agar of New York, as did businessmen, including insurance magnate John C. McCall, department store president Michael Friedsam, and state Chamber of Commerce president Alfred E. Marling. Felix Adler, the Ethical Culture leader; Dr. Henry Dwight Chapin, a physician active in child welfare; Charles P. Steinmetz, scientific chief at General Electric; and Buffalo newspaper publisher and Democratic party committeeman Norman E. Mack were the sole representatives of their fields. There were five women: Sara A. Conboy from the State Federation of Labor; Katharyne A. Steele, president of the Federation of Women's Clubs; Ella Hastings, a Democratic county leader; Alice Chamberlain Chanler, the wife of a former Lieutenant Governor; and Alice Campbell Good, a clubwoman active in charitable and civic organizations. In general, wealthy capital interests and social respectability dominated.[16]

On the advice of Shientag and Perkins, Smith named Moskowitz as the Commission's executive secretary. Perkins recalled thinking that she was the best choice. The commission, after all, had been her idea. "Anybody that has so much devotion, dedication and conviction about an idea always promotes it better than someone you go out and hire, even if you submit them to the most severe civil service tests," Perkins said. She then praised Moskowitz, who "proved a perfect mine of ability," especially in organization and strategy. "Almost as soon as [an] idea was broached," Perkins said, "she could think clear through to the end to see what the possibilities were, what the hazards were and how you could do it."[17]

Moskowitz's ability to foresee effects and work out appropriate tactics in advance was one of many skills that attracted Smith to her.

As the commission's linchpin, Moskowitz needed someone else to oversee the commission's most challenging task, state retrenchment. She had in mind someone hard-driving, ambitious, and brilliant. She chose Robert Moses. Although eleven years her junior, Moses had much in common with Belle Moskowitz. Moses's mother, another "Bella" and an equally forceful personality, worked as a volunteer staff member at Madison House, to which she made large annual contributions. Moses himself had worked for Henry Moskowitz on an ill-fated efficiency rating system for the municipal civil service. Up to then, Moses's career had not been brilliant, but the Bureau of Municipal Research vouched for his high standards and passion for work. In the fall of 1918 Moses was out of a job. When Moskowitz called to ask him if he wanted to work for Governor-elect Smith, he responded enthusiastically, in spite of his mistrust of Smith, whom he considered a "typical Tammany hack," and whose cocked brown derby, gold-filled teeth, and cigar offended him. But working on official programs and having his own research staff offered Moses a proximity to power which up till now had eluded him. He thus became chief of staff for the Commission's Committee on Retrenchment.[18]

Both Smith and Moskowitz were confident that the state legislature would approve the budget request to fund the commission, a modest $75,000. As early as December 1918, Moskowitz began to hire staff. The city loaned the commission a suite of offices and some furniture in the Hall of Records. Moses too selected a staff, many from the Bureau of Municipal Research. After his inauguration on January 2, Smith announced the Reconstruction Commission plan. Two weeks later, he disclosed the list of members. Democrats and most state newspapers hailed the governor's move as a bold initiative, but Republicans were skeptical about the worthiness of the program, and suspected that a commission, if successful, might bolster the career of the new governor, whose reelection they hoped to forestall. Marshalling their forces, they opposed Smith's request for funding.[19]

Smith's goals for reconstruction were moderate,[20] but Republicans interpreted the program as radical. Thaddeus C. Sweet, the newly elected speaker of the Assembly, condemned talk of extensive social and civic reform as "Socialistic and Bolshevik propaganda," and argued that reconstruction should be confined to providing jobs for returning veterans through public building projects. His solution for the state's social problems was to abolish liquor.

During the ensuing weeks, the Speaker's colleagues found more reasons for opposing Smith's plan. First, they resented his failure to consult them about the commission's membership. That he had named prominent Republicans on his own had shown even less tact. Realizing that Smith was trying to set up an independent, bi-partisan base to help him win reelection, Republicans looked at reconstruction as a political ploy. Second, they feared Smith's hints at fiscal retrenchment through government reorganization, which to them meant eliminating patronage posts. Though both parties would be hurt, the Republican party would lose more because a Democratic governor would do the cutting. Republicans therefore had a number of reasons for stopping Smith. Then, in February, the Reconstruction Commission, operating on private funds while awaiting the allocation they still expected, began to issue policy statements and legislative advice. Republicans fumed, calling the commission a "super-legislature" with "unlimited powers." By mid-March they were so aroused that, with a clear conscience, the Republican majority voted down the commission once and for all.[21]

By that time, few Democrats were surprised, but they took it hard. The amount of funding had been reduced to what Senate Democratic Minority Leader John J. Boylan termed "a measly little $60,000 for [a] highly important nonpartisan commission." Yet Republican opposition was based on political not monetary concerns. Smith knew this, and tried to make the best of his first major defeat. In a statement issued March 19 he said, "There was an opportunity for the Republican majority to help do a great work for the State, and it is sad to say that the State and its interests, and the interests of all its people, were subordinated to the interest of partisan politics." On the same day, Abram Elkus announced that the commission members, convinced of "the great importance of the work and the necessity for its complete performance," would raise their own operating funds. Smith would never forget this extraordinary gesture.[22]

Republican opposition to a reconstruction plan masterminded by Democrats did not end with their refusal to fund it. They continued to oppose the legislative ideas that emerged, forcing Smith to mount massive publicity campaigns and to make countless compromises to get the ideas into law. Smith eventually succeeded with many of them, but only in his third two-year term and after much struggle. One wonders how much more he might have accomplished had he been more diplomatic in presenting the commission to the legislature. One also wonders why, after so many years in the Assembly, Smith bungled his very first act as governor.

Smith's biographers have never asked this question. They treat the Reconstruction Commission as the best thing that ever happened to Smith, and condemn Republicans for being bullheaded about its program. The fact remains that Smith's tactical judgment failed him. Before 1919 he was reputed to be adept at getting fellow legislators to cooperate with him. After the Triangle fire, he had warned Frances Perkins that an investigatory body made up of experts alone made politicians nervous, and advised her to be sure to allow legislators a role in naming appointees.[23] Upon becoming governor, he ignored his own advice. Belle Moskowitz and her crowd had provided him with an impressive program, complete with the personnel to accomplish it. Dazzled, he perhaps lost sight of the tricky politics involved. He forged blindly ahead, assuming that no one would object to a program that every newspaper in the state praised. He then named his commissioners without consulting Republican legislators, thereby alienating them and arousing doubts about his motives.

More caution might not have made a difference. Still, the first few months of a new executive's term is usually a "honeymoon." Minority Leader Boylan reminded his Republican colleagues that, during the war, Democrats had lined up with Republicans to vote emergency funds. Since some funds were left over, and Republicans had promised Smith at his inauguration to help him any way they could, funding the reconstruction idea seemed a good way to keep their promise. But Smith's tactical errors soured relations from the start.

Had the Reconstruction Commission funding bill passed, even for a "measly" $60,000, Belle Moskowitz and Abram Elkus would have rejoiced. As it was, their own fund-raising efforts netted only $44,000, a paltry sum for the projects they had in mind. And this amount trickled in slowly, leaving the commission in arrears much of the time. During the spring of 1919 the staff went unpaid for six weeks. Nonetheless, the group, young and committed, stayed on, living off relatives or borrowing from friends.[24]

The commission's leadership kept the staff enthusiastic. Moses, chief of a staff of ten, drove himself and others with prodigious energy. Governor Smith sometimes dropped by the Hall of Records offices to cheer on the staff. Frances Perkins recalled that Smith came to see the commission as "the key to all kinds of problems," and his support for it deepened. He was also beginning to notice that Belle Moskowitz was a source of consistently reliable information and advice.[25]

Commission staffers referred to Moskowitz as "the boss." Moses reported to her daily. During these sessions she instructed him in practical politics, tutelage he did not enjoy. Each day he arrived full of projects, all conceived according to an ideal plan of good government—a department sliced here, professionals replacing hacks there. Moskowitz then quietly but firmly reminded him that departments contain personnel who vote, and that legislators fight the loss of patronage. Moses must become more sensitive to political realities. Fuming, he would pace the floor, arguing his case anew, to no avail. "The boss" always won. Outside her hearing, Moses continued to rage, but soon accepted her cautiousness as the right approach.[26]

Tutoring Moses was only one of Moskowitz's tasks. She engineered every aspect of the commission's work, coordinating its eleven committees, participating in their hearings and debates, and publicizing their work. She saw their reports through the process of drafting, editing, printing, and distribution. She organized fourteen meetings of the whole commission in several of the state's major cities, and in weekly executive committee meetings apprised the commission's leadership of the latest progress.

Moskowitz's special expertise was publicity. She developed new techniques to let the public know how Smith, advised by the Reconstruction Commission, was solving state problems. She persuaded him to be filmed and recorded discussing reconstruction. The Industrial and Education Department of the Universal Film Company, which later hired her, produced footage which she sent throughout the state, arranging with local chambers of commerce to show them from the back of trucks. Announcing this "novel approach" to publicity, Elkus explained that the motion pictures would reach the state's many non-English-speaking residents. Moskowitz probably had an even wider audience in mind. Long aware of the power of visual information, at last she had the resources to bring messages directly to the people.

One of Moskowitz's films showed "Mr. Citizen" coming home with his pay envelope. He finds his butcher bill—his pay envelope shrinks. He finds his gas bill—it shrinks again. Then comes his grocer's bill, his wife's dressmaker's bill, and after his envelope has shrunk to almost nothing, his income tax bill. The question is then asked: how was New York spending its citizens' hard-won funds? wisely? efficiently? humanely? Every taxpayer could identify instantly with Mr. Citizen's need for answers.[27]

The culmination of the Reconstruction Commission's work came in twelve published reports that appeared between April 1919 and March 1920 and varied in length from a few pages to more than four hundred. They covered five broad areas of public concern: labor and industry; the rising cost of living; public health; education; and government reorganization. Except for the education reports, which had little impact on policy, the rest defined Smith's future programs in these areas. All the reports were a team effort, combining drafts provided by staff, committee heads, and outside consultants; but Belle Moskowitz supervised each one and wrote some parts herself. The reports bear the stamp not only of her own ideas about the interaction of society and government, but also those of an entire generation of reformers.

The Reconstruction Commission's most immediate concern was to ease industrial conditions, a complex area involving problems of demobilization, industrial readjustment, and labor unrest. The commission began by allaying fears about the effects of demobilization. "Apparently," Moskowitz reported, "every soldier in the Union was determined to pass through New York City on his way home and to stay there if he could." The commission's committee on unemployment coordinated the efforts of relief agencies, deciding that the problem was not so much a shortage of jobs as of information. Early in 1919 the federal government announced it would curtail its employment services. In March, the unemployment committee urged Smith to ask the state legislature for an emergency appropriation of $50,000 to enlarge the State Industrial Commission's employment offices, and to increase that amount to $400,000 the following year. After extensive lobbying organized by Moskowitz, the legislature granted these requests.[28]

In April, the committee reported on the relationship between unemployment and the slow pace of business readjustment. A survey of industry in the state indicated that relief was not imminent. High and unstable prices, unsettled international trade, and lack of money for loans discouraged business expansion. Stimulation from a carefully planned public works program seemed essential. The committee surveyed officials at all levels of government to determine the status of public improvements under their authority. Finding that $155 million was still available for projects suspended during the war, it urged their resumption to reduce unemployment.[29]

These recommendations covered short-term issues but later were linked to the committee's report on chronic unemployment, which emphasized the need for a "more rational organization of the market." While acknowledging that government could do nothing about general business conditions, the committee felt that state government might alleviate some causes of unemployment, such as seasonal fluctuations and the time lost in finding new jobs. The committee suggested that the state supplement public employment services, strengthen the regulation of private employment agencies, and study new ideas such as centralized planning in the labor market, vocational training for youth, and annual conferences of public works authorities to coordinate building projects with employment needs. In conclusion, it begged the governor not to suspend public works in bad times, as his predecessors had done, but to use public works to spur the economy. Should this or other remedies fail, the governor should even consider unemployment insurance, an innovation as yet untried in the United States.[30]

Today such ideas appear tame. Even though they represented views long held by many Progressive Era social reformers, most legislators of the 1920s still saw them as threatening to American free enterprise. Moskowitz recognized that the Reconstruction Commission program was ahead of its time. She hoped, however, that it would establish "standards and ideals of progress to which the State might grow for years to come."

The Reconstruction Commission was also concerned about labor unrest. Drawing upon her experience in the garment industry, Moskowitz urged the commission to recommend that the state act aggressively in settling labor disputes. In August 1919, over Elkus's signature, she wrote the governor a letter summarizing the recent labor strife. There had been two transit strikes in a period of two weeks in New York City, traction worker riots in Olean, a city in the southwestern part of the state, resulting in deaths, and work stoppages in the building trade. She recommended a state-wide conference "of representative employers, workers and public-spirited men and women for the purpose of preparing a program . . . to prevent strikes if possible, and to bring about arbitration and mediation of differences and disputes between employer and employee by men and women who are willing to serve the State voluntarily."[31] All the follow-up correspondence on the conference and the Reconstruction Labor Board that emerged from it shows Moskowitz as the fulcrum of Smith's labor policy.

Smith called the conference for September 16 in Albany. About eighty men and women representing business, labor, the legislature, the Reconstruction Commission, state government, and the public attended. In a "personal and confidential" letter to Smith, Moskowitz outlined what she thought the conference should do. She hoped that it would confine itself to one topic, "the maintenance of production and the prevention of strikes," and that it would "attain . . . an agreement on both sides that there will be no strikes or lock-outs for a period of a year and that you will appoint a Reconstruction Labor Board, similar to the War Labor Board, to whom all of these disputes can be referred. This, of course, would have equal representation on both sides." She hoped further that the board would stimulate the formation of similar boards in all the state's leading industries. Working in cooperation with the Industrial Commission, the boards would "study these industries and have constantly in preparation data upon which wage adjustments can be intelligently made." Finally, she told Smith that she was preparing a list of invitees to the conference and an agenda. She was particularly anxious that the conference not be a "talk-fest" but result in "constructive" work.[32]

On the back of her letter, Smith scrawled with the thick-nibbed pencil he used for all his informal notes: "Mrs. M to send to me list of names of people to attend Conference—Confirm on Sept 16th." He or his secretary, George Van Namee, referred all later matters relating to the conference to her. She drafted the letter of invitation, suggesting whom to telephone instead of write; developed the publicity crediting the Reconstruction Commission with the idea for the conference; offered to come to Albany "to help get this material out"; and kept a tight rein on the invitation list, for instance scratching Women's Trade Union League organizer Rose Schneiderman, "pending word from [James] Holland," president of the State Federation of Labor. "I think it might be unfortunate if she were invited," Moskowitz wrote. She preferred Mary Dreier to represent the WTUL. It was essential to have practical people, she wrote Van Namee, not "industrial doctors."[33]

The Albany conferees may have thought everything unfolded spontaneously, but the meeting went exactly as Moskowitz planned. Smith opened with a brief analysis of the causes of the present unrest—production lags, the rising cost of living, a dearth of necessities, all exacerbated by strikes. Then James Holland and other labor leaders spoke of the need for more vigorous government policies to prevent lockouts and achieve agreements. They also spoke in favor of an investigatory board similar to the War Labor Board. At last Smith unveiled "his" idea:

> I have an idea that we could create a committee of nine . . . made up of three representatives of employees; three . . . of employers and three to be named by the Governor, who would be men and women in this community about whose fairness and justice there could be no possible question, and who would . . . represent the whole State.

This committee, housed in Albany, could "determine ahead of time whether it is possible to avert a crisis." Such a group, Smith believed, could have stopped the recent transit strike before it began. Concluding, Smith expressed the hope that, "[after] the majority of that committee has spoken and they are right about it, I propose to put the power of the State behind them as far as I am able to do it. And that after all is what will probably compel the decision." His audience applauded warmly.

The conference then debated how the board would operate. Responding to concerns that a Reconstruction Labor Board would duplicate or interfere with the State Industrial Commission, Henry Moskowitz spoke in favor of the "fresh blood" and "inspiration of the outsider," the roles both he and his wife had played during the protocol. Another speaker from the garment industry called for the new board to hear grievances as under the protocol system. Smith, reiterating his plan to throw the "full power of the state" behind the board, never clarified exactly what he meant, nor was he asked. After several hours of discussion, the conference split into groups to submit lists of potential members. On September 25, Smith appointed the Reconstruction Labor Board and made Belle Moskowitz its executive secretary. The usually hostile state legislature, perhaps contrite over its treatment of the Reconstruction Commission, unanimously passed a $25,000 appropriation for the board's expenses.[34]

As secretary of Smith's Labor Board, Moskowitz finally had official sanction for the labor policies of progressive reformers. Even if the extent of state support was vague, the board's official status distinguished it from previous boards on which she had sat. She still believed in the voluntary cooperation of society's diverse elements, a theme that recurred in other Reconstruction Commission proposals. But failing that, she wanted the state to help. State help would come, in good progressive tradition, through the work of unpaid public servants and a small staff, who would collect data, make rational, scientific proposals, and educate the public to endorse them.

The Reconstruction Commission reports on the rising cost of living received wide attention. The reports focused on two fronts, food and housing, blaming the rising cost of both for current strikes. They recommended state-funded housing incentives to the building industry, and a total revamping of the state's marketing networks. These recommendations profoundly influenced Smith, who fought for years to implement them.

The commission issued two reports on food costs, one arguing for modernized terminal markets, the other for new transport systems. In the commission's view, without better means to get produce to markets, all market modernization would be futile. It therefore urged the governor to promote rural motor truck express lines. Only trucks, with their flexible routing, pick-up at farms, and deliveries on time, would reduce costs. During the war, when congested railroad lines hindered food distribution, rural truckers had demonstrated their "usefulness and economy." The report advised Smith to set up a temporary, non-salaried "State Highways Transport Committee" to "stimulate the development of motor truck routes that will serve rural communities." Hastening to add that the state would not thereby be entering the trucking business, the report concluded: "The function of the State in this connection is largely educational. In the advancement of any new enterprise of a public service character the interests of the public should not wait upon the private interest or initiative of individuals." This last phrase reflected Moskowitz's view of government as a spur to the private sector. The National Automobile Chamber of Commerce was so happy with this report that it bore the cost of printing ten thousand copies.[35]

Early in 1920, the Reconstruction Commission published its final statement on food. It began by condemning laissez-faire policies that resulted in "profiteering, waste and excessive costs." Shortages or oversupplies, losses due to poor transport, and price uncertainties, all caused by faults in the system or individual malpractice, could have been avoided if the state had taken action. The commission advised enlarged licensing, regulatory, and arbitration powers for the Department of Farms and Markets. Pointing again to wartime experiences, the commission cited the New York Federal Food Board and how it had settled thousands of disputes among suppliers, distributors, and retailers, thus preventing the ruin of perishables. Finally, it recommended making milk a public utility, an idea discussed at the beginning of the war and raised again because of a crisis in milk distribution that occurred in 1919. Charges of corruption in the Farms and Markets Department, and later of impurities in milk, led to investigations of the dairy industry and eventually to a call for a State Milk Commission. Smith tried to implement the idea but failed. Responding to upstate dairy interests, the legislature defeated a Milk Commission bill in April 1920.[36]

Relieving New York's housing shortage seemed urgent not just to the governor but also to Mayor Hylan and the state legislature. All three addressed the problem, the Governor through the Reconstruction Commission, the Mayor through his Committee on Rent Profiteering, and the legislature through a joint legislative Housing Committee chaired by Senator Charles C. Lockwood. As in the labor field, Belle

Moskowitz influenced the development of Smith's housing policies and his later efforts to get them written into law.[37]

Social reformers had long pressed the government to establish a centralized housing policy. A population explosion and rising land values forced workers to live in areas that were increasingly crowded and run-down. Developers did not build low-cost housing. Tenement house laws enacted after the Civil War subjected urban housing to safety and sanitation codes, but these had driven up costs and thus discouraged investment. By the turn of the century, most reformers were asking government to offer inducements to builders. The idea of using state resources to interfere in a private enterprise aroused deep conflicts. Before the Great War, only Massachusetts had experimented with tax exemptions and low-cost loans to limited-dividend corporations planning to construct homes for workers. While a short-lived federal housing program in 1918–19 established a precedent for public assistance, the numbers of actual homes built remained small. Opposition to increasing the tax burden to help house the poor continued throughout the 1920s.[38]

During that period, urban areas faced critical housing shortages. Housing construction fell sharply during the war; afterward, inflated building and financing costs kept home-building in a slump. Urban rents rose dramatically. In an atmosphere of complaints of tenants against landlords and speculators, and of warnings from charitable institutions that they could no longer care for the homeless, the Reconstruction Commission's committee on housing attacked the problem.

John Alan Hamilton, an eloquent lawyer from Buffalo, chaired the housing committee. In a speech given to welfare workers at a meeting organized by Moskowitz, Hamilton began by observing that the state had been regulating housing for a long time. In the present shortage, however, the state had to move toward providing housing. This did not mean putting the state into the housing business but providing tax exemptions on mortgages and cheap loans to builders of decent, low-cost housing. It also meant centralized planning on the relationship between dwindling land and rising populations, and on how best to save on building costs through large-scale projects. Hamilton admitted that some might view such proposals "with alarm, as smacking of socialism, or as visionary and theoretical, or as opening the door wide to State and municipal partisan patronage, waste and corruption." He therefore urged his listeners to convince the state's "enlightened opinion" of the wisdom of a coordinated, government-fostered housing policy. In a stirring finish he called up a vision in which every child would be assured "healthful and reasonably decent physical surroundings," every dwelling would comply with simple "sanitary and esthetic requirements," every man would rent a home at as low a cost "as co-operative endeavor can effect," and all people "shall have learned that it is more profitable to share the rise in value of their city land jointly than to scramble for it individually."[39]

Moskowitz devoted a major portion of her energies during the life of the commission and afterward to fulfilling this vision. She began by fact-gathering, issuing a call for volunteers to make a block survey of housing conditions. She discussed the results of this survey both in a preliminary report issued in May 1919 and in more detail in the commission's final statement on housing of March 1920.[40]

The survey, covering a thirty-four-block area in various sections of Manhattan, the Bronx, and Brooklyn, exposed the seriousness of the crisis. Volunteers found residents enduring levels of filth, darkness, and overcrowding that made a mockery of tenement house laws. Yet, even in buildings that lacked heat, garbage disposal, and washing and toilet facilities, less than 1 percent of the apartments were vacant. In newer buildings, rents in some cases had doubled. Often, professional lessees guaranteed owners against losses from vacancies and repairs, and then jacked up the rents. The commission believed that laws against rent-profiteering would not remedy the situation. Maintenance and managerial costs would still go up, pushing rents up, too. "The rising rents are merely a symptom," the first report declared. "The disease is lack of sufficient houses," the only cure "building more houses at once."

But declining building rates were unlikely to turn around. The Reconstruction Commission argued that the state had to foster large-scale building projects on cheap land near existing industry. This building should be of a "better type," spacious, light, and well-ventilated. Practical but imaginative designers should advise the state on its construction. To discuss this further, Moskowitz asked the Governor to call a conference on May 16, 1919, of "representative men and women" interested in relieving the housing shortage.[41]

During the May 16 conference, the Reconstruction Commission temporarily skirted the issue of public funding for housing by unveiling a private plan. Elkus announced that a semi-philanthropic "Housing Reconstruction Corporation" had been formed to raise money to loan to builders at modest rates. Well-known financiers, such as J. Pierpont Morgan and Cleveland Hoadley Dodge, had already subscribed to the fund. Governor Smith praised the "women and men of the Reconstruction Commission" for the "good sound common sense" they had brought to the question, all "without publicity, without looking for the friendly lines of the newspapers." The state cannot build houses, Smith said, and laws cannot "make houses grow on empty lots." Only the continued patriotic service of the private sector could remedy the crisis.

In the weeks that followed, Elkus and Moskowitz issued sanguine statements about how well their "limited-dividend corporation" was doing.[42] But keeping up a sufficient flow of money proved impossible. Early in June, in cooperation with Senator Lockwood, they urged Smith to ask Congress for tax relief on mortgages over $40,000 and on bonds of the State Land Bank to make these securities as attractive as tax-exempt Liberty Bonds. Moskowitz reiterated these requests in the final housing report of 1920. By then, building costs had risen so high that a limited-dividend corporation putting up apartments in Brooklyn had already raised its rents from an original nine dollars a room per month to fourteen dollars, just to guarantee a 5-percent return. The situation was desperate. There were hardly any vacancies in apartments "fit for human habitation." New apartments, costing between fifteen and twenty-five dollars a room, were beyond even average wage-earners. New York City alone was forty thousand apartments short; upstate the situation was similar.

To meet the crisis, the Reconstruction Commission asked the state to pass laws that would reduce building costs. After analyzing all the factors that influence housing costs—land, taxes, money supply, planning, building and zoning laws, and management—the commission made three requests of the state: that it extend credit to builders of moderate-priced homes; that it grant permission to municipalities to purchase and develop lands for low-cost housing; and that it establish local and state boards to study and coordinate regional housing policies.

A six-member minority of the commission's housing committee opposed two of these actions—extending state credit and allowing municipalities to develop lands. The majority, however, argued that American private enterprise had failed to house the labor needed for its industries and that immeasurable suffering had ensued. As Moskowitz observed in her final report, the only persons deriving benefit from rising land prices were the speculators. The community as a whole got very little, and the poor were driven into slums. If, as she suggested, the community owned and controlled "large tracts of land . . . along the lines of the next probable development," it could rent parcels to builders or tenants for long terms at moderate rates. In her view, the benefit of increased land values should accrue to the whole community, not to speculators.[43]

The entire committee endorsed the idea of local and state-wide housing boards. These would collect and distribute information, propose revisions of building and zoning codes, and study means to lower building costs through "better planning." The report noted that almost every other civilized country, except America, had a coordinated housing policy for its industrial workers. America's failure to implement such a policy condemned its cities to inexorable decay.

Smith and Moskowitz had to wait three years for legislative approval of a housing board. Their only immediate successes in 1919 were a few laws, passed in the legislature's extraordinary session, that relieved some minor aspects of the crisis.[44] Moskowitz's housing policies had political consequences for Smith, however. By endorsing them, he distinguished himself from Tammany and its New York City leader, Mayor Hylan. Hylan's housing committee had not cooperated with Moskowitz. In fact, both Hylan and his committee's chair, Nathan Hirsch, had been openly hostile to the Reconstruction Commission. Hirsch scoffed at the limited-dividend corporation as a "drop in the bucket," saying the governor looked ridiculous coming all the way from Albany to announce it "with a flare of trumpets." Hylan had told Senator Lockwood he would not object to one interior, hence windowless, room in remodeled tenements. When told that the Charity Organization Society opposed such a plan, Hylan retorted, "Reformers oppose all sorts of legislation."[45] Since Governor Smith had agreed with the reformers, Hylan decided that Smith had sold out to the impractical idealists.

Up to this point, all the commission reports shared certain themes: government should spur business interests to cooperate voluntarily; when voluntarism failed, the state should step in. Thus the Reconstruction Labor Board, State Highways Transport Committee, expanded powers for the Department of Farms and Markets, and a State Housing Board, while differing in particulars, were all of a piece, inspired by policies long cherished among progressives and, more recently, by the success of similar institutions during the wartime emergency.

In the realm of public health, the role of the state, within certain limits, was already more widely accepted and therefore less controversial than in areas traditionally dominated by private enterprise. The commission report on the subject came out in October 1919 without fanfare, yet it had a major impact on the development of rural health facilities and services in the state of New York. Written by Dr. Henry Chapin, the report began with the axiom that in modern society the health of individuals was a community concern that could no longer be left to "chance" or "individual prudence." After holding public hearings, Chapin made a series of broad proposals covering health issues for all age groups within the population. He recommended active educational campaigns and state grants for maternal and infant welfare, stricter training and licensing of midwives, foster homes instead of institutions for destitute children, and state-wide pasteurization of milk. Since physical defects preventable in childhood had rendered ineligible almost 40 percent of New York's draft-age youths, he urged local governments to extend child welfare services, including milk and hot lunch programs for schoolchildren, and to make physical education compulsory. He called for the combination of state health services in small communities, a larger staff for the State Health Department, an increase in the number of public health laboratories, and cooperation with federal authorities in venereal disease campaigns. He noted with alarm the growing death rate from cancer, heart disorders, and arteriosclerosis, recommending that the state continue to fund research facilities. Finally, he concluded with a request that had long been a component of Progressive Era programs—compulsory health insurance for industrial workers.[46]

Almost point for point, the essence of this report, including the request for health insurance, reappeared in Governor Smith's annual message of 1920. The legislature cooperated with the Health Department in carrying out many of the proposals, even when Smith was out of office during 1921–22. When Smith returned in 1923, he noted this progress but urged the legislature to do more, especially in the area of providing matching grants to counties improving their health services. This legislation was enacted, but workers' health insurance, an idea too far ahead of its time, did not resurface.[47]

New York was ready for administrative reorganization. Yet the task proved to be the Reconstruction Commission's hardest. The committee designated to draw up a plan had its own staff who worked long hours for many months preparing a report. Most of its proposals eventually became law, but only after an arduous, time-consuming, and frustrating process. Interests opposing change were so powerful that only the combined efforts of Smith, his close advisers, a state-wide lobbying team, a bi-partisan State Reorganization Commission, and Smith's successors in office could fulfill the Reconstruction Commission's reorganization plan.

In New York, as in other states, neither master plan nor rational order determined how governmental agencies related to one another or to the executive. By the turn of the century, the state had become a hodgepodge of agencies, boards, bureaus, councils, departments, or commissions, many with overlapping or redundant jurisdictions. Some of these bodies were permanent, others temporary, with vague termination points and unclear lines of authority. Elected or previously appointed officials sometimes served longer than the governor, or worked under authorities beyond the governor's control. When investigators found the head of the Department of Farms and Markets guilty of corruption, for example, Smith had no authority to remove him. Yet, when trouble arose in that department, all looked to Smith to resolve it. Worse, the governor served only a two-year term, which gave him even less power over the executive branch than in other states. Worse yet, the state legislature controlled the budget; the governor had power only to veto line appropriations. Executive weakness, along with an inefficient, wasteful blurring of administrative and legislative functions, gave rise to calls for root-and-branch reform.

From Charles Evans Hughes to Whitman, New York governors had campaigned for reorganization of the government, supported by civic groups such as the Citizens' Union and the Short Ballot Association. A report from the Bureau of Municipal Research influenced the delegates to the Constitutional Convention of 1915, who included a reorganization plan in their new constitution. But the voters rejected the new document, not because of its reorganization plan but because of other problems. Subsequent efforts to reorganize through statutes or constitutional amendments also failed. Amendments needed approval from two successive legislatures before being submitted to the voters, and party bosses, loath to lose the sinecures they dispensed, and civil servants, fearing loss of jobs, joined forces to defeat them.[48]

Thus matters stood when Smith came to power. He favored reorganization, but did not at first see it as the Reconstruction Commission's top priority. Instead he asked it to find new sources of tax revenue for the state's expanding needs. In the spring of 1919 the legislature passed a state income tax law that solved the problem of new revenue sources. When the commission's Committee on Taxation and Retrenchment turned its attention to economy, it decided that reorganization was the key. By then Smith had discovered some of the weaknesses of his new office. A powerless governor presiding over an array of uncoordinated agencies staffed by political hacks could never implement the reforms the Reconstruction Commission was proposing.[49]

After many months of work, the staff responsible for drawing up a reorganization plan released a first draft on September 22, and two days later held a public hearing. Heads of state agencies and civic organizations attended, offering only minor criticisms and high praise for the job the commission had done. A final version appeared on October 10, both in a summary format readable in one sitting and a book-length volume complete with historical surveys and illustrative charts for reference.[50]

While the reorganization plan borrowed much from the proposals of 1915, it boasted many unique qualities. Its principal author, Robert Moses, relied on Bureau of Municipal Research work and personnel to bring it together, but commission staff and members made further contributions. Thus the plan reflects broadly the whole commission's ideals and prejudices.

The plan's main feature was a reduction of the number of separate agencies in the executive branch. With 187 of these, no unified planning was possible; moreover, without a centralized budget system, 187 "streams and rivulets" ran unchecked out of the state treasury. The commission recommended sixteen groupings of agencies, each under a single head or board. The plan's second feature was a strengthening of the governor's powers with a four-year term and control over all department heads and vacancies on boards, subject to Senate confirmation. There would be a short ballot. New Yorkers elected seven state officers: a governor, lieutenant governor, comptroller, secretary of state, treasurer, engineer, and surveyor; reorganization would reduce the list to the first three. The final feature of the plan concerned the budget. The governor would prepare it, retaining a veto that restricted the legislature's ability to make appropriations.

The commission's reorganization plan reflected a larger hope that government become not only more efficient and economical, but more accountable to the people and responsive to social needs. To ensure that these needs would be addressed, the commission applied ideas contained in its earlier reports to the internal structures of reorganized departments. The new Labor Department, for example, would have a

"Bureau of Employment" functioning exactly as described in the commission's report on unemployment. In addition, the department would administer minimum wage and health insurance laws, should they pass. A reorganized Health Department would emphasize county districting for rural health services.

The plan also incorporated reform goals long of interest to progressives. The new Education Department would administer an Americanization program. Boards and councils in the Health Department would hold a number of places for women. The Civil Service, changed from a commission to a professional department, would receive enough funds to guarantee the best supervisory personnel. After identifying irrational and unfair civil service practices, the report recommended major changes, including the reclassification, regrading and equalization of positions, removing non-policy-making jobs from exempt status, repealing fixed salaries for employees other than department heads, and establishing a uniform pension system.[51] The report presented all of this in a thorough and clear manner, without jargon. Convinced of the ultimate wisdom of an informed democratic process, commission staff had produced a report that any educated person could easily fathom.

Almost immediately upon the report's release, party pols attacked it. Warning civil servants of impending doom, they charged the commission with trying to make the governor a "czar." Hoping to forestall such attacks, the commission included a survey of similar reforms across the country, showing that its plan was not radical but in the mainstream of contemporary policy. Moreover, Moskowitz had held back on radical changes. In planning the departments of education, labor, and agriculture and markets, for example, Moses had argued for single heads. Moskowitz insisted that staying with the traditional boards would least endanger the entire reorganization plan.[52] Such precautions failed to prevent major opposition to reorganization.

The process by which Smith, Moses, Moskowitz, and others achieved reorganization was tedious and frustrating. It can only be summarized here. In 1920, Republicans offered their own plans, which omitted the executive budget and four-year term. When Republican Nathan Miller was governor from 1921 through 1922, the legislature passed statutes to streamline state departments, but constitutional changes never emerged. Back in office in 1923, Smith repeatedly conferred with Republicans to hammer out compromise amendments, but the four-year term and executive budget remained obstacles. Finally, during Smith's third term, and only with the aid of the state's "federal Republicans" and the support of influential newspapers, a reorganization amendment won both legislative and voter support. The new plan, essentially following the Reconstruction Commission blueprint, with some minor changes, became effective with the 1926 legislative session. An executive budget was approved the following year, but Smith lost his battle to institute a four-year gubernatorial term with elections held in non-presidential years. When the legislature insisted on a four-year term elected in presidential years, Smith campaigned against it, winning voter support two-to-one. New York governors would not enjoy a four-year term until the constitution was revised ten years later.[53]

Throughout this struggle for reorganization, Moskowitz kept a close watch on its twists and turns, supervising publicity, advising the governor on tactics, and helping with his speeches as well as giving her own. She reached out for support especially among women members of civic, welfare, or reform organizations, and addressed women's organizations. She stressed with them the non-partisan side of administrative reform. "All citizens of a state must live under its government," she wrote in a press release announcing the formation of women's committees on reorganization; "and as cost and efficiency are paramount," everyone must take the subject to heart. Then, appealing to feminine solidarity, she provided the name and affiliation of every woman serving on the Reconstruction Commission's Retrenchment Committee.

Women's groups that set up reorganization committees included the Council of Jewish Women, Women's City Club, New York State League of Women Voters, New York City and State Federations of Women's Clubs, National Civic Federation, Consumers' League, Women's Municipal League, and United Neighborhood Workers. Women's City Club members made a special effort in behalf of reorganization, taking part in a special study class, endorsing the plan, and then appropriating $75 to print and distribute the endorsement. The board called the plan "the only basis on which economy and responsible government, consistent with good service, are possible." In March the club sent a delegation to Albany to attend the Senate hearings on reorganization bills. Belle Moskowitz's name never appeared in these matters, but her hand surely guided them.[54]

When she spoke to welfare workers, she emphasized the need to spend the state's revenues more wisely. She was not promising simple economies. Eschewing a piecemeal cut-and-trim approach, the Retrenchment Committee had sought changes that allowed more spending in areas underfunded in the past. "Only a week ago I heard Governor Smith say, that he for one was not satisfied with the way in which the work was being done in the departments that had the care of the State's dependents; . . . [and] while it cost fourteen million dollars last year to run those departments, it ought to have cost at least twenty millions, if the work were adequately done." Continuing, she cited poor equipment and salaries so low that not even "the most unskilled type of labor" would want a state welfare post. The state could never broaden its activities along "progressive lines," she warned, unless it had more funds at its disposal.[55]

In other publicity Moskowitz tried to answer Smith's foes. Reorganization would reduce patronage, she admitted, but not just of legislators. Governors, too, would lose gifts they had traditionally bestowed. Instead of making the governor a "czar," however, greater control over officials' behavior would make him more responsible. To critics who charged there would be no economies from reorganization, she reiterated Smith's pledge to welfare workers that public funds would be more efficiently and wisely spent. To those who argued that reorganization was too radical, she noted the many states that had reorganized successfully. These points found their way into speeches of Smith and those of other members of the Reconstruction Commission who went out to promote the plan.[56]

Once, in the heat of the legislative battle for reorganization, Moskowitz allowed herself to be directly quoted. She was angry. On April 20, 1920, Senator Henry M. Sage, an upstate Republican, had introduced reorganization bills that reduced the governor's appointive powers and gave them to the legislature. Moskowitz denounced his bills the next day. "The Legislature has made a political football of the Governor's reconstruction plan," she fumed. Now that they were worrying about how to account for their actions, they were throwing "dust in the eyes of the people, pretending to sympathize with a general principle and killing its application." Playing "cat and mouse," they conceded the consolidation of departments, but instead of centralizing the administration in the governor's office proposed "regencies," department heads elected by the legislature or removed only with the Senate's consent. The result, she said, was a system "more diffuse and irresponsible than ever."[57]

This was one of the rare times Moskowitz vented her anger. Otherwise she worked quietly throughout 1920 to realize Reconstruction Commission goals. She continued to work even after she lacked the funds to keep a staff. As secretary of Smith's Labor Board she had a salary to pursue some issues. She also had Robert Moses, who became secretary of a "Citizens' Committee on Reorganization in State Government." Moskowitz had hoped that the Reconstruction Commission would evolve into a "permanent organ of research and consultation for the guidance and assistance of the Legislature and the Executive," an idea first expressed by journalist Walter Lippmann.[58] As head of such an agency Moskowitz could have influenced much of future state policy.

While this dream was not to be, other successes consoled her. Smith had adopted as his own the Reconstruction Commission's solutions to New York's problems. Thus armed, he made ready for his second term. The concrete proposals of the commission were widely respected. Some of New York's most knowledgeable professionals were advising him. Belle Moskowitz, to whom he owed much of his arsenal, was staying on to continue the fight. There could be no outcome but triumph.

In the long run, he did triumph. But not immediately. In 1920, the voters turned him out of office. He came back stronger than ever two years later, riding in part on the continued strength of the reconstruction idea. By opposing this idea in so partisan a manner, state Republicans made themselves look foolish. In the mid-twenties, the "federal" Republicans joined with Smith to reorganize the state, thus rescuing their party from total debacle. But there would be no more Republican governors for another two decades.

CHAPTER 8

A Power for Good

[Public relations] is an enthralling occupation for the woman who has constructive ideas. . . . She may be a power for good. Her goal is of her own choosing.

Doris Fleischman (Bernays)
Outline of Careers for Women, 1928

When the Republican landslide of 1920 swept Al Smith from office, Belle Moskowitz was not without resources. By then she considered herself a "public relations counsellor." Public relations was a new field that had arisen in the latter years of the nineteenth century, when large firms began to use press agents to get them free publicity. Most firms were content with the short-term results that followed. But during the late nineteenth century, with big businesses and monopolies coming increasingly under attack, some firms began to take a more systematic approach to publicity in order to influence voters. Three men helped develop the new field of "public relations"—Theodore Newton Vail, head of the Bell System, and Ivy Lee and Edward Bernays, both early-twentieth-century publicists.

Credit for having coined the term "public relations" usually goes to Vail, who used publicity to expand the telegraph and telephone system. As publicity director for the Pennsylvania Railroad from 1902 to 1910, Lee invented the news release, an account of his firm's policies sent regularly to the press. His reliability helped counter the generally bad reputation of the "press agent." Bernays took publicity even further. He developed a theory of propaganda and professionalized the propagandist's status. In 1913, at the age of twenty-two, he became editor of the *Medical Review of Reviews.* Actor Richard Bennett planned to produce *Damaged Goods,* a play by Eugene Brieux that attacked public ignorance of the consequences of venereal disease. To help Bennett raise money for the production and prepare public opinion for it, Bernays set up a "Sociological Fund" of the *Medical Review.* The Fund's executive board consisted of well-known social and moral reformers, including his friend Mrs. Charles H. Israels, then already well known because of her campaign for dance hall reform. The success of the Fund convinced Bernays that propaganda, which he defined as the "conscious and intelligent manipulation of the . . . habits and opinions of the masses," could be used as a "power for good." That it could also be used for evil purposes became clearer to him later on.[1]

To Progressive Era social reformers, the "conscious and intelligent manipulaton" of public opinion was nothing new. Many of them, including Belle Moskowitz, were adept at the art long before the field had a name. At the Educational Alliance, she had used exhibits and entertainments to win neighborhood goodwill; as a reformer, she had molded public opinion and influenced legislators with staged events. By the time she went to work for Al Smith's campaign and later the Reconstruction Commission, she could claim she was a "public relations counsellor."

In a friendly feud, Moskowitz and Bernays disputed who had coined the name of this new vocation. Moskowitz claimed priority, but Bernays, while acknowledging her as a pioneer, took credit. According to him, after finishing war propaganda work for the Committee on Public Information during World War I, he and his wife Doris Fleischman opened a "publicity direction" firm. Between 1919 and 1923, the couple came to call themselves "counsels on public relations," a term that avoided the connotation of the aggressive if not mendacious "press agent," while sounding more professional than "adviser."[2]

The Bernays-Moskowitz feud aside, Moskowitz's pioneering role is beyond doubt. Early in the development of the field, she not only recognized its potential, but made a special contribution by developing the public relations motion picture. Few non-fiction films were made before 1920. Most dealt with scientific or educational topics, or were made by the government for propaganda or instruction during World War I. Some larger industrial firms, such as United States Steel and International Harvester, also made films. But thirty-five-mm film was both costly and flammable. It was not until after 1923 that cheaper, non-flammable sixteen-mm film was introduced. Other technical improvements, including sound, came only in the late 1920s.[3] To use film at all in the early twenties was innovative. In March 1920, when Belle Moskowitz became director of the Industrial and Education Department of the Universal Film Company, the step caused comment in the press not only about her high salary but also the novelty of the work she would be doing.[4]

Moskowitz's first venture in public relations film-making had been for the Reconstruction Commission, the short film of Smith explaining the benefits of government reorganization. Two years later she produced a second film for the New York-New Jersey Port and Harbor Development Commission. This commission had been created in 1917 by the legislatures of both states when, after New Jersey had asked the Interstate Commerce Commission for preferential shipping rates over New York, the ICC told the two states to work jointly on the port's future.

The port district was a huge area bounded by Yonkers and New Rochelle on the north, Great Neck and Far Rockaway on the east, Perth Amboy and Sandy Hook on the south, and Paterson, Newark, and New Brunswick on the west. Serving the immediate needs of eight million people, it also handled 50 percent of the nation's foreign commerce. A staggering quantity of goods passed through this gateway: annually,

4 million tons of food for the district alone, and 120 million tons of freight. Twelve railroad trunk lines (nine in New Jersey, three in New York) and over 8,000 steamships functioned in the area. Stretching along 800 miles of waterfront, the district affected the lives of 105 municipalities.[5] Anyone hoping to change the interaction of all these elements—manufacturers, shippers, railroads, truckers, longshoremen, wholesalers, retailers, consumers, government agencies—had to proceed with great sensitivity toward many powerful interests.

Merely laying the groundwork took almost four years. On the advice of Eugenius Outerbridge and Julius Henry Cohen, the Port and Harbor Development Commission proposed that the two states create a Port of New York Authority modeled after England's London Bridge Authority. Directed by a commission consisting of an equal number of members from each state, the authority would plan, fund, and execute all port improvements. Many commercial leaders approved this plan, but it had strong opponents. In 1920, because of complaints about dock and harbor congestion, New York City and Newark had both launched independent renovation projects. They now refused to work on a joint plan. But expensive delays continued. Trade was being diverted to other cities, such as Philadelphia and Baltimore. New York City's commercial leaders urged action, but neither New York nor Newark would give way.

Governor Smith favored a port authority, but because of his close relationship to New York City had refused to sign an enabling act over the city's objections. Ironically, his defeat in 1920 helped break the deadlock. The new governor, Nathan Miller, had no debts to Tammany Hall. In his first post-victory utterance, he assured New York's commercial interests of his support. Encouraged, the Development Commission pressed its case throughout 1921, engaging speakers such as Smith and Cohen to address clubs, organizations, and chambers of commerce. Press publicity explained and justified a coordinated port plan. At the end of March, the New York legislature passed a bill to enable the two states to sign the treaty creating the Authority. The New Jersey legislature passed a similar act over the governor's veto. Although opposition continued, the following August Congress passed and President Warren G. Harding signed a resolution authorizing the treaty. Early in 1922 both state legislatures authorized the Port Authority, an act President Harding approved on July 22.[6]

A publicity campaign directed by Belle Moskowitz helped determine this long-awaited outcome. The campaign's centerpiece was a film she produced that stressed the link between inefficient transport and rising living costs. Early port development publicity had told stories about fish caught up north, brought down to the city through the Bronx to the Fulton Fish Market in lower Manhattan, and then redistributed in the Bronx at inflated prices. The turnip offered another example. If a Long Island farmer sold an entire bushel for twenty-five cents, he considered himself wealthy. By the time one turnip reached a Manhattan grocery stand it alone cost twenty-five cents.

Moskowitz's film focused on the potato. "Picture the potato leaving the farm, unwrinkled, round, and with bright eyes," suggested a news article announcing the film's release. Tossed into a barrel, it travels by freight car to the Jersey Meadows. A food dealer in Manhattan is waiting for it. He waits over three full days. The potato sits twenty-six hours in the Meadows, then twenty-two in Jersey City. After two hours, it has crossed the Hudson by "lighter," but then lies on a Manhattan pier for another twenty-four. Finally unloaded, the potato waits for a truck, which worms its way through traffic before arriving at the food dealer, not far away. Even after all this, the potato, now wrinkled, has not reached the local market.[7]

The film was called "Mr. Potato." It began with aerial views of the port district's railroad and steamship facilities. Using diagrams, animated pictures, and printed dialogue, it illustrated proposed changes. Then came the potato. Drawn with eyes, nose, ears, and limbs, the potato wept over the length of its journey. "I've been in this train twenty-six hours. It is terrible!" The film then represented the passage of ninety-one hours. The potato, at a cost of .6895 cents, finally reached the retailer. The film ended by claiming that the port plan, which would replace lighterage with automated trains running through tunnels, would reduce the time and cost of commerce.[8]

Throughout 1921, in private gatherings and movie theaters, viewers watched "Mr. Potato" lament its fate. It was shown at the ceremonial signing of the treaty, as well as to a party of over one hundred congressmen and state officials who toured the harbor by boat and then had dinner together at the Commodore Hotel. Clubs, societies, and chambers of commerce used the film as a warm-up for speeches by civic leaders. To coordinate the showings and other publicity, Moskowitz organized an "Advisory Council on Education," a large group of men and women who advised the Authority on how to get its plans across to the public. The Council offered prizes to high school students for essays on the relationship of the port to the cost of living. It set up professional subgroups, such as in real estate or silk manufacturing, and organized meetings at which Authority boosters spoke. Such events ended with motions calling for swift approval of port development. Using the press contacts she had made while publicizing the Reconstruction Commission, Moskowitz arranged newspaper coverage for every meeting, speech, film presentation, and resolution.[9]

Once New York and New Jersey had signed their treaty, each state had to pass a law establishing the Port Authority. At the end of January 1922, in a move Moskowitz probably organized, Smith, Outerbridge, Cohen, and representatives of twenty-five New York City organizations went to Albany to support the Meyer-Mastick Port Authority bill.[10] Smith made a long speech, a *tour de force* that laced a masterful display of facts with vigorous, sometimes caustic humor. He condemned the New York City Board of Estimate for having refused to attend an informational meeting on the port arranged at its own request. He then quipped that, when the board asked not only for eleven more copies of the port development plan but also of the supporting documentation, he voted to comply after recognizing it as a "trucking job which would be well worth while." (Everyone present knew that Smith was then chair of the board of a trucking company, and would have been pleased to give that job to his firm.) Later, instead of recalling the fate of Mr. Potato, he picked the cabbage, a vegetable also common to an Irish family's table. His tale of the cabbage's adventures trying to reach his plate in his old Oliver Street home evoked gales of laughter. He ended by promising that the cabbage that arrived by the port's new routes would be "worth eating!"[11] The bill passed by a wide margin, with Smith widely credited with the victory.

Smith's activities in favor of the Port Authority kept him in the public eye while he was out of office. During that time, Moskowitz continued to serve his political goals, especially through her contacts with New York City women's groups. Invigorated by enfranchisement, these groups were now turning toward political education, lobbying, and supporting women candidates for elective and appointive office. Women's divisions were established in the political parties. Almost immediately, however, women suffered setbacks. Fewer of them voted than had been expected, and barely a handful won elective or appointive office. Further, women were still excluded from the inner circles of their political parties.

Despite these setbacks, even the most pessimistic among former suffragists refused to give up. Through past experience they knew that persistent educational and organizational work paid off. The old National American Women's Suffrage Association under Carrie Chapman Catt became the non-partisan League of Women Voters. A more militantly feminist National Woman's Party formed to achieve women's total civil and economic equality. While this approach eventually proved divisive, it did not prevent women from acting in concert on other issues. On the local level, several women's organizations—the New York State and City Consumers' Leagues, the Young Women's Christian Association, the New York State Women's Suffrage Party, and the New York Women's Trade Union League—formed a Women's Joint Legislative Conference under Mary E. Dreier to lobby for social welfare laws and support women candidates for office. Other women's organizations from the presuffrage era remained active in the 1920s, often taking stands on controversial issues or actively lobbying for reforms. Such, for example, were the National Council of Jewish Women, the Women's Municipal League, and the Federation of Women's Clubs, all of which had local branches in New York City. Belle Moskowitz's membership or leadership in many of these provided her, and consequently Al Smith, with a network of women active across the spectrum of New York politics.[12]

During the early 1920s, the non-partisan New York Women's City Club absorbed a large part of Moskowitz's political energies. Soon after New York women won the vote, the club's membership had soared to three thousand. The increase allowed it to lease, with an option to buy, a mansion on the corner of Thirty-fifth Street and Park Avenue. Designed by Stanford White for his sister-in-law, Mrs. Preston H. Butler, the mansion boasted a marble foyer in the shape of an oval, an Empire staircase, and four stories of large rooms, some oval, for offices, lounges, dining, and meeting. In 1920 the club bought the mansion for $160,000, an investment that doubled in value by 1927.[13] This was the environment in which Belle Moskowitz flourished during the 1920s. It was a stunning place for meeting, socializing, and making contacts.

Not all the club's members took its purposes seriously. Many merely lunched there while shopping in midtown. But the elite who ran the club were determined to make it a "civic center for women." They set up a committee structure to investigate and develop positions on municipal and state issues, and on national issues when these touched women's concerns. The work of the committees eventually became complex enough to justify hiring a professional coordinator, but even so committee chairs and board members bore heavy duties. These members, an elite of an elite as it were, became the club's life force.

Although Moskowitz's place in this top elite did not emerge until the end of 1921, her name, or topics reflecting her interests, appeared with increasing frequency in club minutes. Late in 1916, for example, the club's board learned that its legislation committee would begin to focus on the strengthening of dance hall laws. In March 1917, Frances Perkins, chair of this committee, had taken over this work. By April the club had created a separate recreation committee with a subcommittee on dance halls. After the fall elections, a slate of Hylan appointees replaced the members of Mitchel's Committee of Women on National Defense. According to Perkins, the Mayor's new chair was Mrs. William Randolph Hearst, who planned to run things along "decidedly different lines." This left the old committee, chaired by Mrs. Willard Straight, with important work undone and no headquarters. Perkins moved that the club give the old committee a place to work, renaming it the "Home Defense Committee of the Women's City Club." Since Moskowitz was the committee's executive secretary, she probably did most of her war work out of the club's headquarters.[14]

The Women's City Club may well have been where Moskowitz first approached Frances Perkins about addressing women's issues in Smith's 1918 election campaign, and her later plans for a Reconstruction Commission. During the life of the Commission, Moskowitz used club members to promote its projects.[15] Between 1921 and 1922, she moved steadily toward the club's inner circle. In January 1921 she represented the negative in a debate on whether to restrict immigration, an idea gaining currency in an era of growing nativism. The following month the club's legislation committee asked her to help devise a substitute for a bill that gave veterans civil service preference. Since women lacked veteran status, such a bill hurt their chances for civil employment. In December she joined the club's recreation committee, and the following April was put on the nominating committee, a position she resigned when her name was entered for election to the board of directors. At the club's seventh annual meeting, held on May 15, 1922, she won a place on this board. Reelected in 1925, she served until forced to withdraw by her growing involvement in Smith's presidential politics.[16]

Moskowitz moved toward the center of the Women's City Club in part because she recognized the club's potential and wanted to influence its direction. In part also, club members recognized in her a link to power. Smith was no longer governor, but early in 1922 rumors were rife that he not only would oppose newspaper publisher William Randolph Hearst for the state Democratic nomination that September but had a strong chance of beating Nathan Miller. In that event, Moskowitz's ties to him would give the club inside knowledge of politics. Her use of the club as a forum for Smith's programs was thus not one-sided.

One member of the club's inner core who gained both personally and politically from Moskowitz's active membership was Eleanor Roosevelt. Then in her mid-forties, Roosevelt was beginning to carve out an independent life for herself. She and her husband Franklin were no longer intimate, their marriage having almost dissolved over his affair with Lucy Mercer. But they remained life partners, in business and in politics. After running for the vice presidency on the Democratic party ticket in 1920, Franklin had been paralyzed by polio, and his political future seemed doomed. But neither his wife nor his manager Louis Howe accepted this prognosis. In an effort to keep Franklin's name before the public and maintain his interest in politics, Eleanor joined several women's political groups in New York City. She found that she liked political work, which gave her not only a renewed sense of purpose but a world of friendships with women that filled an emotional void.[17]

Roosevelt divided her political energies between partisan and non-partisan organizations. Her partisan group was the Women's Division of the New York State Democratic Committee, of which Moskowitz was also a member. Founded in 1921, the division formally dissolved in 1923 when a rule change gave women full membership status on the state committee. It remained an informal network, organizing upstate where the party was weakest and publishing the monthly *Women's Democratic News*, which Eleanor Roosevelt edited from 1925 to 1928.[18]

Eleanor's non-partisan work revolved primarily around the League of Women Voters and the Women's City Club. League members Esther Lape and Elizabeth Read taught her how to analyze legislation. As a league officer until 1924, she made speeches and wrote articles on political issues. At the Women's City Club, whose board she sat on from 1925 to 1928, she investigated many controversial issues, formulating club positions and traveling to Albany to lobby before legislative committees. She published legislative reports in the club bulletin, organized and took part in debates, made radio broadcasts, and ran countless meetings. Her first term on the club's board coincided with Moskowitz's second. Thus they saw one another frequently at board and committee meetings and often lunched together in the club dining room. The two women shared more than politics, and exchanged tales about their difficulties in raising their sons. This does not mean they were close friends. Eleanor had overcome many of the prejudices of her class and youth, but still harbored traces of anti-Semitism,[19] and Belle's social circle did not normally include Protestant patricians. As far as we know, the two women did not see one another outside the women's organizational or Democratic party circuits. But within those circuits, they shared many goals in common.

Moskowitz and Roosevelt may first have met through a movement of women Democrats to defuse a "Hearst for Governor" boom in 1922. Since the early part of the century, newspaper publisher William Randolph Hearst had been a sometime Democrat, who in one election supported a Tammany slate, in another a Republican; in other years, he put himself up as a third-party candidate. Apart from a short term in Congress, Hearst had not succeeded in his own bids for office. But the power of his newspapers made a potential Hearst candidacy threatening to Smith.

In 1918, Hearst had reluctantly supported Smith for governor, only to feud with him the following year over the price of milk. In a series of editorials and cartoons, he attacked Smith for bowing to the dairy industry, which by law was exempt from antitrust legislation and thus free to set its own prices. Hearst ignored the fact that the industry was regulated by the State Department of Farms and Markets over which Smith, under the present administrative structure, had no control. When Hearst accused Smith of killing babies by denying them milk, Smith exploded. He demanded not only a retraction but a public debate in which to answer the charges.[20]

Belle Moskowitz organized the debate through her favorite public relations device, a "citizens' committee." This one, called the "Citizens' Fair Play Committee," hired Carnegie Hall for the confrontation. Hearst, announcing he had no time to waste on dishonest politicians, left for California to oversee the building of his castle, San Simeon. On October 29, 1919, Smith appeared on the Carnegie Hall stage—alone. The galleries were packed. Smith went into action with an impassioned speech that enthralled his audience. One by one, he listed Hearst's political tergiversations. He explained why Hearst hated him, painting him as too "white-livered" to appear in person to answer for his charges. If only by default, Smith easily won. A few days later, the citizens' committee issued a manifesto, drafted the week before, that recognized the "right and duty of a free press" to criticize public officials, but declared "wilful misrepresentation of the policies and acts of public servants" to be "inimical to the Republic." It planned to "frame a sound public opinion directed against the insidious and disintegrating opinion of [Hearst's] journals."[21] Belle Moskowitz would be the chief architect of this opinion.

In the ensuing months, Hearst buried the hatchet and supported Smith's 1920 unsuccessful reelection bid. By 1922, however, Hearst decided to try for the governorship himself. Women Democrats, led by Belle Moskowitz, were ready for him. On May 26, 1922, at a luncheon honoring Emily Newell Blair, New York's Democratic National Committeewoman, a "Democratic Union of Women of Manhattan" was born. Moskowitz read a resolution that endorsed for state office only those who had supported the national Democratic ticket in 1920. The nominees on that ticket "represented the highest principles of the Democratic Party and it is only reasonable that this test should be applied," she said.[22] She did not mention Hearst by name, but since he had opposed former president Woodrow Wilson on the Versailles Treaty and League of Nations, he was the obvious target.

Wilson sent the Union a letter of thanks, which gave it wide publicity. Despite the Union's thinly disguised aims, its high-minded principles of party loyalty and support for international idealism appealed to many New York Democratic women. Applications for membership exceeded expectations.

Mayor Hylan, Hearst's chief supporter, retaliated. "The ladies starting this organization headed by Mrs. Moskowitz, asking that only those be considered as nominees for high State office who have been loyal in the past to the Democratic candidates, were opposed themselves to the Democratic candidate for Mayor." Hylan's Commissioner of Accounts, David Hirshfield, claimed that Moskowitz, Frances Perkins, Mrs. George McAneny, and Mrs. Abram Elkus had never supported Democrats. Indeed, he continued, the husbands of Moskowitz, McAneny, and Elkus had received "lucrative places" in exchange for their support of fusion and Republican candidates, and enjoyed a $2 million "slush fund" during Mayor Mitchel's reelection campaign. "Persistent opposition . . . to regularly nominated Democratic candidates, does not qualify these ladies to pass judgment as to who is and who is not deserving a Democratic nomination," he announced.[23]

At first inclined not to respond, Moskowitz protested, "Why does Mr. Hylan think that we opposed him?" As an enrolled Democrat, she had helped at campaign headquarters, she said. She pointed out further that several of the Union's officers—Ethel Stebbins, Ella Hastings, and Anna Naughton—were Tammany officers. Exchanges such as these kept the Democratic Union in the news. Soon the Union idea spread beyond Manhattan. Resolutions of support came in from Mary A. Morse of

Buffalo, Alice Good of Brooklyn, Caroline O'Day of Rye, and Eleanor Roosevelt of Hyde Park, all of whom organized unions in their own districts. Former suffragist and now State Democratic Associate Chairman Harriet May Mills, assisted by field secretary Nancy Cook, oversaw the organization of Democratic women in the state as a whole. Efforts to build a similar structure of pro-Hearst women came to little. In July, when upstate Democratic leaders met in Syracuse, a resolution proposed by William Church Osborn denounced Hearst as a traitor to his party. The force of the Hearst "threat," while not over, was now fairly well depleted.[24]

After the Syracuse resolution, Harriet Mills talked about the growing power of women in the party. She observed that its male leaders had virtually abandoned organizational activity in strong Republican counties. She believed that Democratic women would not give up so easily, and therefore might soon be replacing the men in leadership positions. "Of course," she said, "we . . . have not entered this work in rivalry, except the most friendly, with the men."

> We merely want to do our full duty to the party, and in that spirit we are offering our co-operation. We have had some rebuffs from the men leaders of our own party, some of whom seem to look upon us as nuisances, but the younger and more progressive element . . . appears to welcome this new and militant force in their local party organizations.

She said further that women's success in organizing stemmed from their long experience in clubs and suffrage. "Where the men begin work only about the time election is due, we believe in working all the year round."[25]

Mills's point helps explain how Belle Moskowitz penetrated the male-dominated inner circles of the State Democratic party. Personal skills gave her advantages. But in addition she could mobilize a phalanx of women determined to carry on relentless organizational work. No party politician could afford to ignore such help. And, once again, her relationship to the phalanx was not one-sided: just as she relied on them, they used her to represent their interests to the party's central powers.

On the eve of the state Democratic convention, a meeting of representatives from county Democratic party women's divisions overwhelmingly endorsed Smith. The endorsing resolution was proposed by Marion Dickerman, an unsuccessful candidate in 1919 for the seat of former Assembly Speaker, Thaddeus C. Sweet, from Oswego County. Along with her companion Nancy Cook, executive secretary of the state committee's Women's Division, Dickerman would join Eleanor Roosevelt, Elinor Morgenthau, and Caroline O'Day in developing the division into the party's most effective organizational tool.[26]

The Democratic Union of Women of Manhattan remained in existence throughout the ensuing gubernatorial campaign, working for

Smith's victory. After the election, the union pledged to help Smith get his legislative program adopted, a pledge Belle Moskowitz signed. To stimulate women's political awareness, the Union then organized a "National School of Democracy" to be held between January 29 and February 3, 1923. Moskowitz put together the school's course on democracy and gave one of the lectures. Other organizers and lecturers included Ida Blair, Harriet May Mills, Elisabeth Marbury, Anna M. Kross, and Eleanor Roosevelt. Governor Smith himself gave the closing lecture to an audience of 1,200 participants. The following month, a tiny article buried on an inside page of the *New York Times* whispered, "Women to Start Smith Boom." This time the goal was not another governorship, but the presidency.[27] The whisper soon became a roar.

Smith won the 1922 gubernatorial election against Nathan Miller by 387,000 votes, and his party won a one-vote edge in the state Senate. Republicans retained control over the Assembly, however, as they did for the rest of Smith's years in office. In 1919, Smith had been ready to launch a major overhaul of the state's administrative structure, only to discover more legislative resistance than he had expected. In 1923, his plans remained the same, but this time he moved with more caution.

His Annual Message to the legislature sounded an unimpeachable theme: the preservation and perfection of democratic institutions.[28] Calling again for government reorganization, he highlighted the short ballot, four-year term, developmental consolidation, and executive budget. To achieve greater government accountability, he wanted a more comprehensive direct primary system, popular constitutional initiative, and greater home rule for cities. All "unjust" discriminations against women should be removed, he said, without, of course, jeopardizing laws that protected women in home and industry. In that regard, he favored a minimum wage and eight-hour work day for women and minors. He further showed the influence of women social reformers by asking that New York implement the federal program on maternal and infant health provided in the Sheppard-Towner Act.[29] For labor in general, he urged a more efficient Workmen's Compensation system, and better machinery for conciliation. Current drives to censor the arts and impose loyalty oaths on teachers encouraged "intolerance and bigotry in the minds of the few directed against the many," he warned. He attacked overcrowding in mental institutions, the lack of rehabilitation programs in prisons, low pay for teachers, unsafe highways and railway crossings, and the lack of centralized motor vehicle licensing and coordinated housing policies. He closed with a strong appeal for public development of water power.

Frustration, defeat, and partial triumph followed. None of Smith's victories was easy. All required his constant vigilance of the legislative process and a willingness to trade and compromise. In addition, voters had to be wooed. Smith's opponents blocked not only government reorganization but other programs that involved new spending. As Smith's public relations counsellor, Belle Moskowitz hoped to convince voters that Smith's programs were in their interest and should be supported.

That Moskowitz would serve in this way came as no surprise to Albany observers. Her work for the Reconstruction Commission, Port Authority, and election campaigns had established her reputation as an effective publicist. Still, Smith and Moskowitz were very different from one another. Culturally, religiously, and ethnically, they came from widely divergent backgrounds. Their shared political goals bridged many of the gaps between them, but what of the gender difference? No male politician had ever relied so heavily on the advice of a woman outside his own family. Smith's personal attitudes toward women were traditional. How did he justify to himself and his supporters the incorporation of a female into his innermost councils? Further, the idea that a woman could have direct political influence was new and not yet widely accepted. By according Moskowitz a key role in his affairs, Smith opened himself to the charge of "petticoat government."

Moskowitz was certainly conscious of the risks. Early in their partnership she decided to stay out of the public eye as much as possible. She declined Smith's offers of state government posts, and rarely allowed herself to be quoted directly on Smith's affairs. Before the 1928 campaign, one had to read the back pages of newspapers, be closely connected with the Democratic party, involved in women's political networks, or be a family friend to know the extent of her influence. This tactic not only protected Smith, it was completely consistent with Moskowitz's view of how a woman should function in society.

The Moskowitz-Smith relationship was not "modern" in the sense that it featured two equals, one male, the other female, working together. Their relationship bore with it an institutionalized inequality: Smith was the authority, Moskowitz the adviser. But a closer look reveals a gender-specific side. Moskowitz framed her relationship to Smith within the boundaries and roles established for women during the era in which she grew up and came to maturity, from 1890 to 1920. Some of these roles were domestic and familial in nature; others, more in the public sphere, took inspiration from women social reformers of the Progressive Era.

The roles Moskowitz played that were inspired by domestic models are vividly encapsulated in a letter Moskowitz wrote to Lillian Wald in August 1923. Wald, the head of the Henry Street Settlement, spoke occasionally in the 1920s in support of Democratic party causes and had agreed to deliver a speech lauding Smith. She asked Moskowitz to pass along some colorful anecdotes about his upbringing on the Lower East Side. Moskowitz obliged with a four-page, handwritten reply sent from her home. Because she was away from her files, she explained, she was thinking out the material "afresh." To be sure, Moskowitz was a public relations expert, a professional who seldom sent out a message without calculating its effect. But, because the letter was written by hand and from memory, it has unusual spontaneity.

Moskowitz presented Smith anecdotes, learned directly from Smith or his family, in four sections, each longer than the one before: his life as a newsboy on Park Row, his love of animals, his youthful triumphs in elocution contests, and the effect on his character of being a "widow's son." In telling them, she used the word "love" seven times:

> [He] loves the Thanksgiving Day dinner of the newsboys and never misses one. . . . He always loved animals. . . . He loved the amateur theatricals at the Church. . . . He loves the East Side dearly. . . . He loved the fire engines. . . . He loved practical jokes of a simple kind. . . . He loved singing and a funny story. He still does.

In describing his leisure activities, she used "fond," "greatest pleasure," and "endless fun." These words reflect her perception of Smith as a man of feeling, but they also express her own pleasure in being around him.

Many people "loved" Al Smith. Moskowitz's letter to Wald helps explain why. In setting out for Wald the objects of Smith's love, she cited newsboys, animals, amateur theatricals, fire engines, practical jokes, and stories—all simple pleasures, signs of gregarious largeheartedness. Who could not love such a loyal and sentimental man, whose very desires showed him to be endearingly childlike? Moskowitz's maternal feelings must have been especially charmed.

In addition to elucidating Smith's character, Moskowitz's letter also established biographical links between herself and Smith. When she told how he participated in elocution contests and theatricals, for example, she may have been thinking of her own experience in drama. A stronger link appeared in this story:

> He tells of the long trip in the horse cars from Chatham Square where his school was located, to Manhattanville College (125th and Old Broadway) where the speaking contests were held. He tells how on a snowy night he and the other boys from his school carried off the prize and then had snow fights all across 125th Street to Third Avenue, with the defeated contestants.[30]

From all the anecdotes she heard, she chose to recount one set in the neighborhood where her own youth was spent.

The letter reveals one final element of Smith's character that endeared him to Moskowitz. She described his "devotion to his mother," whom he always "minded," his "reverence for teachers," and his tremendous "respect for education." In emphasizing these points, Moskowitz drew attention to a side of his character vital to their relationship—his openness to instruction from a maternal source. Without that openness, he might never have welcomed a woman into his inner councils.

To Moskowitz, then, Smith was a loving, simple, large-hearted man. His background bore striking similarities to her own—they were both raised by immigrants, and both had found the stage an appealing means of expression and communication. And finally, as a widow's son, Smith was willing not only to admit the deficiencies in his education, but more importantly to repair them through a woman's instruction. These elements of Smith's character not only won Moskowitz's devotion but made their relationship possible.

Moskowitz's letter to Wald suggests both maternal and sororal aspects of the roles she played in her relationship with Smith. And yet, for Moskowitz, these roles had their limits. At the same time she saw Smith as someone to instruct and develop, as a mother would view a son, two other maternal figures—Smith's wife and mother—were far closer to him. Further, Smith could reject Moskowitz's influence at any time, and without guilt. Thus, while offering him the unconditional love and loyalty of a sister or mother, she had to play gentle instructress, never flaunting her influence in public or in private. The portrait of her drawn by Robert Moses, a close observer and participant in meetings of Smith's "kitchen cabinet," rings true. According to Moses, Moskowitz took part in these meetings from the sidelines, plying her knitting needles and speaking only when asked for her views. As a woman, and especially one whose power depended on the continued favor of the man she advised, she knew her place and kept to it.[31]

In this sense, she was conforming to another traditional female role model, that of the supportive wife who manipulates her husband's career from behind the scenes. To keep from threatening his masculinity, she never revealed the extent of her control. J. M. Barrie's character Maggie Shand, the heroine of his 1908 play *What Every Woman Knows,* is the literary archetype for this role.[32] Shand not only runs her husband's political career but is chiefly responsible for its success. She even risks his infidelity in order to maintain the fiction that he has succeeded on his own.

Frances Perkins, another observer close to Smith, and one whose affections for him were as strong as Belle Moskowitz's, perceived this aspect of the Smith-Moskowitz relationship. She was not uncritical of it. In her view, Moskowitz manipulated Smith, "ran him in [a] subtle way." Perkins qualified this judgment by pointing out that, in her view, Moskowitz never acted out of "any spirit of ill will." It was simply that Moskowitz's competence was "so much greater than anybody else's" that she slipped easily and naturally into a manipulative role.

Perkins remembered that Eleanor Roosevelt assessed the Smith-Moskowitz relationship in the same way. At campaign headquarters, Eleanor's and Belle's desks were "right beside each other." The two women had "daily, hourly contact for weeks together, and about matters of great importance." Eleanor was thus in a position to know. Besides, Moskowitz never concealed her method, except perhaps from Smith.

Perkins herself had observed it. "All of us knew," she said, "that Belle arranged things so that when they went before Al, it was so logical to make the decision that she'd already told you would be made." When Perkins went to Moskowitz with a plan, Moskowitz would reply, with a kind of "supercilious smile," "It'll be all right, Frances. Don't give it any more thought. I'll see to that. This will be done. I wouldn't speak to the Governor if I were you. It will be better if I handle it alone." Perkins went on: "She would prepare it for him so it just seemed as though it were something that he had thought of." Having said this, Perkins again added a qualification. The things Moskowitz thought up, Perkins said, "were good things." She never gave Smith "a bit of advice, or prepared anything for him, that was not good both for him and for the people of the State of New York. I never saw anything done that wasn't, but many things were done into which his mentality and personality did not enter."[33]

Perkins developed her characterization of the Smith-Moskowitz relationship in another section of her memoirs where she expressed herself more positively. In her view, Moskowitz succeeded with Smith because of her selfless, total, and non-threatening loyalty. "She never belittled him," Perkins said. "She never in the slightest way betrayed him." Even more important, she "never grabbed credit for herself. She gave it to the Governor."[34] In these four pithy phrases, Perkins captured the essence of Moskowitz's strategy: hatch ideas, but give all attention to Smith. Thus, as secretary of the Reconstruction Commission, she allowed herself press coverage only on a few select, and usually minor, items. In 1923, in a speech to New York State social workers, she credited Smith with having thought up the idea for the Commission. She adopted the same tack when Smith took credit for her labor arbitration or housing commission ideas. And when friends spoke of Smith as her protégé, she corrected them by calling him her mentor.[35]

One could argue that any adviser of a political figure would have taken this approach. Remaining "invisible," as it were, eschewing publicity and public office, is the best way to influence a leader who lives in the public eye. As Oliver Garrett expressed it in his *New Yorker* "Profile" of Moskowitz, "To modesty her friends attribute her comparative obscurity; her critics call it shrewdness."[36] Whatever the case, Moskowitz certainly seemed aware not only of the limits gender placed upon her but of how to use those limits to her advantage. If she were to influence events, she would have to do so subtly, even at the risk of being perceived by friends and enemies alike as domineering and manipulative.

The care with which she interacted with Smith and his male advisers is illustrated by the following anecdote. Actress Aline MacMahon, who had met the Moskowitzes during the 1920s through her husband, architect Clarence Stein, was on her way to Hollywood on *The Chief* in the spring of 1933, shortly after Belle Moskowitz died. The last time she had taken that train, she had traveled with Moskowitz. The dining car waiter, who recognized her as Moskowitz's friend, told her that he used to see Moskowitz on the New York-to-Albany run. What he remembered most about her was the way she handled herself during discussions. She would remain silent until Smith asked her for an opinion. She would then say, "Gentlemen, the ideas you have presented are all very interesting, and should certainly be considered, but let's look at it from another point of view." With this simple, non-threatening phrase, she changed the perspective of the discussion without making a single criticism.[37]

Perhaps a male counsellor could have been as diplomatic. But the story is consistent with other aspects peculiar to the Smith-Moskowitz relationship. Perkins said that Moskowitz never took advantage of her access to Smith by allowing others to "work personal preferences through her." When a post-seeker called her up, she would say, "No, I never recommend people to the Governor. If you know someone who you think would be good to be the health officer at XYZ, I suggest that you make the recommendation straight to him. He'll be very glad to hear of it." On the other hand, she frequently nominated people to Smith at his request and on her own intiative, and Smith usually took her advice. Perkins thought this policy "very honorable and very shrewd."[38] It established Moskowitz as someone pointedly not engaged in the empire building usually associated with ambitious men. Moreover, by refusing to intervene with Smith on behalf of others, she assured him she was not representing the interests of others and was loyal only to him.

Sister, mother, wife; controlling her environment through subtle manipulation; loyal and selfless, or at least to all appearances not self-interested—by following the models women had traditionally observed in the private sphere of the family Belle Moskowitz remained in safe territory. Although her sphere was now public, she conformed to roles that made herself and Smith feel comfortable.

Her conformity to tradition is hardly surprising. Her career developed out of social reform in the early years of the century when few professional opportunities were open to women. Service in a cause might bring a feeling of self-fulfillment, but it was seldom pursued for personal gain. Women's suffrage changed things to some extent, opening new opportunities. But Moskowitz was not sure that society was ready for women to be as ambitious about their careers as men were, or that women should relinquish their historic place in the scheme of things. As she said in an interview in 1923, women with political ambitions may err in seeking office or "political pie." Often discouraged by the political process, they conclude they are "too good for politics." To make her point, she cited the previous legislative season in Albany. The "petty tricks, deceit, underhanded scheming and mean personal motives" she had observed there had disgusted her. Had she been a seeker of office or of other preferment, she might have given up politics.

Instead, since she followed the "service ideal," she was able to endure temporary discouragement and even defeat.[39]

Many women and men of Moskowitz's generation shared this view of women's political roles. According to it, men were society's natural leaders, with the intellect, ambition, and drive to lead. Women worked best as supporters, bringing their instincts and compassion to bear on social problems. Moskowitz expressed this sentiment in a controversial remark she made in 1926 at a forum at Columbia University. The topic of the forum, attended by almost a thousand women, was "Women's Part in Politics." "Women have qualities of mind peculiarly feminine," Moskowitz said, "but they are not the intellectual equals of men. Their intuitive sense is the biggest thing that they bring to politics. Combined with the thinking ability of men, this makes a splendid working team. It would be better for women to use the abilities they have and not attempt to do what they can not do."[40]

Other prominent women of Moskowitz's day also believed in the complementary but distinct natures of male and female brainpower. Eleanor Roosevelt, for example, echoed them in her first book, published in 1933. In it, she adhered to the view that women succeeded in politics best if they followed a "service ideal." "[I]f a woman wants to work and can prove her ability and is not too anxious and insistent upon recognition and tangible reward," she wrote, "she can be part of almost any party activity except the inner circle where the really important decisions in . . . politics are made. . . ." On the issue of sex equality, Roosevelt argued that the physical differences between men and women gave women a special perspective on social problems. She envisioned a future society built not only "by the ability and brains of our men, but [one] which also represents the understanding heart of the women."[41]

At various points in her Columbia Forum speech, Moskowitz made it clear that, in assessing women's roles in politics, she was describing the present reality in which men did not listen (at least not openly) to women's views. Only the radical political parties believed in sex equality. Neither of the two major parties did so, with the result that women had not been influential in forming party policies. "The reason for this," she said, "is that men are realists in politics, while women are individualists. Masculine efforts are devoted toward one thing only, *success at the polls* [my emphasis], and men are not yet sure that women can demonstrate that solidarity which means success at the polls."[42] Similarly, Eleanor Roosevelt wrote in 1933, "Women have made no great changes in politics or government and that is all that can be said of the past and now for the present."[43] While asserting women's limitations, then, both Moskowitz and Roosevelt conceded that women's roles might expand in the future. Moskowitz's self-effacing behavior as Al Smith's adviser may have been intended, in part, as a tactic for protecting Smith from criticism. But it was also in keeping with long-held beliefs about the proper roles of women.

Ironically, for all the conventionality of Smith's views on women, he was more prepared than she to consider women as intellectual equals. After Moskowitz's death, Smith wrote a fifteen-hundred-word tribute to her. In it, he emphasized not her instincts but her brainpower. She had, he wrote, "one of the finest intellects" he had ever known. He gave her full credit for the idea of the Reconstruction Commission, saying further that, during the Commission's work, he had found himself calling upon her advice with increasing frequency. This advice he invariably deemed "sound." The "keenness of her mind" impressed him, "especially her ability to get at the essentials of a complicated question and to explain it briefly and simply." Finally, he avowed that he "saw no difference between the processes of her mind and the way the best men's minds work." And yet, he admitted, in a group of counsellors she was "very loath to express her opinion," and he "frequently had to ask her for advice before she gave it." Here, Smith seemed unaware of her reasons for reticence. He concluded: "A discussion of a problem with her served to clarify my own mind." And although on the purely political side of things "she did not pretend to have expert judgment," he usually found her "political sense sound."[44]

At the same time as Smith praised Moskowitz's intellect, he also affirmed her "essential feminine qualities." In his view, her thought processes were no different from a man's, yet unlike other women in politics who tried to imitate men, Moskowitz always kept her sights on "the true values of life." To Smith, this meant that she remained "essentially a wife and a mother." She frequently spoke "of the woman's point of view," and was a "womanly woman" whose "mother heart" impelled her toward making the state an "instrument for the welfare of the people." After citing the legislation for which she had worked, including government reorganization, housing laws, and state aid to education he then paid her his ultimate compliment: "Though she was tender, as only a fine woman could be, she was a fighter; she was a hater of insincerity and bunk . . . of sham and hypocrisy."

Smith gave approximately equal weight to the gender polarities in Moskowitz's character. He used a form of the word "woman" nine times: "service" appeared four times, "mother" three, "wife" twice, and "feminine," "modest," "heart," and "tender" once each. Interspersed with these attributes of traditional femininity, however, he used terms associated with maleness: "men" and "mind" appear three times each, "sound," "tireless," and "hater" twice each, "intellect," "organized," "interesting," "true," "administering," "keen," "ideal," "industrious," and "fighter" once each. Smith's view of Moskowitz was a distillation of the mixed-gender image she had been conveying since late adolescence, an image which combined external feminine and internal masculine attributes, and which had evolved out of the transition of the young Belle Lindner from "James, the Tailor-Made Girl" into Mrs. Charles Henry Israels.[45] Smith perceived Moskowitz exactly as she wished him to.

Nonetheless, there is a disjunction in the ways Moskowitz and Smith understood their partnership. Moskowitz denied important aspects of its dimensions, and in both her view of Smith and her behavior toward him she emphasized traditional female roles. One might argue that she had no choice: as a woman in a male-dominated society, any pretensions she made to intellectual equality would not have been well received by Smith or his male friends. Smith's view was not only much more balanced but had distinct political advantages for him. By saying her mind was as good as a man's, Smith established her advice as smart and reliable; and, by affirming her womanliness, he avoided any suggestion that he associated with a woman society might deem "mannish," if not actually deviant.

Smith was not alone in his mixed-gender perception of Moskowitz. Robert Moses, intending to pay her a compliment, called her "a real woman with the brains of a man." Oliver Garrett, in his *New Yorker* "Profile," said that he had heard others describe her that way and found it amusing: "It is part of the curious conceit of most men that those who know Mrs. Henry Moskowitz claim her mind, and thereby her unique political success, for the male sex. 'Practical,' they call her, in the complimentary sense, as if that were a prerogative of males." To this view, Garrett riposted: "Her mind is no soil where may be flung out a flag marked 'Man.'"[46]

On the other hand, Garrett recognized that the way Moskowitz chose to work for Smith expressed conventional femininity. Throughout her association with him as governor, she deliberately refused all offers of government posts. According to Smith, he was prepared to offer her "any position within my gift."[47] But she always turned him down, to the bafflement of Tammany Hall. From Tammany's point of view, patronage was the essence of politics. For Moskowitz, it was service. Garrett said this sounded like "hokum," but it was not. "The truth would appear to be that she is a woman of the Jewish tradition," he wrote, "whose career must be a part of and no greater than that of a man."[48]

Garrett's interpretation is off the mark. Moskowitz was interested less in a career than in getting things done. By presenting herself as a self-effacing Jewish woman who had no interest in outshining the man she served, she made it possible for other men to tolerate her influence. Moskowitz's daughter Miriam remembers the moment when her mother decided to refuse all offers of posts from Smith. As the work of the Reconstruction Commission was winding down, Belle had come to recognize the extent of her growing influence on Smith. Sensing the imminent offer of a post, she and Henry discussed how they should react and came to the conclusion that she should refuse. In her view, working for Smith "without portfolio" would in the end result in her having more impact on him.[49] She was right. High office can mean power, but it is power circumscribed by the limits of duty. A bureaucrat in charge of excise, for example, cannot advise on housing or hospitals.

As general adviser and confidante, Moskowitz achieved a wider effectiveness.

That Moskowitz was interested in, and enjoyed, power is clear. Away from Smith and his coterie of male advisors, in Democratic party offices, women's political groups, and dealings with outsiders, she functioned with executive aplomb. In these contexts, she knew and felt personal ambition, showed interest in the progress of her own career, and derived personal satisfaction and pride from her work. She may even have reveled in power.

In describing her at work, Garrett profiled a tough executive who seldom expressed her feelings in public. No matter how aroused to anger, she kept "her vibrant emotions locked behind . . . a steady control," her mouth in an even smile, her temper unshaken. "She reveals herself so little to anyone," Garrett mused, "that a dozen or more men and women to whom she has shown as little of herself as most disclose to acquaintances, boast: 'I am her only confidant.'" When she did lose her temper, the effect was unforgettable: "Her eyes blazed and her great, slow-moving body seemed to tighten as if it were about to perform the miracle of abrupt action. But the guarded lips gave forth no more than a few quiet words, sharp-edged but passionless." Other journalists called her a "high powered machine" that performed miracles without ever making a noise, or referred to her "cool-headed, unruffled self" or thorough mastery of her "thoughts and actions."[50] A capped volcano, Moskowitz commanded respect.

Her pride in her work is illustrated by a letter she wrote to *New York World* columnist Walter Lippman, a friend since the early twenties. She was commenting on his praise for the Women's Democratic Union's plank on prohibition. Moskowitz wanted him to know she was its author: "Your editorial . . . leads me to point with pride to the fact that I am a vice-president of the Association and a member of the Platform Committee which drafted the plank and I was very largely instrumental in having this one inserted instead of the absolutely colorless one which had originally been urged."[51] This is not the self-effacing Belle Moskowitz who, in working with Smith, never took credit for anything she did. Within a women's political group that she had helped found, she wanted and sought credit.

At the same time as Frances Perkins described Moskowitz's modest self-effacement, she also described her "grasp for power." According to Perkins, Moskowitz "did love the recognition and the realization that she, herself, had power, that she could make you a promise and it would be fulfilled. She certainly basked in that. She basked in the fact that the great and powerful . . . called her up first. She arranged that." Perkins went on to Moskowitz's tactic of letting someone "stub his toe" when he brought up some matter with the Governor. Having failed to get action from the Governor, the man would finally come to her. She would smile, again, "in a supercilious way," go on with her knitting, and say, "Well, it's too bad. I'm sorry. You should have seen me first. I think I could have arranged it." This was a masterful manipulator at work, someone who took pleasure in watching the less adept twist in the wind. Perkins believed that Moskowitz "trained" supplicants to see her first. "The result was that they did see her first, not only the public officials, but the newspaper editors. She had them all dependent upon her for news, for information, for advice and for introducing their idea [to] the Governor. She was as good as her word. She really could get things done."[52] Thus, as irritating, even exasperating as her mode of operation was, in the end Perkins redeemed her, assuring posterity that Moskowitz's influence was always "excellent." "I think that those of us who worked with her when Smith was Governor had great respect for her judgment about the substance of matters that we wanted the Governor to take up."[53] To Perkins, then, as well as to other contemporary observers, Moskowitz used her public relations skills as "a power for good."

There are many routes to power. The one Belle Moskowitz chose—to remain out of office and work behind the scenes for a male politician—seemed right to her. Certainly it was consonant with the tradition of Progressive Era women social reformers. No other Smith adviser chose that option. Robert Moses became Smith's secretary of state and then parks commissioner. Having clashed with Franklin Roosevelt over parks, he never climbed the federal ladder after Smith left office but carved out a base for himself as a builder of public projects in New York. Smith's other close associates—Abram Elkus, Joseph Proskauer, and Bernard Shientag—all accepted judgeships. Frances Perkins was one female associate who made a choice different from Moskowitz's. She accepted a post on the State Industrial Commission and later moved into Franklin D. Roosevelt's state and federal cabinets. But, loyal to the values of her time, she disclaimed careerism. "It may be true that unlike many other women I have had a life of my own and a career," she later recalled, "but I had no intentions of having that. I never, never dreamed of being Secretary of Labor. I never dreamed of being Industrial Commissioner. I never had any notions in my mind—never, never, so help me, God! I'm not a different kind of woman than most of the women in this country. . . . Those who have a career have been thrown by a series of circumstances and their own energies into situations where they had to assume responsibility, did assume it, were asked to assume more, did assume more. Before they knew it they had a career."[54]

By refusing posts, Moskowitz increased her effectiveness when Smith was in power but lost it when he was eclipsed. She could not have known how the choices she made in the early 1920s would affect her in the long run. As her mentor prepared to take up his second governorship, she felt herself sailing on a rising swell. She foresaw only the crest.

A portrait by Lewis Hine, early 1930s. *(Courtesy Miriam Israels Gabo)*

The only surviving photograph of Belle as a child. Visiting her father's parents in East Prussia, she is examining a family album. *(Moskowitz Papers)*

A "Poster Party" at the home of Belle's friend, Anna George (DeMille), seated in the center, dressed in black and smiling at Belle, who (far left) is wearing a basket on her head and a dress decorated with magazine covers. *(Courtesy Agnes DeMille Prude)*

Belle in 1896 as "James, the Tailor-Made Girl," her monologue character. *(Author's collection)*

A year later she posed for the same photographer. *(Courtesy Alice Long Goldsmith)*

Charles Henry Israels. *(Author's collection)*

Belle as a young matron. *(Courtesy Miriam Israels Gabo)*

Charles, not too long before his death in 1911, reading to his children, Carlos and Miriam. *(Author's collection)*

Belle with her youngest child, Josef. *(Author's collection)*

The Moskowitz-Israels family, 1919. Belle had christened the submarine *Sutermco*, built at the plant where Henry was personnel manager. A few years later, Joe would sign on as the ship's radio operator. *(Moskowitz Papers)*

Henry Moskowitz. He inscribed the portrait to the family of Carlos Israels's first wife, Irma Commanday. *(Author's collection)*

Above left, Carlos in the early 1920s. *Above right,* Alfred E. Smith in a portrait sent as a gift to Carlos and Irma Israels. *Below,* Miriam, with Caesar, a gift from Governor Smith, who loved Great Danes. *(All: Author's collection).*

New York Democrats meeting in Eleanor Roosevelt's house to plan the 1924 state campaign. Roosevelt is facing Louis Howe; Belle Moskowitz is to her left; Moskowitz's son, Carlos Israels, is standing on the far right. *(Jessie Tarbox Beals, Courtesy Franklin Delano Roosevelt Library)*

Above, A snapshot of Belle from the early 1930s. She does not look in the best of health. *(Courtesy Miriam Israels Gabo). Below,* After her accident in December 1932, the Moskowitzes sent this card to their friends. *(Author's collection)*

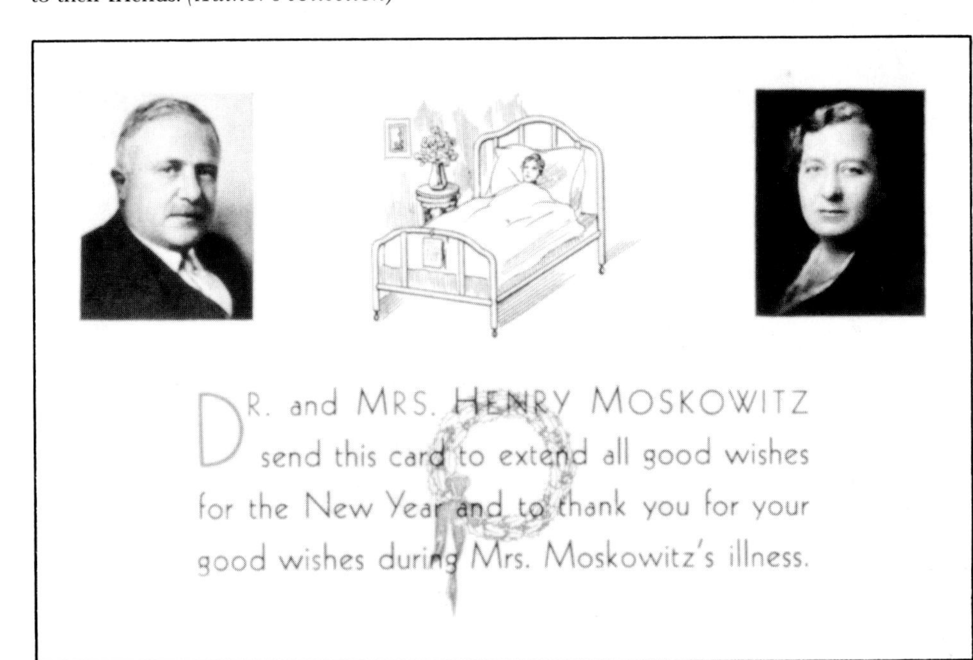

DR. and MRS. HENRY MOSKOWITZ
send this card to extend all good wishes
for the New Year and to thank you for your
good wishes during Mrs. Moskowitz's illness.

CHAPTER 9

Policy and Publicity

Politicians cannot afford to be visionaries. They must deal with actualities recorded in terms of votes. They must deal in candidates elected or defeated. Social workers are prophets—the cranks of one generation moving the next to action. They see in terms of eternal aspects of great principles.

Belle Moskowitz
Brooklyn Daily Eagle,
June 8, 1923

Belle Moskowitz took only one post from Smith's hands, that of Publicity Director for the Democratic State Committee. She started early in 1923 at a salary of $4,000 a year, less than she had earned in the garment industry. But the post gave her more power than she had ever known. She performed a wide variety of services for Smith—editing his public papers, writing speeches, monitoring legislation, ferreting out information, finding the right person for a vacant post, all services that made her a central figure among the governor's advisers. In addition, she had access to all information going in and out of the party organization. As Smith reported to Joseph Proskauer in 1926, when "Mrs. Moskowitz . . . had no news there isn't any around."[1]

George R. Van Namee had founded the publicity bureau in 1914. Van Namee, then secretary of the state committee, became Smith's secretary during his first term as governor, and later a member of the Public Service Commission. He remained active in the State Committee throughout the 1920s. During Smith's second term, the bureau's rising workload prompted Van Namee to suggest Moskowitz become permanent director. Under her leadership, the bureau launched the Committee's first fully developed public relations program, sending out weekly news releases, planting favorable editorial commentary in newspapers, and distributing copies of Smith's speeches and messages. The publicity bureau had three major goals: to win voter support for Smith's policies, to reelect the governor, and, after 1924, to promote him as the Democratic presidential nominee. As Moskowitz herself explained, her office was "a sort of repository for information concerning Governor Smith and the significance of this or that political or legislative activity of the party." Until mid-1927, Moskowitz accomplished this work with a budget of approximately $12–15,000 a year.[2]

In helping to win support for Smith's legislative policies, Moskowitz provided him with a unique and lasting service. New York's government had become increasingly complex. Its growing number of citizens required more and better organized public services—hospitals, highways, bridges, parks, schools, transportation, and utilities. Bureaucrats, politicians, and citizens did not always agree on how best to develop such services. In some fields, water power for example, fortunes could be made. Should private companies be allowed long-term, low-cost leases on natural water resources in order to develop hydroelectric facilities and sell electricity? How should government regulate projects that benefited the public? To what extent should technicians and planners map out development strategies? How should projects be funded?

Whatever policies Smith adopted, he would need wide popular support for them to be realized. To win this support, Moskowitz brought into politics the same public relations methods she had used in selling the concept of a Port Authority. In May 1923 she summarized these methods in a public relations proposal she wrote for another group, called "Committee on the Regional Plan of New York and Its Environs." The proposal provided a model for what she envisioned for the Democratic State Committee.[3]

Public relations, Moskowitz wrote in her proposal, began with good press relations. The publicity director must develop an informed approach to the press, recognizing its diversity and creating material custom designed for each publication. The director should supply editorial comment in advance, calling a conference of editors or having a friendly publisher give a luncheon for the presentation of material. Since reporters had many interests and little time to read, Moskowitz advised furnishing them with brief summaries in question and answer form. "Brevity is seldom an error," she wrote. She recommended a series of brochures "simply and entertainingly stated." "More social ideas have been sold" by reiterating the name of the enterprise and a series of catchwords than by "weighty treatises and ponderous maps," she observed.

Drawing on her Port Authority experience, Moskowitz suggested forming an "education council" of members from organizations. In addition to maintaining a "first class" speakers' bureau, the council should plan conferences and determine how best to contact each organization. Schools were "the real cradle for a long time program." High school students would soon be voters, teachers comprised a "large army of thinking citizenship," and "pupils reach parents at home." Her approach to them included meetings with civics teachers and essay and drawing contests, which provided multiple opportunities for news articles. "Each year," she concluded, "new groups of pupils and their parents are interested and the nucleus of an intelligent and informed public opinion is thus founded." Finally, Moskowitz recommended graphic means of expression, including radio, and short motion pictures, an "expensive but unexcelled medium for education and publicity."

Applying these principles to politics, Moskowitz brought to the Democratic State Committee a systematic program for achieving Smith's legislative goals. Although there was a Republican majority in the legislature, her publicity neutralized much of its opposition. She did not win all of Smith's battles, but her victories were sufficient to provide him with the impressive record of accomplishment that led to the presidential nomination in 1928.

After 1923, Moskowitz approached Smith according to patterns she had established during the term of the Reconstruction Commission. When suggesting a plan of action, she rarely used the first person singular. Instead she invoked a plural "we" that meant various things—Abram Elkus and the Commission, the Commission and Smith, or herself and Smith. Most of the ideas she presented were hers, but the use of the plural forged an identity that linked Smith with the Commission. Further, she often justified a project by its potential popularity, sometimes implying that votes beyond New York were at stake. She thereby played on Smith's understandable desire for victory at the polls. Envisioning herself as a "social worker/prophet," she also accepted his need to win. Finally, once he approved a project, she gave him full credit for it (as Frances Perkins observed) while developing and controlling it from behind the scenes. At the same time, to avoid all appearance of being domineering, she took pains to remind him that her ideas were only suggestions.[4]

She performed this work out of a two-room office at 331 Madison Avenue, near the Biltmore Hotel, where Smith kept a suite. This placed her close to Smith when he was in the city, and within a few hundred yards of Grand Central Station. From there, once a week, she boarded the night train to Albany. She had telephone access to Smith at all times.[5] Her work touched many facets of Smith's executive policies and campaign strategies, but was most visible in three areas: the administrative reorganization of state government; improvements in the quality of life for New York residents; and issues of justice and morality, such as protective labor legislation and prohibition.

When Smith returned to office in 1923, one of his first priorities was to pursue the issue of administrative reorganization. He stumped the state, sent special messages to the legislature, and held conferences with the Republican majority in the Assembly, all to little avail. In 1924, he appealed to the press by inviting editors from across the state to a conference at the executive mansion. He persuaded them that at the very least the legislature should submit a reorganization amendment to popular vote. The editors pushed the idea in their papers, the vote was organized for 1925, and the result was positive: the amendment passed.[6]

When the legislature set about appointing a reorganization commission, however, divisions over its personnel threatened further delays. Smith wanted a core group from the Reconstruction Commission, including Belle Moskowitz and Robert Moses. The legislature would have none of them. After long negotiations, they accepted ten Smith nominees but rejected Moskowitz and Moses. A battle loomed over who would chair the commission. Republicans wanted H. Edmund Machold, former Speaker of the Assembly and outspoken foe of reorganization. Smith favored former Governor Charles Evans Hughes, a prominent "federal Republican." After persistent lobbying, Hughes was unanimously chosen chair at the commission's first meeting.[7]

Rejected as a commission member, Belle Moskowitz remained active behind the scenes in many of these maneuvers. Some documents showing her involvement appear in Smith's official correspondence; many of them were placed, often haphazardly and probably by Moskowitz herself, in Smith's private files. Not all documents bear her name, but the presence of her handwriting, a notation to "Mrs. M," or actual reference to her establishes her role. In February 1923, for example, she worked on Moses's draft of Smith's special message to the legislature on reorganization. In April she wrote a press release condemning the executive budget proposal of Joseph McGinnies, chair of the Assembly Ways and Means Committee. She then sent this statement to Reconstruction Commission contacts in Buffalo, urging them to make statements to the press. In 1924, when Speaker Machold released a long apologia for his stand on the executive budget and four-year terms, she helped refine the text of Smith's reply. Her handwriting appeared on the final list of Smith's proposed additions to the Reconstruction Commission. No document shows her arranging the March 24 conference of newspaper editors, but, since she dealt with the press all the time, she probably did.[8]

In 1925, she directed the publicity campaign for the constitutional amendments of that year, including those for reorganization. In August, Smith asked her to prepare a list of friendly Republicans. He wanted to send the list to his friend Addison Colvin, a Republican banker from Glens Falls. Colvin planned to send "one of his men" throughout the state to collect Republican opinion on reorganization. The next month, while in England for her daughter Miriam's wedding, Moskowitz continued to dwell on Smith's affairs. Again, the first person plural dominated her correspondence. "Even in far-away England," she wrote, "I cannot forget things that are of importance to us, and especially you." She then quoted parts of transcripts of speeches made at the Economic Club, which she had read on the steamer crossing the Atlantic. In one excerpt, she quoted Republican Ogden Mills saying that there were no "politics" in state reorganization: "Wouldn't it be a good idea just about the time that the State campaign is being opened up to address an open letter to Mr. Mills resuming *our* [my emphasis] correspondence with him, and inviting him to accompany you to various centres of the State and to speak from the same platform with him on the reorganization amendement?" She then told Smith how the English praise him, calling him the "straightest man in American politics, absolutely honest and square." Suggesting a wider field of action, she flattered him, saying,

"You would be surprised to know how much they do know about you, and how they follow every step in your career."[9]

In addition to aiding Smith directly, Moskowitz used her connections with women's groups to foster reorganization. She had been on the Women's City Club board since 1922, and was probably involved in, if she did not initiate, the club's pro-reorganization efforts. In April 1924, the club sent every Assembly member an appeal to pass the executive budget and also four-year term bills. The following year, while Republicans kept reorganization bills locked in committee, the club sponsored a debate on the executive budget. To represent opposing sides, Molly Dewson, the club's civic secretary, had intended to invite Seymour Lowman, the Republican lieutenant governor, and Jimmie Walker, the Democratic leader in the legislature. Governor Smith decided to take Walker's place. Knowing how formidable an opponent Smith would be, Dewson and the club's president, Ethel Dreier, went to Albany to tell Lowman that he did not have to go through with the debate. "It was in the winter," she later recalled, "but great drops of persperation [*sic*] came out on his forehead . . . he was a sport and did not back out."

The event proved hard on Lowman. He accused Smith of trying to act like King John prior to Magna Carta, raising and spending money without consulting the people. He also claimed that Smith rarely advanced plans or policies original with him, but grabbed them from others. The audience hissed, but Smith applauded, staving off embarrassment. When his turn to reply came, Smith arranged some exhibits on the table. Recalling his humble origins on the Lower East Side, he said: "No king from Oliver Street is complete without his exhibits." His unpretentious manner having softened the crowd, he proceeded to show the effect of budget reorganization on government efficiency. Lowman left early. The event packed a large hall and made front-page news.[10]

Moskowitz knew how to make Smith's reforms appeal to middle-class women. Writing in the first issue of the *Women's Democratic News*, Moskowitz promised that reorganization would "do away with waste and duplication." "Women appreciate executive organization," she said. "We know what it is to departmentalize our households, and we do not put the laundress to work at serving tea, and we do not tolerate two people doing the same work."[11] The following issue of June 1925 announced an Albany conference of women interested in the constitutional amendments.[12] From then until the election, the monthly hammered home the reorganization theme.

Moskowitz did not control all aspects of the women's campaign for state reorganization. But, as state publicity director, she was probably consulted at every stage. Eleanor Roosevelt certainly went to see her often. In December 1926, as legislation chair at the Women's City Club, Roosevelt made passage of the final aspects of reorganization her top priority. By then, the reduction of the number of state agencies had won approval. The issue of an executive budget had become moot: the reorganization plan included a budget department in the executive's office. There remained the governor's four-year term. Because Republicans usually polled more votes in presidential years, they wanted the term to end during those elections. Democrats had equally strong reasons for wanting the gubernatorial elections to be held in "off years." Eleanor Roosevelt defended the Democratic party position. She wrote editorials in the *Women's Democratic News* and participated in two public debates against Republican women, one broadcast over the radio. In the vote on the amendment in 1927, the Democratic position won, causing Moskowitz to praise Eleanor for her "logically sound, excellently put and widely quoted" arguments.[13]

Belle Moskowitz was central to many of Governor Smith's programs to improve the quality of life for New Yorkers. The Reconstruction Commission and others had identified some of their needs—better hospitals; elimination of railroad grade crossings; subsidized low-cost housing; cheap electric power; expanded public education, especially in the rural areas; and an accessible parks system. Smith was committed to all of these goals, and more. But how would they be funded? Republicans urged delay, taking a conservative, pay-as-you-go approach. Democrats retorted that delay was uneconomical. In their view, bond issues, which permitted loans to the state against future revenues, distributed the cost of capital outlays among the future taxpayers who would benefit most from them.[14]

In 1923, Smith sought voter approval for a $50 million bond issue to eliminate fire hazards in state hospitals. A fire in February that took the lives of patients and staff in the Manhattan State Hospital had aroused citizen concern. The State Charities Aid Association appointed a "Citizens' Committee on Protection of the State's Unfortunates" which organized publicity to get the bond issue passed. Moskowitz served as liaison between this committee and the Governor's office. She also communicated with women's groups to help them understand the Governor's policy. She urged the Women's City Club to endorse the bond issue, and made a speech at the club's monthly meeting in its support.[15]

Moskowitz's publicity activities for bond issues escalated in 1925. Two bond referenda came up that year. One asked for $10 million worth of bonds issued annually over ten years to pay for general public works improvements, the other for the elimination of railroad grade crossings. To promote passage, Moskowitz organized a campaign with the following elements: a source book containing the central facts of public works and the views of state department heads and prominent citizens; speakers addressing organizations, which then passed favorable resolutions; and the appointment of key publicity people in upstate cities, especially those likely to receive bond proceeds. She herself prepared and distributed leaflets. One, called "Plain Facts About the Permanent Public Works Constitutional Amendment," received wide distribution. Placards appeared in public transportation throughout the state and near polling places on election day. As election day neared, newspapers received two-page supplements. State departments and institutions stamped their correspondence: "Vote Yes on the $10,000,000 Bond Issue." Letters were sent to relatives and friends of State Hospital patients to urge support. And finally, speeches and articles emphasized the absence of log rolling in the way the previous bond issue for $50 million was being spent.[16]

To implement the publicity campaign, she helped form a "Citizens' Committee on Bond Issue Amendment" consisting of over two hundred prominent men and women from the state. A stream of press releases and pamphlets poured forth under its name, all emphasizing that a "pay-as-you-go" method of financing public works led to the accumulation of hazards which, in the end, hurt citizens and cost more to repair. Having received this message in dozens of easily understood ways, including conferences and debates between prominent leaders, voters approved the bond issues. The elimination of grade crossings passed overwhelmingly, 1,032,109 to 859,702; public works by only 22,000 votes. The closeness of this latter vote convinced Republicans that fiscal policy would be Smith's weak point in the next election.[17]

Moskowitz hoped to strengthen Smith's position by proving how well the earlier bond issue of 1923 was being administered. "What do you think," she wrote the Governor early in January 1926, "of putting in a special message to the Legislature when the next allotment of the Fifty Million Dollar Hospital Bond Issue is asked for, showing in detail what the results accomplished to date have been, etc? It might even be illustrated by pictures." She accompanied the suggestion with a draft of the message. By May she had taken her idea further, issuing an illustrated pamphlet called "What the State of New York is doing with the money realized from the fifty million dollar bond issue."[18]

In addition to creating publicity campaigns on select issues, Moskowitz played an especially important role in developing housing policy. In 1923, the legislature terminated the joint legislative (Lockwood) committee, replacing it with a bureau of (later, commission on) housing and regional planning. The agency was a version of what the Reconstruction Commission had advocated several years earlier. Its chair was Clarence S. Stein, architect, regional planner, and the commission's housing adviser. Stein told Smith that New York's shortage of workers' homes was critical. Real estate tax abatement and limited-dividend companies had produced some relief, but it was insufficient. In his view, the state had to use public funds to encourage building.

Smith turned to Julius Henry Cohen for advice. Cohen, using the Port Authority as model, proposed a state housing authority that would raise capital by floating bonds, using the revenue to spur private building of low-cost homes. Both political parties sent housing authority bills to the legislature, and eventually a Republican measure passed. Smith and Cohen were disappointed because it lacked a housing bank, but Smith signed it so as not to lose "the main thing." In the end, the authority sponsored fourteen housing developments (six thousand apartment units) built by eleven limited-dividend companies. Again, hardly a dent was made in the problem.[19]

Moskowitz helped prepare Smith's housing messages and sent him a steady stream of ideas throughout the 1920s. In one letter of early January 1925, she presented a bold plan to expose how few savings banks made loans to individual owners who wanted to buy small homes. She suggested that Smith ask the superintendent of banks to compare the number of mortgages awarded for one- and two-family homes to the number awarded to office buildings. The purpose would be to show that, while the banks were "a depository for the poor man's funds," they make it "almost impossible for him to borrow from this source when he needs it."[20]

The following year, she acted as liaison among Smith's office, the bill drafting department of Columbia University, and the Commission on Housing and Regional Planning. At the end of April 1926, for example, she urged Smith to delay signing the new housing bill so as to keep the old commission in existence a few more weeks. On the verge of completing its report, it needed more time before being replaced by the new Board of Housing. In the meantime, she forwarded a list of nominees for the new board. In the end, four of her candidates were chosen, including the board's chair, Aaron Rabinowitz, whom she recommended not only because of his reliability and sound business sense but because of his long association with settlement houses and social reform.[21] The Rabinowitz appointment was one of many occasions in which she connected Smith to her own network of Jewish professionals from the Progressive Era.

Many of New York's women civic activists were interested in housing. Moskowitz persuaded Eleanor Roosevelt, then chair of the Women City Club's housing and city planning committee, to endorse a bill that would extend state credit to housing. Roosevelt subsequently planned a model exhibit at the club of an Astoria housing development being erected by the City Housing Corporation, a limited-dividend company. Moskowitz also arranged a debate between architect Clarence Stein and Walter Stabler, comptroller of the Metropolitan Life Insurance Company, about the use of state money to finance mortgages on low-income housing. In 1926, she spurred the women's club to call a conference involving all organizations interested in housing, and the following year arranged several housing presentations at the club, including one in April by her husband Henry, who spoke on housing projects in Vienna. Finally, she wrote on housing topics for the *Women's Democratic News*. Calling housing for wage earners "a quasi-public service," Moskowitz pointed out in a lead article how the Governor hoped to remove it from the field of private speculation.[22]

Water power was another area the Governor hoped to keep in public hands. In 1921, when Smith was out of office, Governor Miller had established a Water Power Commission to grant licenses to private enterprises. The next year Smith ran on a platform calling for the agency's abolition. A stalemate followed. Again drawing on Julius Henry Cohen's "authority" idea, Smith proposed a water power authority. The legislature would hear none of it. In 1926, Smith relented somewhat, agreeing that the state should only own water resources but lease transmission rights to private companies. Still, year after year, the legislature called the plan "socialistic."

According to the state reorganization plan that would go into effect in 1926, the Water Commission would cease functioning on December 31, 1925. That fall, with only a few months left to act, the commission's wholly Republican membership began procedures to grant hydroelectric power licenses to private companies. Opponents exposed their plans to the press. The Commission postponed action until the November elections had passed, knowing that afterward they would still have two months in which to act. To stall them further, Smith hired special counsel Samuel Untermyer to threaten an injunction. The power companies applying for licenses "lost their nerve" and dropped away.[23] In the end, while preventing private exploitation of water power, Smith failed to convince the legislature to develop a water power authority. The New York State Power Authority would not come into being until the next gubernatorial regime.

Moskowitz played two roles in water power controversies, those of liaison and publicity. She brought lawyers Cohen and Untermyer together when the latter needed a background on the "authority" concept. In July 1927 she was a guest at Untermyer's estate and impressed him with her grasp of a complex transit plan he showed her.[24] After this visit, he turned to her often for information and contacts on water power. Later, Untermyer expressed continuing pleasure with Smith's views on water power. Because Republicans were taking a stand in favor of federal development of the water power resources of the St. Lawrence Seaway while arguing against state development for New York, Untermyer thought Smith had a "big and appealing national issue." Cohen and Moskowitz also discussed Republican inconsistency on this matter. Cohen could barely contain his amusement when sending her extracts from the report of the Federal Power Commission, whose "Bolshevist" stand for federal hydroelectric power concurred with Smith's plans for New York.[25]

In her publicity role, Moskowitz performed the usual tasks of supplying journalists with materials and of writing press releases. When the Water Commission was threatening to grant licenses to private companies, she organized a large dinner at which Smith made a well-publicized talk on state ownership of hydroelectric power.[26] She also wrote on the topic for the *Women's Democratic News,* describing the state's water power as "the legitimate property of the people." Unless the state accepted Smith's water power authority, she argued, the new power commission would have members as ill-fitted to develop water power as the retiring commission. They would be the Attorney General ("the most harassed lawyer within the bounds of New York"), the Commissioner of Public Works ("whose duties are as manifold and all absorbing as the Attorney General's"), and the head of the Conservation Service Department ("in no way equipped to devise or meet technical problems of engineering"). She explained: "Such an organization as a Power Authority equipped with the most competent engineering brains and the soundest business talent, completely divorced from the State in the sense of its being politically controlled, and yet profiting by the State's backing to the extent of raising money at a low rate, offers all that is best in both private and public enterprise."[27]

In adopting this point of view, Moskowitz expressed her deep faith in the skills of planners and the objectivity of a state agency. That laissez-faire capitalists called her views "Bolshevist" is not surprising. The following spring, the *Women's Democratic News* published a water power study guide which, while unsigned, bore many similarities to Moskowitz's views on the subject. Beginning with a history of water power in New York, the guide ended with a warning: if hydroelectric power fell into the hands of a small group of men, they might win "absolute control over industry and through industry control of the country itself." Such fears, probably raised by the ongoing controversy over the water power resources of the Muscle Shoals section of the Tennessee River, were perhaps overstated, but they reflected growing sentiment among former Progressives throughout the 1920s.[28]

Moskowitz affected the development of Smith's educational policy. New York State faced three challenges in this area: to raise teachers' salaries, equalize resources across the state, and consolidate rural schools. Throughout the 1920s, conferences of educators, citizens, and legislators had met to address them. Pressure on Smith to act mounted during his third term. Everyone seemed to agree that public education needed massive infusions of money, but few favored the concomitant rise in property taxes. At the end of 1925, Moskowitz organized a conference of educators and civic leaders that recommended the funding of a special Commission on School Finance and Administration. Smith accepted this idea, appointing Republican Michael Friedsam, former Reconstruction Commission member, and president of B. Altman and Company, as chair. To facilitate the commission's research, and perhaps avoid the funding embarrassment of the Reconstruction Commission, Friedsam donated $7,500 himself to defray research expenses.[29]

Moskowitz recommended the commission's membership and otherwise monitored its work. Its charge was to study how to finance public education in an era of expanding need, to identify sources of revenue, and to secure "the most effective educational administration." It concluded that state aid to education should be increased immediately by over $18 million, and by $5.5 million annually for three years. It also advised an increase in taxes on gasoline, inheritances, corporations, and personal incomes, and urged more studies of ways in which local boards of education might attain fiscal independence. The commission's report resulted in the passage of the Dick-Rice Law of 1927, which authorized a large increase in state funding for education. In an otherwise lackluster legislative season, the law stood out. "The curtain has fallen, the comedy is over and the 1927 Legislature has passed into history," Moskowitz wrote in the *Women's Democratic News.* In her view, the Dick-Rice law had been the only major accomplishment of the season. During the 1928 presidential campaign, when Smith's enemies charged that he would never champion public schools, Moskowitz reminded them of his support for the Dick-Rice Law.[30]

Another facet of Smith's plan to improve the quality of New Yorkers' lives involved the building of a state park and parkways system. The idea came from Robert Moses, Belle Moskowitz's protégé from the Reconstruction Commission, and proved much more controversial than Smith's plans for the educational system. Nonetheless, Moskowitz gave Moses her support and publicized his programs for him. If she had any criticisms of his methods, she kept them private.[31]

Moses had become aware of deficiencies in the state's park system while serving as executive secretary of the New York State Association, a non-partisan "good government" group. In December 1922 he published *A State Park Plan for New York* which asked for a $15 million bond issue to acquire, improve, and provide access to a unified parks system that would be run by regional, unpaid park commissions, whose presidents would form a Council of Parks to administer the whole. Moses took the plan to Governor Smith. Parks for mass recreation meshed well with Smith's other programs and were sure to be popular with voters. In April 1923, Smith sent to the legislature a parks message that led to a bond issue appearing on the ballot in 1924.

Moses planned to develop the Long Island segment of the state park plan himself. In the summer of 1923, Smith and Belle Moskowitz drove out with him to look at the sites he had selected, and gave him their approval. Moses then set about drafting legislation to establish the Council of Parks and Long Island State Park Commission, the presidency of which Smith promised him. Moses was an unexcelled bill drafter. Aiming for the strongest possible council, he granted park officials the right to "enter and appropriate" lands before the legislature had set aside funds for compensation. Although some doubted the legality of the right, no one challenged the clause in Moses's bill. He also gave broad powers to the future president of the Long Island State Park

Commission. Introduced during the legislature's adjournment rush of April 1924, Moses's bills passed both houses unanimously. Smith signed them, appointed Moses president of the Long Island Commission, and picked the rest of the commissioners from Moses's list. At its first meeting, the State Council of Parks selected Moses as its chair.

Conflict and bitterness followed. Through a combination of negotiation and threat, Moses began to "appropriate" Long Island land. The bond issue passed in November, but the legislature had not yet set aside funds. When landowners refused to sell, Moses claimed that the law empowered him to "appropriate" in advance. Small farmers were powerless against him. But a small group of wealthy landowners, who wanted to protect a wooded parcel known as the Taylor Estate that they used for hunting, went to court. Moses then asked Governor Smith to sign an appropriation form attesting "that there is available a sum representing fair compensation for the land to be entered and appropriated." Smith was torn. Parks were popular, and had helped win him reelection in 1924. But he was worred about legality. His advisers, including Belle Moskowitz, thought he would not sign. But then the wealthy Long Islanders came to tell him their side of the story. One of them, in explaining why he opposed a park in the town of East Islip, expressed the fear that the town "would be overrun with rabble from the city." "Rabble?" Smith queried. "Why, that's me." A few days later, he signed the appropriation.[32]

As the court fight dragged on, the Republican-dominated legislature tried to amend the previous session's parks legislation by putting parks under the Land Board. Smith vetoed the amendment. In response, the legislature refused to set aside funds for Moses's Long Island purchases, and adjourned. Smith's only recourse was to call a special session. He waited until the dog days of June 1925, when everyone longed to get to a beach or park. Then, in his first state-wide radio hook-up, he announced the special session and gave a long speech on the park issue. Moskowitz organized a conference of the regional park commissioners, fifty-four of whom met a few hours before the start of the special session. Although the commissioners packed the visitors' gallery and cheered Smith's opening speech, which was also broadcast, the legislature refused to budge. It passed again the parks bill Smith had already vetoed, forcing Smith to veto it again.

In the meantime, Moses was losing his options on the Long Island land he had hoped to buy. Since he still held an option on the Taylor Estate, Smith suggested he find a philanthropist to loan or donate the quarter million dollars necessary to buy the land. Moses worried about asking someone who might refuse and then spread the story that the Council of Parks lacked the funds to buy what it had "entered and appropriated." Moskowitz suggested the name of someone who would not refuse—philanthropist August Heckscher, a recreation enthusiast.

Offering to telephone him herself, she obtained his assent. In return, the park was named after him.[33]

The next year matters improved for Moses. As soon as he had secured Heckscher's gift, and before the court had rendered judgment, he began to turn the estate into a park. He lost the first judgment but won on appeal. In the trial of the appeal, Governor Smith himself, accompanied by Belle Moskowitz, took the train to Riverhead to appear as a witness. On the stand for fifteen minutes, Smith recounted the "rabble" and related stories, was not cross-examined, and then had lunch with the judge, who ruled that the first appropriation of the Taylor Estate had been illegal but that, even though appropriation could be made only with public funds, the second purchase, with private funds, had been made in good faith and was hence legal. The plaintiffs appealed to the Supreme Court. But in January 1929, Justice Louis Brandeis ended the case by refusing to hear it.

According to Moses's biographer Robert Caro, in 1924 the only state park on Long Island was a useless two-hundred-acre tract on Fire Island. By the end of the summer of 1928, New York owned fourteen parks there, totalling 9,700 acres. Because two-thirds of this acreage had been acquired through gifts, the parks had cost the state only $1 million when, at 1928 land values, they were worth more than $15 million. One need hardly wonder why Smith, after balking at Moses's quasi-legal actions, eventually defended them. He even accepted Moses's high-handed treatment of two other state park commissioners, Ansley Wilcox from the Niagara Commission and Franklin D. Roosevelt from the Taconic, and of "federal Republicans," such as Henry Stimson, whose support he needed.[34] Moses always managed to allay Smith's concerns or to convince him that the disgruntlement of a few was worth the large ends achieved.

Moskowitz also accepted Moses's methods and developed the public relations aspect of his schemes. The job was not difficult. Most New York newspapers (with the exception of the *Herald Tribune*) heaped derision on the "few selfish golfers" trying to keep working people from getting fresh air. In addition to sending out press releases publicizing the benefactors who had given land to the parks system, Moskowitz also wrote about park controversies in *Women's Democratic News*. Here she confined herself to the legislative and administrative, not legal, issues.

Two of her articles appeared in the summer of 1925 under the single title "The Truth About Parks." They explained why Smith had vetoed the legislature's amended parks bill. Its faults were administrative, she said: the legislature placed supervisory power over parks in the Land Board, a group of elected officials without the experience, time, or staff to oversee parks and parkways. Worse, under pending state reorganization, the Land Board would soon disappear, its functions assumed by the secretary of state. In her view, giving the board power over parks went against modern administrative ideals of efficiency and accountability. Further delay would have dire results, she warned. Options on lands would expire, "playing into the hands of greedy real estate operators and others who are eager to sell property to the state at excessive values." She advised women to contact legislators and express concern about placing control over parks into the hands of other agencies.[35]

Smith and Moses liked one another. Moskowitz also liked Moses, and the feeling was mutual. In private she expressed doubts about his ruthless methods, but retained her faith in his talents.[36] In her view, his programs were models of government action undertaken for the benefit of the masses. Rationally conceived and professionally run, they also reflected her ideals of cooperation between private and public forces. Moses's achievements came at a cost, however. The wealthy few and the displaced farmers whose lands were divided or appropriated would never forget how Moses had treated them, though to Moskowitz their objections paled in importance when measured against the larger good that was served.

An advocate of a strong state role in achieving industrial justice, Smith was committed to legislation that protected employees from dangerous, unhealthful, and exploitative working conditions, and believed that when employers failed to provide decent conditions the state should step in to set standards and help settle labor disputes.[37] Moskowitz of course agreed, and after 1923, although her primary responsibility was publicity, she continued to advise Smith on labor issues. She took interest in a number of them—unemployment insurance, expansion of workmen's compensation, and reforms in the Labor Department. But two issues received her special attention: the settlement of labor disputes and legislation to set maximum working hours for women and minors.

She was particularly active in efforts to avert strikes in the garment industry. In June 1924, contracts expired between the union and the associations of manufacturers and jobbers in the cloak and suit trade. Negotiations broke down, and a major strike appeared imminent. New York City officials feared the impact of fifty thousand striking workers on the forthcoming Democratic national convention. Smith won agreement from both labor and management to abide by the rulings of a special advisory commission chaired by George Gordon Battle, an attorney. In the end, the parties came to an agreement, and the strike was averted. Two years later, a strike loomed again, and Smith called upon the same commission for help. This time it was less successful, issuing a proposal that the unions, themselves torn by dissension, did not accept. After striking on July 1, 1926, the unions emerged with terms less advantageous than those the commission had recommended.[38]

Moskowitz knew the commission members well—Bernard L. Shientag, an intimate from the Smith campaign and a member of the industrial commission; Lindsay Rogers, a Columbia University professor and one of her son Carlos's mentors; Herbert H. Lehman, a well-known philanthropist active in Jewish welfare circles; and Battle, who was prominent in probation reform and recreation movements. Moskowitz, who may have suggested the commission's personnel, coordinated the distribution of the 1924 settlement report so as to maximize its publicity effect. When Smith received a flood of letters of gratitude for his intervention, Moskowitz organized the responses that went out under Smith's name. The following year, Battle communicated directly with her about averting another potential strike in a tailoring company. Moskowitz also followed up on the development of an unemployment insurance fund, an element of the 1924 settlement, first delivering a letter from Smith to a meeting at which he was supposed to hand out personally the first unemployment check, and then writing Smith of the "great ovation" his letter had received. "Everybody" felt that his interest was "the most important contributing factor" in maintaining peace in the industry, she reported.[39]

In addition to fostering government arbitration, Moskowitz also influenced the outcome of maximum hours legislation for women. Women's groups and unions had long been lobbying for a forty-eight-hour work week for women. The National Woman's Party (NWP) opposed special legislation for women, arguing that it put women at a disadvantage. Faced with providing special conditions for women, employers tended to hire men. Also, those women who wanted (or needed) to work longer hours, to work at night, or to get jobs in high-paying but hazardous industries lost the freedom to do so. Finally, protectionism of only women perpetuated stereotypes about female weakness and inferiority; in the judgment of the NWP protection should apply to men and women alike. Their opponents pointed out that legislators and courts were unwilling to go that far. The ideology of the free marketplace remained sacred. Lawmakers occasionally acted upon cases of industrial injury to the "weaker" sex and to minors but would not concede the protection principle. Reformers knew that protection of women and children was only half the battle, but they preferred that half to none.[40]

Moskowitz proved a formidable opponent of an equal rights bill sponsored by the NWP and almost singlehandedly kept it from being reported out of committee. When Jane Norman Smith, New York State Chair for the NWP, asked Senator Salvatore A. Cotillo to sponsor a bill, he replied he would do nothing without consulting other women's organizations, "and particularly Mrs. Moskowitz." Complaining to NWP head Alice Paul, Jane Smith characterized Moskowitz as "a League of Women Voters woman who is . . . 'the whole works' in the Smith administration." She charged Moskowitz and Mary Garrett Hay's "group" (probably the League of Women Voters) with trying to "dictate to us and to Senator Cotillo as to what bills should be introduced."[41]

A year later, Jane Smith complained about Moskowitz in a letter to the state organizer for the NWP. Although she could not prove it, she believed that Moskowitz was working to "keep our bills in committee again." In this letter, she recalled her 1923 meeting with Moskowitz, whom she visited after hearing of her opposition to "our bills." The letter illustrates how Moskowitz interacted with other women and wielded her power.

> She was very amusing, stating that the Governor had not intended to ask for the removal of discriminations in his message; practically said that she had helped write the message; said that he had really intended to ask that a commission should be appointed to investigate, etc. I reminded her that the Governor in his message had simply reiterated the plank in the Democratic Party platform, which he had promised me he would do, and the platform said nothing about a commission. She attacked our industrial policy. At that time she was conferring with the welfare groups and the L. W. V. [League of Women Voters], and I am confident that it was she who put them up to the idea of demanding a commission to investigate last year. Because of what she said about the commission, we prepared to meet that situation, and we downed them at the hearing. However, I heard from certain Senators that she was working to keep all our bills in the Senate Committee. Later a Democratic Senator told me that the N. W. P. ought to be very grateful to him for working for our bills; that he had incurred the enmity of Mrs. Moskowitz by doing so![42]

In dealing directly with other women, Moskowitz was hardly self-effacing. She boasted about how she had helped write the governor's message, took the offensive against a policy she decried, and flaunted her power before a state senator from her own party.

Moskowitz engaged Smith directly in her campaign against the NWP. Writing to him in October 1923, she enclosed a page from the NWP's magazine, *Equal Rights,* which claimed that his gubernatorial message had interpreted protective legislation as a legal discrimination against women. She urged him to send a correction to the magazine. "Women read these papers with great interest," she wrote, "and as you see watch your utterances with care. This is a national magazine and they should not be allowed to get away with such a misquotation."[43] She took action herself by composing an article for the *Women's Democratic News* that reiterated the governor's support for "the removal of any unjust discriminations against women in the laws, by specific statutes but not by a sweeping general equality law."[44]

In 1924, both Republican and Democratic state platforms promised a forty-eight-hour work week for New York women. The following year, the legislature considered authorizing the Industrial Board to reduce hours only in those occupations where longer hours would harm female employees. Representing the Women's City Club and League of Women Voters, Eleanor Roosevelt, Frances Perkins, and Sara Conboy campaigned against it. Belle Moskowitz worked behind the scenes. Despite their efforts, the bill passed, prompting Ethel Dreier, president of the Women's City Club, to urge Governor Smith to veto it. She argued that a board empowered to shorten hours could also lengthen them. In announcing his veto, Smith called upon Republicans to stop trying to "hoodwink" the people with fake forty-eight-hour bills. In 1927, he finally signed a bill that reduced working hours to forty-eight but allowed several exceptions. While considering it imperfect, he called it a "step in the right direction."[45]

National prohibition was the most volatile issue of the 1920s. The Eighteenth Amendment, forbidding the sale of intoxicating beverages, was ratified during Smith's first term in office. The Volstead Act then followed, defining as intoxicating any drink that contained .5 percent alcohol. Prohibition had broad national support, especially among rural, Protestant, and native-born populations, but many Americans, especially New Yorkers, opposed it. Smith was one of them.

His chief grounds were legal, moral, and practical. In his view, Prohibition created more evils than it corrected. On the one hand, strict enforcement threatened civil liberties; on the other, lax enforcement caused a decline in public morality. Smith recognized further that most communities were unwilling or unable to pay for the large police force required for enforcement. There were also political considerations. The voters who had put Smith in office—the urban, heavily immigrant, non-Protestant populations of his state—felt that wine and beer were essential to their way of life. All of these arguments notwithstanding, Smith had no choice but to enforce the law. At the same time, he called for an amendment to Volstead that would allow wine and beer.[46]

During Smith's first term in office, the state legislature passed a bill permitting drinks of 2.75 percent alcohol. Smith signed it. But then the United States Supreme Court ruled that states could not act contrary to congressional legislation. During Governor Miller's term, the legislature passed a state enforcement bill, the Mullan-Gage Act, which defined intoxicating liquor just as Volstead did, and exacted more severe penalties for violations. After Smith's reelection, the legislature had second thoughts, and in May 1923 voted to repeal the harsh measure.

Their action proved embarrassing to Smith, who had thirty days to decide whether to sign or veto it. Either way, he faced certain criticism. Felix Frankfurter summed up the views of those who wanted a veto. Writing his friend Henry Moskowitz, Frankfurter predicted that Smith would be "damned" whatever he did; "if he vetoes the repeal, he will be damned for a comparatively brief time but if he signs it, he is damned for good." Further, signing would encourage those who thought the Constitution was only "a scrap of paper" as soon as it touched their own particular interests. "These latter are not the men who have permanently triumphed in American politics," he warned. Charles Murphy, boss of Tammany Hall, took the opposite view, threatening he would never forgive Smith if he vetoed the repeal.

After a month of discussions and self-searching, Smith signed the repeal, acting less out of loyalty to Murphy than from his conviction that, because Prohibition was wrong, repeal was right. His official argument was a legal one: under Prohibition the states might construct their own enforcement procedures but were in no way obliged to do so. In New York State, a violation of Prohibition would remain a federal crime to be prosecuted in federal courts.[47]

Moskowitz opposed Smith's decision to sign. She had monitored the legislative sessions that voted on the bill and was the one who broke the news to Smith that it had passed. She advised him that a veto would not only show Smith as "presidential" but would be popular. Wet New Yorkers were already in Smith's camp; the people he had to win were the nation's drys.[48] Once Smith had decided to sign, however, Moskowitz tried to turn the event to his advantage, and worked hard on the wording of the repeal memorandum. According to George Van Namee, Joseph Proskauer wrote a brief setting forth the basic line of argument. On May 30, Smith dictated a version to his stenographer, after which he revised it until 3:00 A.M. on May 31. Although he had already made up his mind, later that day Smith went through the motions of an open hearing in Albany on the bill. Final revisions on the memorandum were made that night at the executive mansion until 2:30 A.M., June 1. Four people were present: Smith, his counsel James A. Parsons, Belle Moskowitz, and Van Namee. "The changes are in the handwriting of Mrs. Moskowitz and were made as the document was read aloud by her, on the suggestion of all present," Van Namee recalled.[49] Later, when Smith was running for President, Moskowitz argued that Smith's refusal to compromise with his convictions was one of his most admirable qualities.

Mullan-Gage thrust Moskowitz's relationship with Smith into the limelight for the first time. One newspaper speculated that Smith's decision on repeal would "be determined by the counsel of a woman." When Howard Shiebler of the *Brooklyn Daily Eagle* asked her directly about Mullan-Gage, she denied having influenced Smith. "He made his own decision in that case," she said. Shiebler took her at her word and was right to do so. But he was so impressed by her knowledge of the previous legislative session that he concluded she was playing a central role in framing many of the Governor's speeches and messages and preparing legislation initiated by the executive.[50]

By the time the 1928 campaign was over, Belle Moskowitz had abandoned her dry position and come out for Prohibition reform. In 1929, she helped found the Women's Organization for National Prohibition Reform (WONPR), led by Pauline Morton Sabin, a prominent Republican in the Women's City Club who once had been sympathetic to temperance but now believed Prohibition had failed. "Human experience teaches us that total abstinence through legal compulsion is impossible of attainment," she said; "temperance must come from within, and . . . if that spirit is not there, legislation will be of no avail."[51] Irked by organizations such as the Women's Christian Temperance Union, which clung to prohibition regardless of the results and then claimed to speak for all women, Sabin mobilized women into a reform organization.

In April 1932, Moskowitz addressed a WONPR conference held in Washington, D.C. in a speech that revealed her views on both temperance and prohibition, and revealed also her faith in the power of organized women to bring about change. She began by saying that her early experiences as a social worker had fixed her point of view "at the prohibition angle." At the time, however, she failed "to see the distinction perhaps between temperance, which is a matter of controlled habit and which is a basic quality of character, and absolute prohibition." She saw the fight for Prohibition reform as "essentially a woman's fight," to be waged politically in the same manner as woman has always fought for the social reforms she favored: "She must present it, after careful study and analysis of the facts, to both major parties and recognize her obligation and responsibility to sponsor those candidates, local, state and national who come nearest to realizing her ideal."

In making this appeal, Moskowitz underscored the special concerns of mothers, who in her view bore the prime responsibility for providing moral standards for the young. Youths of high school and college age "have been through fire and many have been badly burned," she said. But what of those even younger? They saw the law flouted everywhere—mysterious gentlemen delivering large and bulky packages, pet or flower shops selling nothing of the kind, cocktails and wine served at family dinners—all acquired illegally. "Just what does a child think and at what point do we explain that all this is against the law?" she asked. Tenement children were particularly vulnerable. "Almost as soon as they can run about they are sent on various errands concerned with the traffic." In their play, they never acted out the revenue officers who did the punishing, but rather the "life of the teacher of crime and terror with which our newspapers, magazines and motion pictures are filled." To Moskowitz, children were women's greatest reason for action. When women "cease to guard and to protect their homes and their children they cease to function altogether," she said.

Her concluding statement called for political action. A social agency that believed in a reform and sponsored it got "nowhere" without achieving its enactment into law. She pinned hopes on the Women's Organization for National Prohibition Reform. "Intelligent and able," it would convince politicians that "women insist that this country shall return to . . . freedom of individual choice." Free choice alone, she claimed, would "bring with it a moral standard based on strength of character and not on the lash of the whip."[52]

Throughout her years of service to Al Smith, Moskowitz never lost sight of the power of organized women. Women's organizations had groomed her for her public roles. Equally important, they had inspired many of the reforms she thought essential if working-class Americans were to lead decent lives. She had opposed the Equal Rights Amendment, believed women were not the intellectual equals of men, and rejected the label "feminist." But she never saw women as marginal to society. Their roles may be primarily private and domestic, but women must also work for society's larger welfare. In the modern era, moreover, traditional "lady bountiful" roles were insufficient. If women reformers did not battle injustice in the corrupt and competitive world of politics, they would fail.

Constructing a delicately balanced compromise with the gender inequalities of her time, Moskowitz built her career on this dual sense of woman's purpose. She rejected a purely private existence not because she wanted a career for herself, but because she saw involvement in the public sphere as her duty and obligation. Moskowitz's statement on prohibition reform illustrates a feminism special to the 1920s, rooted in the social reform tradition of the Progressive Era and yet ready to make use of the political power that enfranchisement had at last given them.

Smith's gubernatorial campaigns were run by citizens' committees that Belle Moskowitz helped put together. The committee memberships of "federal" Republicans and noted independents indicated the broad base, extending beyond Tammany and even beyond the Democratic party, of Smith's support. As executive secretary, Moskowitz played a leadership role in all of Smith's state campaigns.[53]

Each campaign revolved around some issue that required special handling. In 1920, for example, Republicans attacked Smith's close connections with Tammany. Despite the obvious rancor between Smith and Maylor Hylan, Republican propaganda heaped the New York mayor's sins upon Smith. Moskowitz put together a pamphlet of speeches made in December 1919 by Smith and Republican Charles Evans Hughes at the City Club. Hughes was chair of the club's committee on State Retrenchment and Reorganization. Moskowitz felt that an expression of support from him and the anti-Tammany club would help separate Smith from the Tammany taint.[54]

During the post-war "Red Scare," the New York State Assembly expelled five recently-elected Socialist legislators on the ground that they advocated the overthrow of the government. Outraged by this act, civil libertarians made it a campaign issue in 1920. After the expulsion, twelve of Smith's advisers met in Albany to discuss what the governor should do. Moskowitz, the only woman present, was alone in urging a special election to fill the vacant assembly places in conjunction with an extraordinary session called to deal with the housing crisis. She gave this advice even at the risk of seeing the Socialists reelected, which they were. She later wrote the governor that she wanted to make his candidacy that of "a rational, sane liberal, who through the best possible methods of party administration can put over a progressive program for the people of the State."[55] Smith adopted her advice, and the Socialists were reelected. But Moskowitz's advice did not turn out well: Smith lost the 1920 campaign.

In the 1922 campaign, Moskowitz told the leaders of the Women's Division of the State Democratic Committee—Marion Dickerman, Nancy Cook, Eleanor Roosevelt, Caroline O'Day, and Elinor Morgenthau—that the committee would make its first extensive use of them in this campaign. Smith won the contest, but Republicans retained a majority in the legislature. Upstate Democrats were deeply discouraged by the continuing strength of the GOP in their regions. The Women's Division leaders decided to visit every county organization or political club once a year in order to inspire upstate workers and keep them better in touch with the state committee. This was a demanding commitment, requiring weeks on country roads, gracious attendance at hundreds of luncheons and teas, and repetitive speech-making. All the Division officers took turns at these tours, which became the backbone of state-wide party unity in the 1920s.[56] Democratic party men admired the women for making the tours, but none of them had the time, inclination, or temperament to go out on the road themselves unless it was election time.

In 1924, when Smith ran against Colonel Theodore Roosevelt, Jr., the Women's Division made headlines. Roosevelt had served in the state assembly, and then became Assistant Secretary of the Navy under President Harding, a post he held during the current oil scandals. Harding's Interior Secretary had leased oil reserves at Elk Hills and Teapot Dome to private oil companies without receiving bids. When outsiders questioned the propriety of the leases, Colonel Roosevelt sent a detachment of marines to Teapot Dome to protect one of the companies.[57]

Although Theodore was a cousin of Eleanor Roosevelt, her Democratic partisanship took precedence over family loyalty. To call attention to her cousin's failures, Eleanor designed a white canvas teapot to sit on top of a large Buick touring car, along with a device for creating steam, which the car's passengers sent pouring from the spout when they pulled into a new town. During its two-week tour, the "singing teapot" created a sensation. Eleanor Roosevelt later regretted this "rough stunt," but it helped the Smith ticket win in another Republican year,[58] and also improved morale among upstate Democrats, who felt at last that the state committee recognized their needs. In the town of Chateaugay, for example, one Democratic woman organizer told Eleanor Roosevelt that the touring "teakettle" of 1924

> . . . was the first notice which had ever been taken of them in the campaign. They could not get any good men or women speakers in campaign time, and what was the use of working for the state or national ticket which paid so little attention to you, even when you went Democratic locally?[59]

Belle Moskowitz did not directly take part in this "trooping for Democracy," but she approved it, followed the work of the Women's Division closely, and conferred frequently with Eleanor Roosevelt on tactics and strategies. As a reflection of her commitment to the Women's Division, she wrote many of its newspaper's campaign articles.[60]

Moskowitz wrote the Democratic state platform for 1926, probably her most important contribution to the campaign. The document began by endorsing Smith's "record of constructive achievement, earnest endeavor and businesslike management." It continued by contrasting this record with Republican recalcitrance. Democrats had presented programs of "State reorganization, of welfare legislation and protection of natural resources." Assuming that in opposing these they would reap partisan advantage, Republicans never offered "honest and clearly conceived" alternatives. The document ended with a wide-ranging denunciation of Republican policies.[61] The platform not only showed how Moskowitz perceived Smith's achievements but summarized the core ideas of her political philosophy. As might be expected, she reserved special praise for the Smith policies she had helped shape, including government reorganization, the bond issues, social welfare programs, and reform of the Labor Department, "crippled" under the Republicans. "This record of progress," she concluded, giving Smith full credit, "was established by the driving force of Gov. Alfred E. Smith. . . ."[62]

After cataloguing Republican sins, the platform then offered Democratic pledges on state and national issues. These reflected Moskowitz's loyalty to Progressive Era ideologies. Critical of selfish elites, she favored protections for the working masses; at the same time, she hoped to keep taxes down in order to encourage "legitimate business." Democrats, she said, promised to "guarantee perpetual ownership and control by the people of the State-owned water power resources," and to oppose "their exploitation by private interests for private profit." They pledged continued support for protective, child welfare, and housing legislation. They opposed granting temporary injunctions in labor disputes without hearings, and supported further liberalization of workmen's compensation. The platform also promised completion of land purchases for parks, despite the "combined opposition of selfish and partisan interests."[63]

The rest of the platform brought no surprises, but by its end it had risen in pitch. It asked for a more equitable distribution of taxes, continued progress in food regulation and market facilities, and a "liberal policy" on health, education, labor, child welfare, highway building, and public utilities. It expressed opposition to voting for governors in presidential election years. And on the ground that the Volstead Act had resulted in "widespread contempt and violation of the law, in illegal traffic in liquors and in official corruption," the party asked the people to vote "yes" on a referendum to modify it. The platform concluded with a ringing attack on Republicans, who thought that "governmental protection is intended only for big business and great wealth," and who have consistently opposed measures "for the relief of the laboring people of this State."[64]

Moskowitz took pride in this document. A month after the state convention, she summarized it in a lead article for the *Women's Democratic News* entitled—surely to the irritation of feminist readers—"We Give Strong Men a Strong Platform."[65] As for the election results that year, Moskowitz called them "a gratifying as well as an extraordinary expression of public intelligence." Democrats had carried every state office except that of attorney general. And the state now had a new United States Senator, Robert Wagner. In Moskowitz's view, Wagner would go forward to enact "broad liberal policies of social progress."[66]

Reviewing Moskowitz's efforts, only a summary of which has been provided here, one is inevitably impressed by the breadth of her influence on Smith.[67] She was not the only person advising him, but, except for his personal secretaries, she alone had the time and motivation to devote herself fulltime to his political needs. The documentary evidence is necessarily incomplete, but what exists of it is huge. In addition to her handwriting on speeches, messages to the legislature, articles that later appeared in print, and letters that went out over Smith's name, there are dozens of brief communications between her and the Governor. Between 1923 and 1928, Smith's personal secretary George Graves retained copies of close to a hundred typewritten exchanges. These included requests to her to research an issue, get information about an individual, draft or rewrite a message, answer a letter, call on a legislator, compose a press release, confer with Smith in Albany or New York, or contact Moses, Proskauer, or Julius Henry Cohen for advice. As time went on, these requests became so frequent and routine that Smith sent them out increasingly as orders, without even a "please" or "thank you." Myriad tasks, small and great, many involving tact, diplomacy, discretion, or caution—all entrusted to Moskowitz with the tacit understanding that she would never refuse or be too busy to respond.

Without formal training in any area of administration, economy, or law, Moskowitz made herself an expert on whatever issue was at hand. She gave Smith programs herself or connected him to others who could bring him programs; she found the people to make the programs work; and then she presented the programs to citizens in ways they could understand. When they went to the polls, they not only kept returning Smith to office but, in increasing numbers, voted in his party cohorts as well.

In 1924, Al Smith had been a strong but unsuccessful contender for the Democratic presidential nomination. Two years later, his fourth gubernatorial election made his nomination in 1928 virtually unstoppable. Moskowitz's publicity work had played a critical part in bringing Smith to this moment of triumph.

CHAPTER 10

The Selling of Al Smith

Mrs. Belle Moskowitz, we realized after her death, had been Al Smith's tutor and mentor to an extraordinary degree. His labor program had been sound because Mrs. Moskowitz was thoughtful, well informed, and incisive. My favorite metaphor for the pair has always been: Mrs. Moskowitz was Al Smith's tent pole.

Molly Dewson, "An Aid to the End"

The 1920s were not a tolerant decade. Extremist groups contended against one another, usually with bitterness, sometimes with violence. Rural and small-town Americans resented the spreading influence of the large cities. Nativists sought to restrict the entry of the foreign-born. Fundamentalists demanded a literal interpretation of Scriptures to stem the tide of moral and intellectual liberalism. This was the era of the "Red Scare," the Scopes "monkey" trial, the Sacco and Vanzetti case, and the Ku Klux Klan, whose membership had soared into the several millions by mid-decade, using intimidation and violence in their attempts to keep America white, Protestant, and dry.[1] In a nation torn by such polarities of ideology and place, it is a wonder that Alfred E. Smith—urban, Catholic, and wet—received the presidential nomination of a major political party. It is an even greater wonder that his expert in public relations, Belle Moskowitz, thought he could win a national consensus.

Of the two major political parties, the Democratic weathered the divisiveness of the times less successfully. The party's strength had lain in a coalition of the South and West. Woodrow Wilson's failure to win acceptance for the League of Nations and his stroke in 1919 dealt serious blows to party unity. At the 1920 party convention, without any involvement of Belle Moskowitz, Al Smith's name was tossed into the ring, but since he seemed to many delegates to represent only the industrialized, urban, and wet Northeast, he won only a few more than his home state's ninety votes. Republican Warren G. Harding won a landslide election that year. Two years later, the GOP fared less well, losing its hold on local and state offices. Smith's reelection in this turnaround enhanced his chances for the presidential nomination in 1924. But the party convention that year was a disaster. Many forces contended for hegemony, among them an urban, ethnic-dominated, and wet North and East; a dry, Klan-ridden, and religiously fundamentalist South; and a militantly farm-oriented and dry West. In the words of historian Robert K. Murray, "It was an impossible combination."[2]

Two major candidates sought the nomination: William Gibbs McAdoo, Woodrow Wilson's son-in-law and former Secretary of the

Treasury, and Smith. Each had his taint. McAdoo, though not directly involved in the oil scandals, had defended the interests of oil magnate Edward L. Doheny, who had been involved. As for Smith, no one questioned his personal integrity, but many recoiled from his wetness and doubted his ability to act independently of the Catholic Church and Tammany Hall. Early in 1924, Smith's chances looked good. At the end of April, there was a setback when Tammany's powerful boss and Smith booster, Charles F. Murphy, died. Belle Moskowitz put a different slant on the event when she told the press that, in her opinion, Murphy's death "was the best thing that could have happened for the Governor," since he could now establish his independence from Tammany.[3]

Moskowitz was in the thick of the 1924 contest. Both she and Smith felt that a convention in New York would not be to his advantage. They tried to get the Democratic National Committee (DNC) to choose Cleveland or Chicago but failed. Deciding to "make the best of it," Smith kept his concerns quiet on this score. Hoping to free Smith from the Tammany stamp, Moskowitz and Joseph Proskauer asked independent Democrat Franklin D. Roosevelt to head Smith's pre-convention campaign. Roosevelt was not only a national party figure but dry, Protestant, and aristocratic, all qualities Smith lacked. Moskowitz and Proskauer feared that, since Roosevelt was still in therapy to recover use of his legs, he might refuse. When they called on him, they promised to do the work if he would only lend his name. At first he demurred, not because of his therapy but because of his discomfort with Smith's stand on prohibition. But his ardor for McAdoo had cooled, and Smith's 1922 victory had impressed him. He not only accepted the job but later surprised Smith's advisers by issuing a mountain of publicity and correspondence in behalf of Smith's candidacy. Smith's camp began to think they had misjudged him. It would not be the last time.[4]

Moskowitz herself was not idle. Working closely with George Van Namee, she churned out publicity and handled all correspondence on Smith's candidacy. In accordance with her theory of good public relations, she issued twenty-five thousand copies of a pamphlet, "What Everybody Wants to Know About Alfred E. Smith." It contained biographical information, summaries of his accomplishments, and personal and editorial comments. To prepare for the convention, one enthusiast in her office checked every delegate's fraternal, civic, or club affiliations, then arranged contacts with New Yorkers of similar bent. Every "straw, every newspaper clipping, every letter from anybody—was held as most important and significant," she later wrote.[5]

Her office made a special effort with women delegates, more of whom participated in this convention than in any previous one. In 1920, 93 women attended; four years later, there were 199, or almost 14 percent of the total. There were also about 300 female alternates, and women served in numbers equal to the men on the national committee. But few women had access to the party's inner circles. Robert Murray characterized Democratic women delegates, who were necessarily less politically experienced than the men, as also less flexible and compromising. He claimed that McAdoo (whom he characterized in the same way) relied heavily on their support to keep his forces from ceding. Smith, on the other hand, took "little notice of women, which was ironic since one of his chief advisers was female."[6] But Moskowitz's deep commitment to women's political organizations raises doubts about Murray's judgment. Smith himself may have given little time if any to women delegates. Moskowitz, using her close ties to women in the Democratic organization, courted them as diligently as she did the men.

Moskowitz wrote later that she had never thought 1924 would be Smith's "year." When, a few days before the convention Smith asked her to assess his chances, she told him "there were great gaps in our armor." The Smith group went ahead anyway and, according to Moskowitz, "played the game." Smith received delegations, governors, and senators; he talked to reporters and listened to the proceedings over the radio. Moskowitz read the delegates' faces on opening day. "How do you feel?" her husband asked. "Not so good," she replied. The couple went home, returning the next day to their places in a small room at the back of the hall. Belle spent her time noting the influence and effectiveness of personalities. She made friends, for example with the Wyoming delegation, which would become a pivotal group four years later. She watched state leaders hold their delegations in line, noted their names, and never forgot what they did. She wandered through the boxes and sat on the platform to consider individual loyalties. She never changed her first prognosis.[7]

The convention opened in New York's Madison Square Garden on Tuesday, June 24. Squabbles over the number of tickets and seating arrangements marred the early proceedings, a foretaste of worse things to come. Tempers flared from the heat and humidity; the din was maddening. McAdoo's name went before the convention on its second day, Smith's on the third. When it was "Al's" turn, partisan spectators jammed the galleries, shouting and cheering, sometimes so rudely that Smith floor leaders had to get them under control. Finally, Franklin Roosevelt, leaning on his son James, walked to the rostrum to nominate Smith. His speech, the product of a month's work, had been drafted by Joseph Proskauer; journalist Herbert Bayard Swope and Roosevelt himself had edited and polished it. It ended with the famous Wordsworth couplet:

> This is the Happy Warrior: this is he
> Whom every man in arms should wish to be.

The speech was a triumph. In the end, both speech and convention did Roosevelt more good than Smith.[8]

After settling some thorny platform issues, the delegates began to pick their nominee. The balloting, which began on June 30, did not end until the wee hours of July 9. A deadlock between McAdoo and Smith emerged early, with McAdoo leading, but several hundred votes shy of what he needed to win. Although both candidates swore they would fight to the end, neither could. After 103 ballots, the exhausted delegates chose John W. Davis, lawyer from West Virginia, for president, and Charles Bryan, William Jennings's brother, for vice president.[9] At the end, Smith asked for the floor. The delegates, many of whom had never seen him, greeted him warmly. But his speech went badly. As Robert Murray described it, instead of focusing on "the broader concerns of the party," he not only "dwelt too much on his New York achievements" but was "aggressive and contentious." Non-New Yorkers concluded, presaging the response of the nation four years later, that Smith was "merely a local chieftain and not a national leader."[10]

The Davis-Bryan ticket did not heal the rifts in the Democratic party. Neither did the election. Calvin Coolidge beat Davis easily, and Republicans retained control over both houses of Congress. After this debacle, two former McAdoo supporters talked about the future. Attorney Thomas L. Chadbourne told Daniel C. Roper that he was switching to Smith. Chadbourne predicted that in four years Smith would be "the best advertised man in the United States." In his autobiography, Roper mused, "The events of the next four years were to show how successfully their work was done."[11] At the center of this work was the publicity bureau of the Democratic State Committee, and its chief, Belle Moskowitz.

At first, her task was not easy. Until Smith won his fourth term as governor, newspapers outside New York showed little interest in him. In 1925 she visited Denver to attend a board meeting of a philanthropic institution, where she met the publisher of the *Rocky Mountain News*. He was more eager to write about her than about Smith. She also called on Colorado's governor, William E. Sweet, who thought Smith was too wet to be nominated. Attitudes began to change only after 1926.

In spring 1927, Moskowitz went abroad to visit her daughter. Her return trip by steamer gave her time to think things through. William Jennings Bryan had died, and McAdoo's support was fading; Albert Ritchie, governor of Maryland, had only a limited following. Southern Democrats, out of power for ten years and "anxious once again to occupy the seats of the mighty," would look for someone who could control the northern and eastern states with their large electoral votes. She began to see Smith's nomination as inevitable. Securing it, she now thought, would be a question of party management.[12]

Although Smith refused to campaign for the nomination, the volume of his mail and requests for publicity kept growing. The executive office in Albany could no longer handle it. Smith's friends agreed that Moskowitz's office should take it over, and Van Namee undertook to raise money for a larger staff. He paid Moskowitz only $350 above her normal salary for three months—"very little," he admitted, considering the amount of work she did. She coordinated the answers to Smith's mounting number of out-of-state letters. She also sent out copies of Smith's speeches and messages, and decided how to answer attacks from the Anti-Saloon League and Ku Klux Klan. To counter charges that Smith had Tammanyized Albany or filled posts with Catholics, she asked Robert Moses to compile a list of the governor's appointees, with their religious backgrounds and party affiliations noted. By late spring, she had distributed two thousand copies of the list in pamphlet form. In addition, she sent out reprints of magazine articles, buying ten thousand at first, then another ten thousand if requests warranted. Finally, she updated her 1924 pamphlet, "What Everybody Wants to Know . . . ," distributing twenty-five to thirty thousand copies.[13]

By April 1928, Moskowitz's operation had outgrown its quarters and moved to a six-room suite at the Biltmore. At that point, pre-convention publicity had cost $100,000. She herself worked for so little, she said, because "We do not any of us work for Governor Smith for any other reason than that we like to do things for him. Just how much salary we will get is not the question that is relevant in our minds with most of us." Besides, for her it was a matter of pride to keep costs down. Primary campaigns usually cost too much, she later explained, because politicians fail to apply "common sense and business methods" to them. They worry that, without petty patronage and constant circularizing, the big politicians will disaffect the lesser ones, "until the very minor ones at the bottom of the heap who turn out at six o'clock in the morning on election day will not deliver enough ballots."[14] She planned a thrifty campaign.

In addition to publicity work, Moskowitz also took part in strategy decisions. One of these concerned lawyer Charles C. Marshall's open letter on Smith's Catholicism. Appearing in the April 1927 issue of the *Atlantic Monthly,* the letter claimed that as president Smith could not keep church and state separate, particularly on matters of public education. It was an old canard, one that Smith had thought his record as governor had laid to rest. Ellery Sedgwick, editor of the *Atlantic,* sent Franklin Roosevelt an advance copy, suggesting that Smith or someone else respond. Smith refused. Marshall's theological arguments eluded him, and he did not want to stir up the religious issue. Moskowitz thought he was making a mistake. She argued that the issue was alive enough to require resolution before the election. Proskauer concurred, and finally persuaded Smith. Proskauer drafted the reply, and Moskowitz got Sedgwick to hold space in his next issue. Proskauer's draft received careful readings from various individuals, including Judge Irving Lehman, Father Francis P. Duffy, and Moskowitz. "Catholic and Patriot," which appeared in May, received praise from around the country. By the end of the month, Moskowitz's office had sent out twenty-five thousand copies and ordered thirty thousand more to meet the predicted demand.[15] This success may have lulled the Smith camp into an illusion of security about the response of the national electorate to his faith.

In participating in strategy, she held herself apart from Smith's all-male "War Board," the group most actively pushing his candidacy. The members included her husband Henry, Proskauer, Van Namee, Franklin Roosevelt, George Olvany, Herbert H. Lehman, James A. Foley, James J. Hoey, John Gilchrist, Daniel F. Ryan, and occasionally Raymond V. Ingersoll and Robert F. Wagner. Reiterating her conception of how women should behave in politics, she explained, "I was never invited to attend these meetings, largely because I would have been the only woman present," and because she "had so much respect for the political pride of the men leaders of the party." She was invited to present reports to the group only late in the campaign, and then simply because she had the campaign records in her office. Even so, she "was always kept thoroughly informed of all that transpired." Moreover, she frequently sent matters there for discussion "to secure in that way any needed general support for a program." The powerful "Mrs. Warwick," or "kingmaker," as she was sometimes called, manipulator of party regulars and of the press, seemed thus to accept her exclusion from the center of male power.

In reality, she had not only already penetrated that center but could almost be said to dominate it. There was a smaller, more intimate group within the War Board with whom she met regularly, reporting financial needs and receiving "unquestioning cooperation, money when I asked for it and the most generous confidence." Once, when another $20,000 was urgently needed, "the little group of eight or ten men did not leave the room until it was secure and in the bank. They did not ask me how the money was being spent, or what we were doing with it. . . ."[16] Moskowitz did not confuse the appearance with the reality of power; she was content with the latter.

The War Board's inner group not only trusted her, but since she kept and classified campaign records they relied on her for information. Her records were a clear source of power. Since the days of Mullan-Gage, she had kept a name index of every person who had congratulated Smith or offered assistance. Classified by state, the names formed the nucleus of a mailing list. She kept other records: lists of state primary and convention dates with the election laws or party rules that pertained; of state chairs, national, state, and county committee members and other political leaders; and of the 1924 delegates and, as they were chosen, those of 1928, with whatever she could find out about them. She also made digests of all correspondence concerning political conditions in a particular state, and of newspaper and editorial comment. Thus, at any time, she could ascertain sentiment, favorable or unfavorable, almost anywhere. She carried on a vast correspondence, seldom with party officials, she said, more often with a "cross section" of friendly newspaper reporters and other contacts. She encouraged the War Board to establish similar networks.[17]

One of her contacts, Fred W. Johnson, a young lawyer from Rock Springs, Wyoming, was pivotal in preparing the ground for the convention. In late 1926, Johnson sent her an assessment of Smith strength in the Rocky Mountain region. The Democratic chair of Wyoming, whom she had met in 1924, had given him her name. Early the next year, Nellie Tayloe Ross, who for a brief time had succeeded her husband as governor of Wyoming but was then defeated for reelection, came to Albany for a lecture tour. When the Smiths entertained her, Moskowitz asked about Johnson. Ross told her he could be trusted.

Johnson proposed to secure enough western delegates to show the South that Smith had broad national support. Avoiding "old linc political leaders," he corresponded with hundreds of individual members of county committees, telling them he was for Smith and why, and asking for their views. He compiled the results, predicted how a state convention or primary would go, and sent these to Moskowitz. Early the next year, Johnson called together a group of Rocky Mountain representatives to confer with him informally in Salt Lake City. He worried that too few would come, and so told only Moskowitz of his plans. Shortly before the conference, however, he wrote her that nine or ten states would be represented. She mentioned that tidbit to the *New York Times* reporter who, like others on New York's dailies, often dropped by her office to pick up news. The reporter passed the news to a *Times* observer then traveling in the west, who wrote up the conference. When other papers picked up the story, southern Democrats became convinced that Smith was unbeatable. In Moskowitz's view, a national news story she had sent out "would have been classed as propaganda and had no effect at all."[18]

Smith needed California's twenty-six delegates, who in 1924 were loyal to McAdoo. Now, without warning, McAdoo announced he would enter the primary. To organize California, Moskowitz believed she needed someone there, "on the spot." She wanted two organizers, one to tell her "who was who," the other to represent the Smith group. Given money by the War Board, she hired a reporter who had done publicity work during some of Smith's campaigns. Since his newspaper wanted western information anyway, it paid for his trip while she paid his expenses. At a meeting that she, "as usual," did not attend, the War Board chose one of their number to be the direct representative. When her husband told her who was chosen, she knew the man would create "antagonism and an increase of difficulties." Immediately she suggested to her inner group that both Lehman and Raymond Ingersoll be sent together. Her wish was obeyed. The broad yet different contacts of the two men, combined with the information gathered by the newspaper man, created a harmonious working group. Smith's old enemy, William Randolph Hearst, controlled most of the newspapers. Southern California was dry and anti-Catholic. But Ingersoll and Lehman mobilized the scattered support that existed in the movie studios from old-time New

Yorkers, and among liberal elements in San Francisco. Smith won the primary. Moskowitz was so sure she had prepared the outcome that she did not stay with the "little band of faithful followers" who remained together until 3:00 A.M. to hear the returns.

The South was a different matter. A writer planning a tour of southern states for her own work volunteered to send Moskowitz reports on Smith sentiment among women. Moskowitz told her not to bother, that because of religion and Prohibition, southern women were closed to Smith propaganda. The writer felt sure this was not so, but later wrote to Moskowitz, "You were right." The Klan, strong in the South, issued pamphlets and flyers accusing Smith of nefarious deeds. The material was put out under the guise of Prohibition defense societies, with such names as the "West Virginia Law and Order Committee" or the "State Committee on Law Enforcement." Since it came out in spurts, depending on which state convention or primary was imminent, Moskowitz felt she handled it effectively; during the presidential campaign, when this propaganda inundated the entire country in one massive flood, the task became impossible.

Amid these travails, the United States Senate announced an investigation of all candidates' pre-convention expenses. Although both Smith and Van Namee testified, the Senate committee was unsatisfied. Van Namee had disclosed every expense, but the senators, judging the publicity results huge in relation to the amount supposedly spent for it, remained suspicious. One midnight Moskowitz's phone rang. A *Times* reporter told her that the committee had summoned some of Smith's close friends—contractor William F. Kenny, shipbuilder William H. Todd, financier Herbert Lehman, and his friend from the trucking business, James J. Riordan, and one other person whose name he could not get. About an hour later the reporter called back. "I have found the name of the other person," he said. "It is you."

The prospect of testifying worried her. She thought more attention would be paid to her, the sole female testifying, than to the male witnesses. She also knew that "one word out of place, or one little sentence, can make one [appear] ridiculous if newspaper reporters should choose to be unfriendly." She consulted lawyers, who advised her to "answer every question but don't volunteer information." Henry accompanied her to Washington. She followed her lawyers' advice, and all went smoothly. In her opinion, when Smith's backers made their professions of unconditional financial support for Smith, they provided another public relations "bright spot" for which she had paid nothing.

Moskowitz ought not to have worried about unfriendly reporters. Ever since the days of the Reconstruction Commission, she had enjoyed smooth relations with the press. Columnist Walter Lippmann was a good friend who clued her in on rumors, warned her about hidden meanings in reporters' questions, and gave her inside information on whom to contact at the *World*. In return, she fed him news and material for his editorials. Other New York political reporters respected her "long standing rule" never to quote her directly. "Never, intentionally," she wrote, "has a confidence been violated." She felt she could always talk with them about a situation, prepare them for future events, or provide background material, "without ever needing to fear there would be any slip that would result in untimely disclosures." One day a new man assigned to her headquarters violated her rule: "He really meant to be kind and to give me the sort of credit he thought headquarters people always wanted." Worse, he printed confidential information. In an act of loyalty to her, his colleagues ostracized him.[19]

Henry Moskowitz's name came up during the senatorial investigation. His contribution to the Smith campaign had been to collaborate in 1927 with journalist and editor Norman Hapgood on a Smith biography, *Up From the City Streets.* The book was a rewriting of Henry's earlier work, *Alfred E. Smith, An American Career,* written for the 1924 campaign. The first chapters of this work are fairly lively, but the rest merely regurgitate Smith's speeches and messages. *Up From the City Streets* was more successful. Reviewed by twenty publications, it was reprinted twice and excerpted in newspapers across the country. Fulton Oursler made a radio play out of it that starred Heywood Broun and Helen Hayes. Henry also published a collection entitled *Progressive Democracy: Addresses and State Papers of Alfred E. Smith.* In sending Smith a signed copy, he hoped "that it will help to make a dream come true about another inaugural address—this time not at Albany but at Washington."[20]

The senatorial investigating committee was interested in the relationship between the Moskowitz-Hapgood collaboration and the Democratic publicity bureau. Van Namee assured the senators that the bureau "had nothing to do with it, paid nothing for it and received nothing from the publication." When a senator asked Belle about Henry's connection with the campaign, she reported that he worked solely as a volunteer, "in a friendly way." "[H]e meets people," she said, "and then he knows about what is going on. I live with my husband, and politics is a subject of daily discussion."[21]

In the early months of 1928, a conciliatory spirit moved the Democratic party, whose leaders wished above all to avoid the fiasco of 1924. As a gesture to the South, Houston would host the convention. Neither Smith nor Moskowitz saw the site as friendly, but never discussed their opinion for fear of being considered weak.[22] In any event, by then they were confident of victory. Every tabulation showed that the Rocky Mountains, the West, the North, and the East were secure; there seemed no need to worry about the South.

Smith did not go to Houston. Moskowitz would have gone if asked, but Smith preferred that she stay in Albany. Belle and Henry moved into the executive mansion, while Smith's wife Katie and their sons traveled west on William Kenny's private palace car. Smith wanted her to be seen in public as much as possible. Moskowitz had publicized reports by well-known Democratic women, such as Eleanor Roosevelt, Frances Perkins, and Ida Tarbell, that Katie Smith radiated a simple graciousness quite suitable to the White House. But the malicious had already labeled her dowdy, vulgar, and a heavy drinker.[23]

All went as predicted at the convention. Claude G. Bowers, historian and journalist, sounded the keynote by attacking the Republicans on the oil scandals. Moskowitz would later stress this theme in party publicity. In nominating Smith, Franklin Roosevelt again used a Proskauer draft, but this time added more of his own. The result was another oratorical triumph.

In the upstairs hall of the executive mansion in Albany, Smith, his daughter Emily Warner and family, and the Moskowitzes listened to the balloting over the air. When the climax neared, they went downstairs to greet friends. Outside, local residents had driven up, waiting for the results. At the dramatic moment when Ohio changed its vote to Smith, Belle Moskowitz and Smith looked across the room at one another. Some time passed before they could speak about what they both knew had happened: Smith had been nominated on the first ballot.[24]

There were some sour notes. Many delegates had accepted Smith only for lack of an alternative. When, after Roosevelt's speech, a southern delegation refused to march, a scuffle took place that was heard on national radio. Four southern delegations never voted for Smith, and the nomination was never made unanimous. Worse, the platform supported Prohibition, but the party's candidate did not. In his wire of acceptance, Smith called for "fundamental changes in the present provisions of national prohibition." Roosevelt thought the wire a terrible mistake, and later blamed Belle Moskowitz for it. As he told Claude Bowers two years later:

> The telegram was sent on the insistence of Belle Moscowitz [sic]. It came when the convention was on its last legs, and . . . Pat Harrison was presiding. I showed him the telegram. "My God," he said, "this will cause a riot!" And so we agreed that when all the business was over the telegram would be read, since we had to read it, and then I would immediately move an adjournment and Pat would put the question and declare its adoption.[25]

That Moskowitz played such a role is unconfirmed. She and Proskauer worked on the text of the telegram, but Smith had the final say. A September 1928 *New York Times Magazine* portrait of Smith's "Kitchen Cabinet" declared that Moskowitz had opposed Smith not only on the repeal of Mullan-Gage but "would have had him keep silent in the present campaign on any modification on prohibition."[26]

The convention over, the Moskowitzes left Albany for New York. Belle announced she would take the next day off, then meet with Smith to plan the first stages of the campaign. But he took his first step without her. By tradition, the candidate named the new chair of the national committee. Smith picked John J. Raskob, a wealthy industrialist associated with the Du Pont Chemical Company and with General Motors, as well as a devout Catholic and Republican. Raskob and Smith had met some years earlier and found they had much in common. Smith hoped that their friendship would lead to greater support for Democrats from big business.

Moskowitz opposed the choice of Raskob. Her group as well as Smith's Tammany Hall friends favored Senator Peter Goelet Gerry of Rhode Island, an experienced politician and a Protestant. But Smith was adamant. Judge Proskauer, a guest at the executive mansion the night before the July meeting of the Democratic National Committee in New York, tried to change Smith's mind. The next morning Smith announced, "It's Raskob." Later he explained to Moskowitz, "It's the only thing Raskob has ever asked of me, and I've got to give it to him."[27] Thus, before the campaign had even begun, Smith had alienated his own team.

There were other ruffled feelings. Democrats from outside New York soon felt rebuffed. While Harry E. Byrd of Virginia, Scott Ferris of Oklahoma, Nellie Tayloe Ross of Wyoming, and Florence E. Farley of Kansas joined Smith's friend Frank Hague of New Jersey as national vice-chairmen, power was concentrated not with them but with the executive committee, which consisted of Peter Gerry, James Hoey, Herbert Lehman, Belle Moskowitz, Franklin Roosevelt, and George Van Namee—all New Yorkers with the exception of Gerry. Within this group, even Roosevelt felt excluded. He was already irked at Moskowitz for having refused to distribute his nominating speech as campaign literature. That she printed the speech later in the party's campaign handbook was inadequate compensation. Roosevelt then complained that during the campaign Moskowitz and Raskob never let him get near Smith. In a letter never sent but kept "for the record," he wrote that they treated him as "one of those pieces of window-dressing that had to be borne with because of a certain political value in non-New York City areas."[28]

Within Smith's camp, Raskob's relocation of Democratic Committee headquarters to the General Motors building also caused friction. To Belle Moskowitz, the Biltmore had always been Smith's "lucky place." But her superstitious attachment was not the only reason for concern. Raskob had already stamped "big business" on the party and she feared that the move would increase the backlash. In addition, Raskob announced that Judge Proskauer, who had played a role in every Smith campaign until then, would have no official connection with this one. In talking to the press, Moskowitz smoothed this over by saying that Proskauer was too busy being judge of the Appellate Court, but reporters speculated otherwise.[29]

Finally, women party leaders rankled at their continued exclusion from the inner circle. The DNC consisted of one female and one male member from each state. Yet, on July 11, an informal meeting of male committee members chose the party's vice-chairs without consulting the women. Elisabeth Marbury of New York protested directly to Smith:

> The method of selecting women to office without giving those who are to serve under their leadership any opportunity of approval or of disapproval is engendering discontent and discouragement. The day has passed, as I have previously warned our leaders, when the ommission [*sic*] of women from decisive conferences can continue without prejudice to harmony.

She continued by pointing out that the women, who made up 50 percent of the DNC, were highly intelligent, serious, administratively experienced, and loyal. Many had made long trips to attend the committee meeting. "The absolute ignoring of their sentiments, the entire indifference to their views had a most disastrous moral effect. . . ." Eleanor Roosevelt wrote her husband that she was concerned, too. "It would have been so easy for the men" to have consulted the women, she wrote. "I can't understand why they prefer to stir up this current of discontent!"[30] Throughout the 1920s, Democrats had boasted about how well they had integrated women into the party structure. The New York State Committee, in particular, had come to rely on the steady organizational work that women seemed willing to perform. By the end of the decade, perhaps because the stakes were now so high, the men at the center reverted to their old ways.

There was less dissension over choosing Franklin Roosevelt as the Democratic candidate for governor. Having declined the honor over the summer, he was at Hyde Park when Belle Moskowitz called to ask him to reconsider. He refused to speak with her until told that Smith was on the wire. He talked to Smith, but again turned the offer down. At the Democratic state convention that fall, party leaders insisted that only Roosevelt could carry the state for Smith. Franklin was in Warm Springs, Georgia, and not taking calls. Using Moskowitz and Frances Perkins as intermediaries, Smith persuaded a reluctant Eleanor Roosevelt to put one through, after which she passed the phone to Smith. Smith promised Franklin help, including a large donation from Raskob to his Warm Springs Foundation. Franklin yielded, again throwing himself into the campaign with an energy and drive that took the Smith crew by surprise.[31]

Soon after the campaign started, a *Herald Tribune* journalist contrasted the Democratic and Republican organizations. The DNC centralized operations at General Motors in New York. The Republicans had headquarters in Washington, D. C., but scattered other offices across the country. The DNC functioned with the "suave and slightly glacial efficiency which distinguishes the loftier interiors of American business enterprise." The Republican National Committee (RNC) presented "a singularly formless picture." What the Republicans lacked in efficiency, the reporter continued, they gained in enthusiasm; but the methods of the Democrats, modeled on modern business practices, put them ahead. As far as this reporter could tell, Raskob, "captain of one of America's largest mass producing and selling industries," planned nothing less than to sell Al Smith to America "in the modern manner."

> There will be hundreds of ballyhoo wagons all over the country, adapting the ultra-modern radio to a method as old as the old-fashioned itinerant patent-medicine man. A loud-speaker . . . from the car will take the place of the old-fashioned guitar and burnt cork . . . ; an orator then will try to "sell" Smith as the medicine-man used to sell snake oil.[32]

To accomplish the task, Raskob put every executive committee member in charge of a specific area. Roosevelt's—commerce, industries, and professional activities—required little of him. Disappointed that the campaign's publicity was going to be run by Moskowitz, and not put "under some really big men with imagination and organizing ability," he felt there was little he could do. He organized his division, turned the work over to Louis Howe, and left for Warm Springs. Gerry chaired the Advisory Committee (Eleanor Roosevelt, Pat Harrison, Alben Barkley, Parker Corning, Jouett Shouse, and others) and supervised the speakers' division; Hoey oversaw college leagues, clubs, and ethnic voters (Henry Moskowitz worked with him as a volunteer); Lehman raised money; and Van Namee ran the offices. As publicity director, Belle Moskowitz supervised press, movie, and radio publicity, research and clippings, and the preparation of literature and advertisements. These activities absorbed the major part of the budget. She also directed "Social Work, Public Health, and Education" and "Women's Activities." In this last capacity, Eleanor Roosevelt, Nellie Tayloe Ross, and Caroline O'Day reported to her.[33]

Eleanor Roosevelt had known since April that she would head the Women's Division. She had already directed women's pre-convention activities. After the convention, she had hoped to spend the summer with her children, and in the meantime stay clear of the controversies between the DNC and women party leaders. "I'm just doing what Mrs. Moskowitz asks me to do and asking no questions, the most perfect little machine you ever saw," she wrote Franklin. She was not allowed to retire, however. The DNC asked her to direct the division activities from New York, while Nellie Ross and Mary Norton went out on the road. Roosevelt set up committees to appeal to various women voters—independents, business and professional women, college women, working women, social workers, and new voters, the last group directed by Rosamond Pinchot Gaston. June Hamilton Rhodes ran the Women's Publicity Bureau, creating newspaper articles with a "home appeal." Another department formed "Smith-for-President Clubs" which placed placards in merchants' windows listing reforms Smith had fostered that were favorable to women. Following the pattern set by the State

Women's Division, Eleanor Roosevelt planned automobile tours of women speakers for each state.[34]

In describing the campaign, Eleanor Roosevelt wrote that "Mrs. Moskowitz was a challenge to all of us at headquarters, because she really knew whether we were working hard and achieving results." Although the comment implies she felt judged, she later avowed that she and Moskowitz "worked together in full harmony from April to the end of the campaign in November, and I have always been grateful to her for the opportunity." Roosevelt herself performed with all the poise, good nature, and reliability that New York political women had come to expect of her.[35]

The male side of the campaign was organized along more conventional lines. Three speakers' bureaus (in New York, Chicago, and Salt Lake City) sent out orators, who worked under state and national committeemen. Separate committees were organized for lawyers, teachers and students, athletes, even airmen. Foreign language voters received special attention, with distinct sections for Jewish, French, Russian, Czech, Hungarian, Swedish, Spanish, Italian, German, Greek, and Polish groups. Henry Moskowitz served as liaison to the foreign-language press. Radio would be used but also the "ballyhoo wagons" to broadcast pre-recorded programs through loudspeakers.

Belle Moskowitz's publicity department impressed one reporter as being "run with an alluring, unerring smoothness." Foreign correspondent and family friend Vincent Sheean recalled that Moskowitz was "as busy as a general directing an advance, and far calmer." DNC headquarters were a "madhouse," he later wrote, ticker tapes and telephones, worried people scurrying about, papers lost and found, everyone in a hurry. But in the office of "Mrs. M.," as she was called, all was cool and quiet. "She did not allow people to shriek or get excited in her room, and her own slow movements, her deep, thoughtful voice, acted as an anti-hysteria medicine on everybody who came near her." With her aide, Joseph J. Canavan, she arranged interviews for reporters and sent out press releases on a daily basis to 2,700 newspapers throughout the country. She also worked with Proskauer and others on the texts of Smith's speeches and messages. One of her early tasks was to put together the DNC's campaign handbook, which presented the official party positions on the candidates and issues. The handbook set the tone for the publicity of the campaign.[36]

The handbook emphasized three reasons for voting Democratic: the Republican legacy of graft and corruption; the Democratic legacy of Wilsonian "progressivism" and its quest for international peace; and Smith's qualifications for the presidency. None of these were easy to sell. The first was old news, and Republicans could now boast that Coolidge had canceled the oil leases. Besides, Democrats had their own deficiencies to explain, including domination by political machines such as Tammany Hall. Further, in a time of rising, although not pervasive, prosperity, themes of industrial justice and international peace inspired a dwindling few. Finally, next to Republican candidate Herbert Hoover's bland but soothing middle-class respectability, engineering expertise, and war-related humanitarianism, Smith cut a poor figure. He had been a fine governor, however, and was personally more amiable than his opponent, a fact highlighted by Moskowitz's publicity.[37]

The handbook's portrait of Smith echoed the themes of love and innocence in Moskowitz's 1923 letter to Lillian Wald. Smith was "able, exact, and resourceful," "efficient in the highest degree." But his was not the "efficiency of a machine" (unlike Hoover's). On the contrary, Smith was strong and manly, yet "warm with the love of his kind, touched by the sorrows of ordinary people. . . ." People vote for him because "he is always right, to be sure, but also because he is their friend, whom they love." Devoted to his widowed mother and loyal to the "humble friends of his youth," Smith found happiness not among the statesmen and big business leaders who now knock at his door but among "the simple people he has always known." He "enjoys and loves other men, women, children—yes, even animals." He bubbles over with song, dance, and story: "Gayety and contentment spread around Smith, as naturally as a thrush sings, or a child plays. The big work must be done: why should it not be done in the sunshine, since the only purpose of the work, for him, is to bring sunshine to all those people who labor and hope?"[38]

After this gushing evocation of innocence and charm, the portrait then paid tribute to Smith's wit and "instinct for the truth." In an effort to dispel the Tammany bugaboo, the writers praised his ability to turn down, with "humor and friendliness," all unworthy applicants for patronage. Thus Smith was a leader who knew how to surround himself with experts, not cronies. In the biographical section, his religion received little play, the Catholic Church mentioned merely as an institution playing a central role in his life, as religion did in the case of persons raised in "many a smaller town." Tom Foley's name appeared briefly as the "political leader" of Smith's ward and his first patron, but the name "Tammany" seldom appeared.[39]

While stressing Smith's personal qualities, this *homage* did not completely ignore his policies. It retraced his path to the governorship, giving positive interpretations to all setbacks. For example, Republican blockage of state reorganization forced Smith to wage four campaigns to put through his programs; these campaigns had all educated the New York voter, developing a "splendid relation" between Smith and the people. His commitment during the Red Scare to American "traditions of freedom" also received praise, as did his versatility ("he can sit down and . . . discuss anything, with men from Wall street, from an engineering society, from a Bar Association, from a Trades Union organization, understand them, never be fooled by them . . ."), his flinty self-respect ("people who attack the Governor's probity . . . invariably break their knuckles"), and his avoidance of the "baloney" of most political rhetoric

("... you will search in vain for a single paragraph of the vague verbiage with which most politicians conceal their intentions"). The portrait concluded with an account of Smith's record in office, singling out the following points for praise: his non-partisan appointments (including that of the Reconstruction Commission), commitment to economy and efficiency, expansion of rural schools, winning of higher salaries for teachers, improving the care of the state's unfortunates, saving water power for public control, and supporting (in the face of Republican opposition) New York's acceptance of the Sheppard-Towner Maternity and Infancy Aid Act of 1921—all policies Moskowitz had helped develop.

"So far has the little newsboy gone," the tribute ended, returning to the reverent phrases of the opening:

> The worker in the fishmarket has had nothing to help him but his warm heart, his clear brain, his loyalty to the toiling millions of whom he was one. His will had been to give not noise or exaggeration, but the best that skill, independence, and truth can bring about. The Alfred Smith who left school without a murmur, to work for his widowed mother and his younger sister, is the Alfred Smith who, called four times by the people of his State, had rendered to them the same devotion that in his boyhood went to those around him.[40]

A great man, a "Lincoln of our new age," a "master," yet at the same time human, loving, "up from the city streets," a man one could love, and whom many did.

The DNC handbook did not constitute the entire Smith campaign. There were also live and broadcast speeches by the candidate and others, and hundreds of press releases and articles. But if this portrait of Smith was to "sell" the man to the American electorate, it fell far short of its mark.

Many scholars have dissected the election of 1928. All agree that the results were foregone. The personalities of the candidates played a large part. Hoover's appealed to a broad cross section of America's dominant culture, its very colorlessness making his seem "presidential." His federal administrative experience and compassionate distribution of food after the World War also gave him an edge over Smith. When the silent film *Master of Emergencies,* which told the story of his relief efforts among Germans after the war, was released during the 1928 campaign, Hoover said it would win votes only from morons. But his friend, writer William Irwin, told him that audiences applauded the titles and at the end stood up and cheered.[41]

Smith could not compete against such an aura of accomplishment. New Yorkers and many recent immigrants loved him for the very qualities that turned others away—his caustic, sometimes raucous humor, his derby hat and cigar, even his deliberately mispronounced words, such as "raddio" and "horspital." Historian David Burner summed up Smith as a "provincial New Yorker," insensitive to the social and cultural traditions of the rest of America. "Butte?" Smith asked when his campaign train neared the colorful Montana town. "Never heard of it," he grumbled, perhaps out of irritation at the flaming crosses he had seen as he approached. Nevertheless, the joke—if it was one—became a symbol of the limits of his national experience. Allen Lichtman, in a statistical study of the election, concluded that religious prejudice and Prohibition were the chief causes of Smith's defeat. But he also observed that, despite Smith's humble origins, links to new immigrants, and his progressive record in state politics, his national policy proposals were "scarcely more venturesome" than those of Hoover. Moreover, he failed to define "coherent and persuasive alternatives to Republican policies of the past eight years." Instead, he relied "on his personal appeal to ordinary Americans. . . ."[42]

There were other reasons for Smith's defeat—the general economic prosperity of the times, the difficulty ousting an incumbent party in a presidential year, and the extreme reactions that Smith's candidacy drew from groups such as the Anti-Saloon League, the Ku Klux Klan, and religious fundamentalists. In sum, Smith may never have had a chance, regardless of the publicity Moskowitz developed. Still, the strategy she chose reveals the hopelessness of his attempt.

Moskowitz's publicity revolved around what she considered Smith's best points. But even they made him attractive only to the network of social reform progressives with whom Moskowitz had always associated, and early in the campaign she learned that some members of this group were defecting to Hoover. Advice sent by her old friend Felix Frankfurter, then teaching law at Harvard University, perhaps influenced her to make a special effort to regain their loyalty. "Don't be sore at the progressives," he wrote her in August. "Win them." But in the late 1920s, support for the old progressive ideologies had dwindled. During the reform era a coalition of forces had united to support limited state interference in private enterprise for the benefit of the industrial masses. By the time of the World War, that coalition had already begun to crack, and during the 1920s, in reaction against wartime controls and the triumph of socialism abroad, it had weakened even more.[43]

Thus, the very aspects of Smith's policies that Moskowitz deemed positive kept others away. His reliance on experts, for example, innocuous from our post-New Deal perspective, in the 1920s conjured up images of socialist planning. His fostering of programs such as Sheppard-Towner (whose renewal Congress rejected a year later) seemed an unnecessary governmental intrusion into family life. And other Progressive Era goals that Smith espoused, including voter education and efficiency and economy in government, were equally displeasing to politicians. Voter education often meant the exposure of shady deals and swaps, and efficiency and economy programs cut into established power bases. Further, from the perspective of undecided voters, Smith was not the only able administrator running for president.

No matter how appealing Moskowitz made them sound, some of Smith's policies could never attract conservative and business-oriented voters. Frankfurter had told Moskowitz not even to try, calling such an effort "a wholly idle policy." Smith's defense of traditional freedoms of speech in an era of widespread radicalism alienated those who feared the triumph of socialism. Public funding for the care of mothers and children and public rather than private exploitation of water power were unpopular with businessmen. Middle-class women, to whom Moskowitz also tried to appeal, were repelled not only by Smith's stand on Prohibition but also by his rough manners, lack of formal education, and the rasping of his "whiskey voice" over the radio. A much smaller group, the feminists of the National Woman's Party, believed Hoover more favorable to an equal rights amendment, and so endorsed him. The images of Smith as a strong, loving, humble, and devout family man may have won some female votes. But, as Frances Perkins observed, even after she had reduced an audience of southern Ohio women to tears with stories of Smith's youth, the district went solidly for Hoover. Touring the rural South, Perkins met almost complete indifference toward Smith's social legislation among Democrats, male and female. In the southern cities she was greeted with stony silence or outright hostility.[44]

Moskowitz's publicity failed on yet another level: it did not convince the electorate that Smith had mastered national issues. The only issue on which he took an original and well-developed stand was Prohibition, and the stand he took hurt rather than enhanced his chances. Moskowitz's campaign handbook discussed other national questions—agriculture, the tariff, taxation, regulation of monopolies and trusts, and foreign policy. But for authoritative quotations she turned to other Democrats, not Smith. There were some exceptions—labor disputes, government reorganization, and public ownership of water power. But Smith discussed these only in reference to the experience and concerns of New Yorkers.

Moskowitz was a gifted publicist. She had managed Smith through three gubernatorial races, and turned his major legislative battles into victories. But she appears to have been either ill-informed or naive about attitudes and feelings outside New York. Frances Perkins said that Moskowitz never directly experienced America's "hinterland." "She didn't get out into the country very much in the '28 campaign." Visiting her brother Max Lindner in Cleveland, she tended to draw conclusions from what she heard there. Max's department store had merged with the May Company. As a result of talking with him and his circles, Perkins thought, Moskowitz believed she was talking with "big business men." In reality, they were "small business men in their thinking and estimate of the situation. They didn't touch the great manufacturing or the great mining or timber or farming psychology of the West, or the regional differences." One evening during the campaign, Perkins and Moskowitz listened together to Smith's labor legislation speech broadcast from Helena, Montana. Perkins said, "Belle, I just don't think that will go in the country." Moskowitz replied, "You're quite mistaken, Frances. . . . This is what this country is longing for." Perkins concluded that Moskowitz "didn't get it as much as I got it from tours around the country that I had made. . . ."[45]

Whatever lay behind Moskowitz's choice of publicity strategy, its failure is beyond question. Yet one can understand her reasoning. She knew that Smith could never win the anti-Catholics, Prohibitionists, or others who had decided in advance to oppose him. Thus her sole option was to stress his strengths as she saw them, no matter how inappropriate to the mood and attitudes of the nation. In the end, she could work only with what Smith offered, and what he offered was not enough.

The final outcome of the campaign was of course deeply and personally disappointing to Belle Moskowitz. But even before its conclusion, certain experiences brought special pain. Character assassinations of Smith, based on his religion and views on Prohibition, spread across the country through pamphlet literature and rumor mills. The attacks were generally dubbed the "whispering campaign." There was not much to be done about it. Republicans denied responsibility, accusing Democrats of spreading tales about their own candidate. At a press conference that at Hoover's request was kept private, Hoover said that a "whispering campaign" hurt him as much as Smith, but that rebuttals only "fan the flame" and did more damage.[46]

Moskowitz probably agreed. Yet, since the reply to Marshall's letter about Smith's faith had not backfired, she prepared selected counter-offensives. At the end of July, for example, Kansas columnist William Allen White, once a Smith admirer, called him a "vice king" for having given legislative support to brothels, gambling dens, and saloons. Smith made a personal reply, on which Moskowitz worked, answering the charge "subject by subject and bill by bill," explaining the circumstances in which he had acted in ways that White had misinterpreted. In September, Moskowitz learned that a Reverend Van Nostrand of Albany, New York, had told a large Bible conference in Indiana that, when Smith made his first radio speech after being nominated, he was too drunk to stand. Audience members then spread the story throughout the state and beyond. Moskowitz spent hours trying to track down the minister, finally getting him to deny, in the presence of the governor and two witnesses, that he had ever made the charge.[47]

Even though White withdrew most of his charges, he had thrown the Smith group onto the defensive early in the campaign. And despite Moskowitz's sending of affidavits of Van Nostrand's denial to anti-Smith people in Indiana, the harm was done and could not be repaired. Moskowitz herself suffered a personal backlash from her aggressive counter-attacks. Orville S. Poland, head of the legal department of the New York Anti-Saloon League, accused her of exploiting the very "whispering campaign" she was trying to neutralize. "If the idea wasn't invented by Mrs. Moskowitz," he charged, "it has certainly been her principal reliance." Ever since White's attack, he continued, Moskowitz had convinced Smith to reply to every charge by calling it part of the "whispering campaign." As a result, Poland quipped, Smith's campaign had become a "whimpering campaign."[48]

Walter Lippmann had warned her against the possibility of a backlash. In regard to William Allen White's slurs, Lippmann advised that Smith admit he had done some things as a young legislator that he now deplores. In Lippmann's view, a detailed statement "attempting to defend everything would hurt him," whereas a "good-humored confession would make a fine impression." Moskowitz should "remember that in creating a public picture of the Governor we are not attempting to depict a man who sprang into the world fully armed and perfect, but rather a picture of a man who started under every kind of handicap and gradually became what he is. That's the truth and it's also a very appealing truth." Later, when charges of drunkenness began to fly, Lippmann's advice changed. Since thirty or forty reporters had witnessed the occasion when Smith was said to have been too drunk to stand, Lippmann thought that refutation was easy. He urged, however, that "the evidence be collected in future and given out on the authority of the Democratic Party and not of Governor Smith personally. He should not be put in a position of proving that he isn't a drunkard. His supporters should do that work for him."[49]

In addition to slurs on the basis of religion and drink, race was also at issue during the campaign. At various times, critics accused both candidates of being either too friendly to blacks or not friendly enough. Eva Levy, a Moskowitz friend who worked at the DNC during the campaign, recalled an unannounced visit by two southern Democrats who had seen a picture of Smith dictating to a black secretary. Confusing this person with Moskowitz, they asked to have her pointed out. That done, they left without a word. Although their reasons were wrong, they were right to suspect Moskowitz of pro-black interests. The previous spring, through her husband Henry's contacts with the NAACP, she had asked its leader, Walter White, to help win black voters from their traditional loyalty to the GOP. White declined, but agreed to speak with Smith about what blacks hoped from the campaign. As a result of this meeting, White drafted a statement in which Smith promised not to be ruled by an "anti-Negro south." Smith never signed it. His vice-presidential running mate, Joseph T. Robinson of Arkansas, Senator Pat Harrison, and Joe Proskauer all advised that it would antagonize southern Democrats. Despite Smith's silence, many blacks voted for him on the ground that Hoover was even less open to racial equality. Moskowitz later told White that Smith thought he should have made a stronger bid for the black vote.[50]

Moskowitz did not travel with Smith on his campaign trail. The only speech she attended, following a tradition established during his gubernatorial campaigns, was his last on the Saturday before election day. Thus she never saw the crosses that lit up the Midwestern landscape, nor experienced firsthand the hostile crowds in places like Oklahoma City. She heard them, or thought she did, over the radio. The broadcast of Smith's Oklahoma speech was marred by so much static that she feared for his safety. When the broadcast finished, she left word at his hotel to call her immediately, and when he did was relieved to find him unharmed. Even though he explained that the audience had actually been cheering him on, she urged him not to leave the hotel the rest of the night.[51]

Raskob was another trial. Throughout the campaign, her relationship with him remained cordial but cool. To Frances Perkins, she expressed frequent irritation. He was cranky about petty things, such as the cost of a man going twice to Albany (a twenty-five dollar expense). "I'm exhausted," Moskowitz would say, retiring from a bout. "Raskob is so difficult because he thinks he knows what to do, but he doesn't know anything." Then, with the self-righteous arrogance Moskowitz frequently demonstrated in Perkins's accounts, she said, "I don't know how he ever got on in business. He won't believe those who know how things ought to be done." When the campaign was over, Moskowitz wrote him a letter of appreciation, expressing regret that they had had so little time to discuss matters in detail; all other exchanges between them were perfunctory.[52]

Election day was November 6. In the final tally, Smith split the solidly Democratic south, and in the north won only Massachusetts and Rhode Island. His total popular vote was 15 million. Hoover won over 6 million more, and swept the electoral votes of the larger states, among them New York, which stunned Smith. That the man he considered his protégé, Franklin Roosevelt, would go to the statehouse was a small consolation.

As early as the previous summer, Moskowitz had hinted to Joe Proskauer her awareness of the difficulties Smith faced. One Sunday afternoon, they were driving together to Albany, carrying a revised version of Smith's acceptance speech, the formal words with which he would accept his party's nomination and present his national policies. In the villages they passed through ordinary people were sitting on their porches, rocking and fanning themselves. Belle suddenly turned to him and said, "Joe, look at those people." "Yes, what about them?" he asked. She replied, "Do you realize that those are the people that are going to pass on this document that we're expending all this energy on refining?" Her rhetorical question acknowledged the difficulty, if not impossibility, of educating "the people" to support Smith. As the campaign progressed, the obstacles could only have become more obvious. As she wrote Frankfurter a few days after the election, both she and Henry had been "too realistic not to have expected what happened," but neither had looked for the defeat "in exactly that degree or distributed as it was."[53]

When she analyzed the result, she blamed religious prejudice as the main cause. Economic prosperity and a "resulting lethargy" about social problems also combined to defeat the "liberalism and progressivism" that Smith represented. The numbers of progressives were simply not there, she said in an interview in late December. Even the farmer, once the stalwart of progressivism, was "not much good now" because he "sticks too hard to his old ideas, which mean more to him, it seems, than progress or anything else." In a phrase she would repeat with some bitterness in 1932, she called herself "first a progressive and second a Democrat." In fact, she was "almost a Bolshevik in some ways," by which she meant that she cared more for the welfare of the masses than for rescuing capitalism. Now "the reactionaries are in the saddle," she sighed. Hoover, who to her represented the "mechanizing tendency in modern life," would safeguard business before all. But "growing unemployment, business depression or some false step" will trigger a reaction against him. "Right now there is unemployment. The breadlines are lengthening. The unemployment may spread and become more serious with winter. If it should grow . . . , the big concerns might cut wages. And hunger would bring a new interest in progressive ideas." This interview, given less than a year before the crash of 1929, was uncannily prescient.[54]

On their own home ground, Moskowitz could help Smith win. Within the confines of the Democratic party, she could also get him nominated: that, in her own estimate, had been merely a matter of party management. Working with like-minded people, she had systematically built up a network of supporters from the party's grass roots. Given the lack of competition from other nationally prominent Democrats that year, and the desire for peace in the party, winning the nomination had barely challenged her. But the larger task of making him president was beyond even her considerable gifts. It was, in fact, beyond anyone's. Moskowitz could not make Smith something he was not. She could not make him change his accent or polish his grammar when he extemporized. She could not make him discard the brown derby and cigar that marked him as a "provincial" New Yorker. These failures must have irked her, but the blame for them cannot be laid solely at her feet.

After Roosevelt's inauguration, Smith offered to help him pursue the policies he had initiated over the past ten years. He would keep a place in Albany and come up Sunday nights for the "big legislative days" on Monday and Tuesday. At an earlier meeting, Smith had counseled

Roosevelt to keep Robert Moses as secretary of state. He had also given Roosevelt Belle Moskowitz's draft of an inaugural address. Roosevelt declined to appoint Moses, whom he disliked and distrusted, and he never used Moskowitz's draft. Now Smith suggested that Roosevelt hire Moskowitz as his private secretary. "He said to me," Roosevelt later told Frances Perkins, "'She knows all the plans. She knows all the people. She knows all the different characters and quirks that are involved in everything. She knows who can and who will do this or that. She . . . could really guide and develop these things with every one of the departments.'" Roosevelt barely knew Moskowitz. His wife knew her well, however. So did his friend and campaign manager Louis Howe, who respected her but thought her domineering and a potential competitor for influence with his boss. Roosevelt told Smith he would think about it.[55]

Shortly thereafter, he discussed Moskowitz with Perkins, whom he had appointed Industrial Commissioner. Perkins told Roosevelt that Moskowitz would be a "fool" to work for him:

> In the first place, what she has done for Al she has done as a free gift. She's been very important to him. I don't think she's a woman who wants power for its own sake. I don't think she wants to be in that kind of a relationship with you. She's got no personal friendship for you. She likes you. She's worked hard for you. She's worked in your campaign. She's devoted to Eleanor, whom she knows. I don't think she would like to do this job.

Roosevelt commissioned Perkins to sound out Moskowitz without letting her know he had asked. A few days later, Perkins speculated with Moskowitz about whom Roosevelt would appoint secretary. Moskowitz ventured that he needed someone "who will really do some work and see that great issues and great problems are properly presented." "Who is there?" Perkins asked, innocently. "I'll tell you something if you won't tell it," Moskowitz confided. "Governor Al has suggested to him that he appoint me. . . . I think I could be very useful."

Moskowitz wanted the job. As early as November 10 she had written Frankfurter that she "was going on to help the new State administration." Whether she wanted the job more for her own or Smith's sake was unclear. Smith saw her placement in Roosevelt's office as advantageous to him. He explained to Perkins, "She can see people. She can arrange things. She can keep in touch with me, tell me what's going on. I can tell her what ought to be done. She can get it through to Franklin very quickly. It would be the best thing in the world to have her there." Except to tell Roosevelt that Moskowitz was willing, Perkins kept these conversations secret. Moskowitz later asked Perkins whether she had divulged her willingness to Roosevelt. Perkins lied and said "No." Pressed by Smith to decide, Roosevelt lied, too, claiming he had already chosen another. Making up reasons as he went along, he told Smith he needed a tall, strong man to lean on in intimate situations, and thus had decided to appoint a man to whom he owed a political favor.

The truth lay elsewhere. Perkins claimed that Roosevelt made his decision after asking his wife Eleanor for her opinion of Moskowitz's appointment. "Franklin, Mrs. Moskowitz is a very fine woman," Perkins reported that Eleanor replied. "I have worked with her in every campaign. I never worked with anybody that I liked to work with better. She's extremely competent[,] . . . able[,] . . . far-sighted[,] . . . [and] absolutely reliable. . . . I think a great deal of her and I think we are friends. But I want to say this to you. You have to decide . . . whether you are going to be Governor of this state, or whether Mrs. Moskowitz is going to be Governor of this state. If Mrs. Moskowitz is your secretary, she will run you. It won't hurt you. . . . She will run you in such a way that you don't know that you're being run a good deal of the time. Everything will be arranged so subtly that when the matter comes to you it will be natural to decide to do the thing that Mrs. Moskowitz has already decided should be done. That is the way she works. That is the kind of person she is." Faced with the prospect of subtle domination by a woman loyal to his predecessor, Franklin Roosevelt made the logical and wise choice.

Smith's attempt to retain power in his beloved state failed. Moskowitz was disappointed, too, perhaps as much for Smith and New York as for herself. "Franklin Roosevelt can never run that show," she told Perkins. "Somebody's got to help him, and Al loves the state. . . . It's going to be terrible. He's got that dreadful Louis Howe up there. Louis Howe will poison his mind about everything. . . . He's that kind of a sour person." In the view of a member of Roosevelt's team, Rexford Tugwell, Moskowitz and Smith, whom he considered political "professionals," misjudged Roosevelt with a "carelessness and superficiality" that amazed him. Naively they had thought Smith's retirement would be only nominal. Tugwell attributed their error to the egotistical arrogance of power. They had controlled events for so long that they could not imagine anyone else doing the job.

After what amounted to dismissal by Roosevelt, Smith became "a lost person," in Frankfurter's words, a "most tragic case of unemployment." To fill the void, Raskob and other financiers bought the site of the Waldorf-Astoria Hotel at the corner of Fifth Avenue and Thirty-fourth Street and drew plans for a 102-story "Empire State Building." For a salary of $50,000, Al Smith would run it. Perkins called the job a "phoney." It "gave him an office, a place to hang his hat, and a place to go, but not any really important involvement of his mind." This was true, but until Moskowitz's death, she kept him busy writing and speaking. For Moskowitz, "the battle" was not over. As she wrote Frankfurter, she intended to make Smith a "potent influence."[56]

Eyeing 1932, she developed with him and perfected his syndicated articles, speeches, and memoir, *Up to Now,* negotiating with potential buyers of his writings and serving as his publicity agent. She opened up her own public relations firm, Publicity Associates, taking her son Josef into partnership. When the Empire State Building was ready in May 1931, she moved her offices there, as did Smith. One of her accounts was the building itself, the construction of which, thanks to her suggestion, an unemployed Lewis Hine had photographed. Keeping her contacts fresh, she carried on a large correspondence, gathering information for articles that would help Smith establish a "fighting minority." Friendly DNC officials, such as Jouett Shouse, kept her in touch with their own pronouncements so that Smith could remain apprised of national party policies. She earned a rising income from her work: $6,500 in 1929, $8,300 in 1931, almost $12,000 in 1932.[57]

Throughout the post-1928 period, the Democratic party focused on winning in 1932. To this end, Raskob funded an ongoing publicity drive that was a Belle Moskowitz legacy. Prior to her directorship of the national publicity bureau, the party used to "go to sleep" between elections, maintaining a propaganda machine only for the four months before an election. Someone—whether it was Moskowitz or someone else is unknown—convinced Raskob to keep a national publicity bureau open for the entire period of Hoover's presidency. Raskob hired an experienced newspaperman, Charley Michelson, and gave him free rein to "minimize every Hoover asset and magnify all his liabilities." Since the RNC was complacent with victory, the DNC bureau, without fear of retaliation, proceeded to give Hoover the worst publicity he had ever received. Michelson followed precedents Moskowitz had established in New York State: he stayed in the background, never allowing himself to be quoted directly. Instead he fed statements to prominent Democratic senators and congressmen and arranged for them to be interviewed. That way statements he never paid for became "news." Michelson, a professional journalist, was a more clever writer than Moskowitz; but he was building on traditions she had established.[58]

As the 1932 convention neared, Roosevelt and Howe worried that a challenge from Smith might result in a deadlock. Watching Moskowitz's activities carefully, they interpreted everything she did—probably with good reason—as serving Smith's agenda. Once, when Moskowitz made a suggestion to Frances Perkins about labor conciliation, Perkins told her she would tell Roosevelt that it was Moskowitz's idea. Moskowitz said, "Oh, don't do that. It will just antagonize him against it." Perkins thought not, but when she did credit Moskowitz to Roosevelt he asked, "What's Moskowitz going to get out of it?" "Oh, nonsense," Perkins replied. "That sounds like Louis Howe. She's full of good ideas. Why not take her good ideas? She just hatches good ideas. . . . She's politically inventive and she's well disposed enough to give me this idea." "I don't know," Roosevelt replied. "I'd go slow on it, Frances. She'll sell you down the river some day." Perkins avowed that Moskowitz "never betrayed" her. They "remained good friends to her dying day," even when Perkins declared for Roosevelt in 1932.[59]

In the fall of 1931, Smith opposed a $20-million reforestation referendum that Roosevelt favored on the ground that it "put the government in the lumber business." Their disagreement presaged the final rift that took place five years later. In all of Smith's promotions of government interference in private enterprise during the 1920s, he had carefully distinguished between encouraging an industry to serve larger social purposes, such as the building of low-cost housing, and "putting government into business." Smith was a reformer, but a cautious one. His political philosophy was best expressed by Henry Moskowitz and Norman Hapgood in their biography of him. "There are three general ways of approaching a social evil," they wrote:

> One is the socialistic attitude, which looks upon it as a positive advantage to have work done by governmental bodies instead of by private bodies. One is the complete opposite, which relies on private enterprise for all conceivable activities and would usually rather endure an evil than take public action. The third point of view, which is that of Smith, . . . is that when business organizations and voluntary good will organizations will do a work, it is better to have it so done, but that a serious community disease should not be allowed to continue merely because there is no voluntary cure.[60]

To Smith, Roosevelt's policy on reforestation transgressed this philosophy. When Roosevelt won the referendum on this issue, the gap between the two widened.

Throughout 1931, Smith's former associates asked if he planned to be a candidate the following year. He said no. But on February 8, 1932, two weeks after Roosevelt declared, Smith announced that in the interest of an open convention he would allow his name to be put forward.

Moskowitz's network sprang into action. Throughout the spring of 1932, correspondence, telegrams, and phone calls flew between her office and Democrats across the country. She was in the thick of competing party factions in California and Pennsylvania, checking up on alleged betrayals, sorting out fact from fiction, encouraging loyalists to stick with Smith. In early April she wrote one contact that Smith was "an active and available candidate no matter what anybody else says." She engineered a letter, and worked on a draft of it, from eight Connecticut mayors asking Smith to run. She sent Joe Canavan to Pennsylvania as a troubleshooter and supplier of material to local Smith forces. She sent out copies of Smith's Jefferson Day speech, and tally sheets showing that Roosevelt's vaunted vote-getting ability in New York State was a myth. To one correspondent in Pittsburgh, she sent material on "ingratitude, which perhaps could be used by somebody." Presumably, the ingrate was Roosevelt. This was a slim reed on which to cling.[61]

Smith's prospects never looked good. Even loyal Fred Johnson of Rock Springs, Wyoming, switched to Roosevelt. In 1931, he had suggested to his old confidante that he hold another Rocky Mountain conference of Democrats. Moskowitz, arguing that the economic and international situation was too volatile, advised a "wait and see" approach. By March of the following year, Johnson passed on the bad news: the majority of Democrats from his region hoped Smith would not "encumber the Party" with his candidacy. In their view, "the religious prejudice which resulted in his defeat still exists," and was so unreasonable that people under its influence could not vote for Smith.[62]

Smith's pre-convention campaign angered Felix Frankfurter who saw Moskowitz's continuing efforts to promote a Smith candidacy as harmful to the party's chief goal of ending Hoover's administration. Smith could not win, he insisted. Roosevelt had limitations, but "his general social-economic directions are the desirable directions." "I am interested in what you say," Moskowitz replied, "but my primary interest just now is in Governor Smith and I am not thinking about anybody else." ("Jungle morals," Frankfurter scribbled in the margin of her letter.) She went on: "Many of us feel that the party needs a well-equipped candidate, able to lead, courageous and willing to take responsibility. We do not think that the record of the candidate who at that time was leading the field as we knew him intimately gave that kind of promise." At the end of this letter she made a confession that Frankfurter found "incredible." "Just because I am a progressive and because the Democratic party, as such, means nothing to me any more than the Republican party does, I cannot follow the Albany leadership."[63]

Thus, for Moskowitz, her purpose was less to give Smith another chance than to stop Roosevelt. We will never fully know why. She hinted darkly to Frankfurter of knowledge, gained probably through the Albany rumor mill, that Roosevelt was just as incompetent as she had always suspected. But she would not trust her information "to the channel of ink and paper." Whatever her case against Roosevelt, it was certainly aggravated by jealousy and resentment. Over the previous three years, he had never consulted with Smith on any important policy matter. Out of loyalty to her mentor, she must have felt deep distress at his feelings of neglect. Further, Roosevelt had reaped huge political benefits from the fulfillment of programs, such as hydroelectric power, parks and parkways, and other public works, for which Smith and his team had laid all the groundwork. Finally, he had rejected her personal aid. After a dozen years of power and influence, she now exercised none whatsoever over the course of the events she cared about most. Hence she told Frankfurter that, for her, specific programs and the people who put them over took priority over party loyalties. Frankfurter could not understand. The two remained friends, but for the time being tabled their correspondence.

Moskowitz was involved in an embarrassing moment during the pre-convention campaign. On March 22, at her request, Lindsay Rogers sent her a speech for Smith to give on Walter Winchell's "Lucky Strike Hour," March 31. Smith never used it. Rogers, deciding to "try it in another direction," revised it and sent it to Raymond Moley, who passed it on to

Roosevelt's speechwriter, Sam Rosenman. Roosevelt gave his Lucky Strike radio talk April 7, using only one of Rogers's sentences on international trade. On April 10 Moskowitz asked Rogers to review the draft of Smith's Jefferson Day speech. He performed this task at her home that day, and brought her some new material on the tariff the next. Smith delivered his speech April 13, using Rogers's tariff material. That night, Rogers, Rexford Tugwell, and Moley proposed material for Roosevelt's St. Paul speech, planned for April 18. Nothing was said then about the tariff, but without Rogers's knowledge passages from his original Lucky Strike draft found their way into the St. Paul speech. The result was that FDR quoted directly from Smith's denunciation of the Hawley-Smoot tariff bill. After this experience, Belle Moskowitz wrote a contact in Baltimore, "I have finished with college professors for a while at least."[64]

At the Chicago convention, Perkins told Moskowitz that she was foolish to try to get Smith nominated. Perkins gave her sound advice: Smith should act as the elder statesman, a trader, guide to the younger generation, not someone still seeking nomination. Moskowitz retorted, "But he's entitled to it . . . , it is his right to be run again." His right or not, the prize eluded him. Roosevelt's lieutenants, who had been organizing throughout 1931, outmaneuvered Smith's at every ballot. On the fourth, when Smith's old enemy William Randolph Hearst permitted the California delegation to vote for Roosevelt in return for the vice presidential slot for John N. Garner, Roosevelt slid past the two-thirds majority and won.[65]

Smith left the convention hotel through a side door. He was so angry that his friends feared he might, on arriving in New York, deliver an ill-tempered statement to the press. Sam Rosenman got his wife Dorothy to reach Bernard Shientag, who mounted Smith's train before it arrived and calmed him down. By the time he spoke to reporters, Smith was able to make a quip or two, but his bitterness was obvious. As for Moskowitz, there is a story that after Chicago she suggested to a Hoover aide that he campaign first in Smith territory.[66]

A month later the Smith group reunited for a testimonial dinner honoring their hero. "ALFRED E. SMITH (THE MAN WITH A RECORD—WHO HAD A PROGRAM), *Moral Victor of the Battle of Chicago*—1932," the menu read. It was an evening of satire and acid jokes. As an epigraph, Robert Moses quoted Shakespeare's "Politics is a thieves' game,/ Those who stay in it long enough are invariably robbed." The menu included "Nuts McAdoo," "Branchless Olives Roosevelt," and "Coupe Empire State With Mortgage," a reference to the building's office suites, many of which were still unrented because of the Depression. A spoof was performed, entitled "Off the Record. The Knife in the Back—The Chicago Run-Out or McAdoo's Double-Cross." Belle Moskowitz received credit for the "stage settings." Two "Expressions of the Great," or sayings associated with convention moments, were also attributed to her: "I don't think we can afford it but you might see Mr. Kenny," and

> Maryland casts 16 votes for Roosevelt.—*Radio Voice.* That makes Ritchie Attorney General.—*Belle Moskowitz.* Virginia casts 24 votes for Roosevelt.—*Radio Voice.* That makes Byrd Secretary of the Navy.—*Belle Moskowitz.*[67]

At least the faithful band could laugh at itself a little.

This was not the last time the band worked together. During the campaign that followed, Smith was finally persuaded to speak in Roosevelt's behalf. Despite their own bitter feelings, Moskowitz, Proskauer, and Moses helped prepare the speeches. Henry Moskowitz worked for Roosevelt, too, and remained a supporter the rest of his life. Henry sympathized with Belle's disappointment but recognized Roosevelt's interests in social progress. According to Perkins, Belle did not seem to resent Henry's decision: "Mrs. Moskowitz wasn't so emotional that people couldn't disagree with her. She was a very complex person. Her emotions didn't tear her to pieces. She was a very cool-headed person. She would get powerfully mad, and she did." If Henry wanted to work for Roosevelt, Perkins concluded, Belle would dismiss it as "one of Henry's notions" and refuse to get worked up about it.[68]

After Belle Moskowitz died, Smith's criticisms of Roosevelt deepened. Rupture threatened in 1934, when Smith joined other wealthy businessmen in a non-partisan "American Liberty League." Its purpose was to fight the New Deal policies then leading America into "state socialism," and to foster the American tradition of free private enterprise. Two years later, after a lifelong commitment to the Democratic party, Smith took his famous "walk" from the party and supported the Republican challenger, Alf Landon, for president.

If Moskowitz was Smith's "tent pole," as Molly Dewson claimed, and if she had still been alive in 1936, would Smith have taken his walk? Would she have gone with him? Historian Otis Graham, who studied the attitudes of "old" progressives toward the New Deal, has shown that many disapproved of Roosevelt's breach of the constitutional limits of state power. Since Belle died before Roosevelt's inauguration, Graham did not include her in his study, but we know that she held tenaciously to her identity as a progressive. We also know that in 1936 Henry announced that although he was worried about the bureaucracy of New Deal programs he would take a "walk" from his old friend Al Smith and declare for Roosevelt.[69] Henry had been more disposed toward Roosevelt than Belle had been, perhaps because he was less personally grieved by the events of 1928 and 1932. But it is hard to imagine the Belle who called herself a "Bolshevik" in 1928 adhering to the Liberty League in 1934.

By the time of her death, furthermore, she had already shown signs of a dwindling attachment to Smith. As his star fell, he began to drink more heavily. Whenever Roosevelt's name came up, he swore and raged. In the fall of 1932, Walter Lippmann and Felix Frankfurter commiserated with one another over the decline of their old friend. "I think his hatred and resentment and personal frustration are almost overwhelming," Lippmann wrote. He thought that Smith had developed "what almost amounts to a persecution complex," and was rapidly becoming an "awful human spectacle."[70] Working with such a person could only have become a trial for Belle Moskowitz. She herself was not in good health, as her inability to survive a fall in December made evident. But had she lived to watch Smith excoriate the New Deal, it is doubtful she would have followed quite the same path.

In fact, her influence on him peaked with the nomination in 1928, and began to decline when he named Raskob as head of the DNC. After the election, it was easier to maintain the pattern of dependency than to break it. But her death, which caused him genuine grief, freed him from a tutelage he had first begun to resist when he became the presidential nominee.

In Moskowitz's last months, she knew some small satisfactions. Early in the 1932 pre-convention campaign, a woman Democrat in Washington, D.C., wrote her that, to those who had watched Moskowitz "most admiringly" in 1928, she had "long been an ideal 'woman in politics.'" Within the range of possibilities that Moskowitz's life in politics exemplified, she had indeed inspired others. But by the end she was probably too hurt and spent to appreciate adulation. After the 1932 convention she observed to a contact in southern California, "Politics is a strange game and I confess part of it is a sad disillusionment to me." To a correspondent in Chicago, she called the convention "a most trying and difficult set of experiences for all of us. I have learned much which I hope I may not have occasion to apply in a hurry again."[71]

Betrayal and disillusionment: sad feelings with which to close a distinguished career.

Epilogue

The year has begun wretchedly with Belle's death. I can have a little notion of what a wrench her death means to you, for she and I were friends for some twenty-five years. . . . Her virtues were many but she taught one lesson of transcendent importance—that high public service does not imply public office.

Felix Frankfurter to Joseph Proskauer
January 4, 1933

In the fall of 1932, Belle Moskowitz attended a private showing of a new film, *Silver Dollar,* which starred two of her friends, Edward G. Robinson and Aline MacMahon. Aline and her husband, architect and planner Clarence Stein, had invited her. At one point, Belle said to Clarence, "I am dead tired, and I wish to heaven I could get away." Aline, about to embark for Hollywood, suggested that Belle come along. Belle accepted, and on November 12, shortly after the election, they boarded *The Chief.* Belle planned to spend a week or so with Aline, and then another week with her nephew Clarence Lindner, publisher of the *San Francisco Examiner.*

Aline was younger than Belle, but Belle took her into her confidence. Aline wondered if Belle had been too busy to have many "heart-to-hearts" with other women. At the start of their trip, Belle was pale and drawn, "exhausted by life." "They use me up," Aline remembered her saying.

Who "they" were was not clear. Smith was not mentioned, but Belle had confided to her daughter that, after his "betrayal" that past summer, working with him had become a trial. As for her children, whom she did discuss with Aline, her oldest son Carlos was in his late twenties, well on his way to self-reliance and a promising legal career. Miriam was in England, but not happy. Married to a wealthy man, she had no access to money of her own. For some time Belle had been sending her fifty dollars a month so that she could buy things for herself and the children. Belle feared the marriage might not last. Her youngest, Joe, was still a problem. In an effort to get him started in a career, she had taken him into business with her. But he lacked the maturity, energy, and contacts to perform well. When Belle went to California with Aline, she turned over to Joe the promotion of Clarence Stein's garden apartments project, "Hillside Housing," then in the process of being funded. Joe proved a washout. Clarence wrote Aline that on Belle's return he planned to berate her on the inadequacies of her office, but when she did return she neutralized his anger by arranging everything in three phone calls. "Joe just sat there," Clarence wrote Aline, "like a lump."

Belle stayed a week with Aline, resting, reading, and working on articles. Her color and spirit improved. She met with studio bigwigs, including the playwright, Moss Hart, who was charmed by her. She told Aline that, if offered, she would take an executive position in the studios. Aline was surprised she was ready to leave New York, considering how much it had meant to her as well as to Henry.[1]

In San Francisco, Belle met with friends from the Smith campaigns, including the wife of a political science professor at the University of California at Berkeley, Grace Montgomery, who had fought valiantly for Smith in both '28 and '32. Montgomery had met Moskowitz for the first time the previous April, when she rushed to New York to confer with her about making a "real" primary fight for Smith. When Moskowitz told her how much she appreciated Montgomery's "devotion," saying "it's the same as mine," the recognition had meant "far more" to her than "the highest gift in the presidential pie-wagon to a politician." The two women met for the last time on the Oakland Pier on Thanksgiving Day, and Montgomery wrote Smith about it early in 1933: "She asked, 'How do you feel about it all?' I told her that for the first time in my life I understood the feelings of the youth who threw the bomb at Sarajevo. She answered, 'That's the way I feel.'" Moskowitz tearfully confessed that on election night she had been unable to bring herself to go to the Biltmore to hear the returns. Finally, when Montgomery spoke of the "dreary formality of a re-nomination in 1936," Moskowitz had said, with characteristic confidence, "It may not come to that. Watch."[2]

On December 8, about a week after her return from the west coast, Belle fell down the front steps of her home. No one, not even she, knew what caused the mishap. The steps were icy, and she may have had a dizzy spell. She did not remember falling, and regained consciousness only as she hit the sidewalk. A doctor who lived next door ran out and helped get her to bed. Her own doctor, Joe Girsdansky, set her left arm, which had been fractured above the wrist. After a day or two of rest, her right arm, badly fractured just below the shoulder, was put into a cast and strapped to her body.[3]

Complications began. Pain, thought to be neuralgia or nerve pain, developed in her chest. This passed, and she became more comfortable. She listened to the radio, read, had visitors, and anticipated being up soon. Lying in bed with her door open, she listened to recordings her son Josef played for her downstairs. Wagner was her favorite, "the greatest love music in the world," she had said to Aline. Then phlebitis developed in her left leg. Her circulation had always been sluggish, and now, because she was so heavy and hard to turn in bed, it worsened. An embolism, or blood clot, lodged in a vein. Massage and manipulation improved the condition, and no one seemed alarmed. Then, one morning shortly after Christmas, the embolism broke off and reached her heart, lodging in the coronary artery. For a few days she was in great pain. Her family thought the end was near. Since she had nursing and medical care around the clock, however, she insisted they all continue as usual, going to work and to the seasonal parties to which they had been invited.

New Year's Day was on Sunday. On Monday, because he thought his mother wished it, her son Carlos went to Albany to attend Governor Herbert H. Lehman's inauguration. Knowing how much his mother had fostered Lehman's career, he called this event the last of her "great creations." Henry had been invited to spend the day in White Plains at the home of Howard Cullman, a friend and Smith supporter whose appointment as a commissioner of the New York Port Authority Belle had helped arrange. Joe stayed home. That afternoon the nurse and another doctor, who had been attending Belle daily, summoned Joe to help move her up in bed. Her breathing became labored. Girsdansky arrived to administer oxygen, but it was too late. Knowing that at best Belle had but a few painful hours to live, he gave her morphine, and she died.

"She was lovely to look upon, radiant even in the moment of her death," Joe wrote Miriam, who was in England expecting a child. Joe and Girsdansky, a close family friend, cried for a while, and then Joe summoned Carlos and Henry. Eva Levy, a family friend, broke the news to Smith, who was having lunch at his daughter Emily Warner's house. He said later that he had had a premonition of disaster, and had declined to lunch with Governor Lehman in order to be nearer a telephone. On hearing the news, he wept, and then took the first train for New York.

Over the next two days, hundreds of people came to the Ninety-fourth Street house. Dorothy Rosenman, Sam Rosenman's wife, and Eva Levy kept the house going and answered the phone. Many sent condolences to Smith; the condolences to the Moskowitz-Israels family numbered nearly a thousand. A few weeks before her death, Belle had told Carlos that she considered Joe her "one great unfinished job." Her death matured him. In spite of being the youngest, he arranged the entire funeral.

Remembering his mother's pride in Charles Israels's refusal of elaborate service or mourning, Joe kept the event as simple as possible. It was held at Temple Emanu-El, which Clarence Stein had designed. By special permission, Stephen Wise, who was not rabbi there, spoke the eulogy and gave the prayer. Joe called him "an old faker," but knew he "was a good friend and sincere admirer" of his mother, and that he meant every word he said. "They little understood or understand Mrs. Moskowitz who think of her as a skillful and successful politician," Wise said:

> She was a great citizen, to whom politics was never a game to be won nor an arena in which to achieve success. . . . Politics was the cause which possessed her soul. . . . She served not that she might succeed or that success might crown the hopes of a friend or friends, but solely where she believed that triumph was bound up inseparably with the well-being of the State.

Thus was she eulogized consonant with the image she had carefully cultivated all her life. She was the "ideal woman in politics" because her chief goal was "to serve others," like all true women, before she served herself.

After the service, a spray of orchids and roses that Rose Van Namee had prepared for the dull-silver casket was placed in Belle's hands. She was buried in a plot overlooking a deep ravine at Sleepy Hollow Cemetery in Tarrytown, where Charles also lay. Grace Goodale, in memory of the bouquet of violets Belle had placed on Charles's grave, brought a similar one to put at her old friend's side.

By the terms of Belle's will, Henry received the house and its furnishings, appraised at $24,000, for use during his lifetime. After paying off various debts—a $14,000 mortgage on their home, $27,000 in other debts, and about $1,000 in funeral expenses—her gross estate of almost $57,000 was reduced to a net of just over $14,000. Joe, who planned to keep Publicity Associates and to seek a new partner, agreed to pay $7,100 due in rents to the Empire State Corporation. Belle's niece Louise Lindner, Max's daughter, of whom Belle was especially fond, received a special gift of $500. Belle owned stocks, some of which were found to be worthless; the estate owed money on them. Her personal effects were worth about $1,000; her jewelry, about which she cared little, was worth only $262. Mary Moses, Robert Moses's wife, helped sort Belle's belongings; June Rhodes, who was sailing for Europe at the end of the week, was entrusted with giving Miriam Belle's Persian lamb coat and her wedding ring, designed to represent the theme of Wagner's "Ring Cycle."

Henry survived his wife by only a few years. He had had a productive career in Jewish philanthropy. He founded and for many years ran the American branch of the Organization for Rehabilitation Through Training (ORT), a group that helped Jewish refugees from Europe retrain for useful occupations in their new homes. Throughout the 1920s he had worked as a volunteer in Smith's campaigns and in the years following Belle's death in the campaigns of other Democrats, such as Herbert H. Lehman. When added to Belle's earnings, his own from work as an impartial chairman in the textile, leather, and other industries, as well as from his publications on Smith, had allowed the couple to purchase waterfront property on Lake Oscawanna in Putnam County. Henry, his friends, his step-children, and later his grandchildren, spent many weekends there in refuge from the stresses of city life. At the end of Henry's life, he was executive director of the League of New York

Theatres and Managers. Both Joe and Carlos had married by then, and were making common household with him. The sudden loss of Belle's income at the height of the Depression had forced the family to keep their expenses down. In addition, without Belle's nurturing influence, the three men seemed to need one another to carry on. "We are like the three Musketeers—one for all and all for one, with the vision of Belle's life to inspire us," Henry wrote Miriam's in-laws. In 1936 he developed cancer of the jaw and died on December 17.

Neither Joe's nor Carlos's marriages lasted, but both remarried later, only Carlos with more success. After a career in public relations, which included representing Ethiopian Emperor Haile Selassie in the United States, Joe, who had inherited his mother's tendency toward obesity and his father's weak heart, died in 1954. He was only forty-five. Carlos, a corporate lawyer, published several textbooks, taught as an adjunct professor at Columbia University Law School, and was active in Jewish philanthropy. He died in 1969. Miriam's marriage ended in 1936; she later married Naum Gabo, a Russian émigré sculptor, and returned to America. She now divides her time between homes in London and Connecticut.

The outpouring of grief, the numbers who attended the funeral and wrote to the family, the long obituaries, editorial tributes, and feature articles on Belle's life all overwhelmed her sons. Joe wrote Miriam that "it opened my eyes anew to what a really great woman she was." Carlos was more emotional, perhaps a bit intimidated: "If I can only live up to that!" he wrote. To the end, even her own sons could not find the words to describe their mother adequately. To them as to others, she remained a baffling mixture of "the brains of a man" with the nurturing, loving instincts of a woman. In planning her funeral, both sons had wanted the organist to play Wagner. Carlos chose the entrance of the gods from *Götterdämmerung.* But Joe had said, "that is for a hero—she was that but she was a woman first, feminine." The organist played the "Liebestod" from *Tristan und Isolde.*

Appendix

Belle Moskowitz's letter to Lillian Wald about Alfred E. Smith's upbringing on the Lower East Side:

Tuesday, Aug. 7th, 1923

Dear Miss Wald,

Here are some random notes. I am away from my files and have to think it out afresh:

1: He sold newspapers on Park Row; and loved the Thanksgiving Day dinner of the newsboys and never missed one.

2: He always loved animals; dogs, horses: as soon as he could afford it he bought the children a little goat carriage and a pair of goats and had endless fun driving them around in it. Today his greatest pleasure is an afternoon all to himself at the menagerie. He owns a monkey, a fawn, a pair of ponies, a baby raccoon and some fat lambs to say nothing of dogs.

3: He was a good student and especially fond of elocution. He used to compete in prize speaking contests. He tells of the long trip in the horse cars from Chatham Square where his school was located, to Manhattanville College (125th E. and Old Broadway) where the speaking contests were held. He tells how on a snowy night he and the other boys from his school carried off the prize and then had snow fights all across 125th Street to Third Avenue, with the defeated contestants. Contrary to general belief he can recite parts of the poetry he learned then and all the poetry the other boys recited besides. He loved the amateur theatricals at the Church and sat in the front row making audible remarks to the actors until he became old enough to be one of the actors himself—as he frequently was, playing leading parts in the Shaughraun and Hazel Kirke and other Irish plays.

4: His father died when he was about thirteen and he left school to go to work; errand boy and general helper to a fish dealer in Fulton Fish market. He has always been "a widow's son" in his devotion to his mother.

He loves the East Side dearly and tells many stories of the long rides he took to get uptown. Of course he went swimming off the docks. He knew all about horses helping his father with them.

He was an altar boy at St. James'.

Has a great reverence for teachers and a tremendous respect for education. He appreciates what was, as he puts it, "denied him" by circumstances.

His mother says, "Alfred was always a good boy and minded me." He loved the fire engines and being around the engine house day and night for a chance to run with the machine. It is today his favorite branch of the city service and he knows by heart every engine house and remembers every horse and engine in the neighborhood.

I can't think of any specific naughtiness. He loved practical jokes of a simple kind and played them on his friends and his family. He loved singing and a funny story. He still does.

If I rake up some more tomorrow through Mrs. Smith I'll send it on. Thank you so much. I know the Governor appreciates your doing this for him. He is like that.

Sincerely,
Belle Moskowitz

Tribute to Belle Moskowitz by Alfred E. Smith, Ex-Governor of the State of New York. *Jewish Chronicle,* 17 Feb. 1933.

"My Closest Advisor. Belle Moskowitz as I Knew Her." [Alfred E. Smith, best known political figure of contemporary America, gives his estimate of the late Mrs. Belle Moskowitz who was regarded by many as the most influential woman in our public life. This article was written exclusively for this publication.—THE EDITOR.]

Women in politics sometimes make the mistake of not being themselves and imitating men. In Mrs. Belle Moskowitz the very opposite was true. Though she had one of the finest intellects I have ever known, though she cooperated very actively with men in political life, she was essentially a wife and a mother. She never lost her sense of the true values of life. She demonstrated that participation by women in public life does not involve any sacrifice of their essential feminine qualities. She dovetailed her life as wife and mother with her public services because she knew how to organize herself. She was one of the best organized persons I have ever known.

I met Mrs. Moskowitz for the first time in the gubernatorial campaign of 1918. She worked very closely with my colleague, the Chairman of the Committee, Judge Elkus. She had a desk in the office of the headquarters, but she was so modest that I knew nothing about her until towards the end of the campaign when she asked me to speak before the Women's University Club. I saw very little of her after that, but when the campaign was over she came to me with an interesting proposition and suggested that my administration concern itself very largely with the problems of reconstruction. I was inaugurated seven weeks after Armistice day and the whole country was busy with reconstruction problems.

I then asked her to draw up a memorandum as to the type of reconstruction she would advocate. The memorandum so impressed me that I called together a group of social workers and my friends at the City Hall at the Office of the President of the Board of Aldermen. As a result of the discussion on this memorandum, I resolved to recommend the appointment of a reconstruction commission and asked her to be its secretary. The work of this commission is now history. Its recommendations cover not only the reorganization of the state government but fundamental questions affecting labor, education, unemployment and other vital problems as well.

As secretary Mrs. Moskowitz was responsible for administering the work of this commission which consisted of some of the most distinguished people of the State. I frequently met her as secretary of the commission and I learned to admire the keenness of her mind, especially her ability to get at the essentials of a complicated question and to explain it briefly and simply. Many questions of policy had to be considered in connection with the work of this commission. In the determination of such policies, I frequently called upon her for advice. I invariably found that advice to be sound.

She frequently spoke of the woman's point of view. She was no doubt a womanly woman, but I saw no difference between the processes of her mind and the way the best men's minds work. In a group of counsellors she was very loath to express her opinion and I frequently had to ask her for advice before she gave it. A discussion of a problem with her served to clarify my own mind.

In the course of the eight years of my administration, Mrs. Moskowitz worked closely with me. Her aim, like mine, was to make the State an instrument for the welfare of the people. At the first meeting of University Women held at the Cosmopolitan Club to which Mrs. Moskowitz brought me, I stated: "I know what is right. If I ever do anything that is wrong, it will not be because I don't know it to be so, and you can write it down as being wilful and deliberate and hold me to account for it. I want to do what is right."

I made up my mind to give the State the best that was in me and Mrs. Moskowitz helped me to realize that ideal. During those eight years, we managed to pass some very fundamental legislation. It involved the structure of the State government and simplified the government by converting 187 departments, boards, bureaus and commissions into 19 divisions or departments. In this work of reorganization I received considerable help from the industrious labors of Mrs. Moskowitz who had a tireless energy. I seldom met a woman with such a capacity for work. She never seemed to tire.

She was vitally interested in housing legislation. There was something about her mother heart which impelled her to fight for slum clearance and to try to establish decent housing for poor people to live in. In all this work, in the work of passing rent laws and providing a constructive housing program, Mrs. Moskowitz worked closely by my side.

Another vital piece of legislation establishing an historic precedent in the State of New York, was the result of the work of the Friedsam Commission. This commission studied the problem of State aid in education. As a result of its labors, one of the most scientific provisions for providing the teachers and the pupils of the educational system of the State of New York with adequate financial aid was established. Mrs. Moskowitz rendered great service to the teachers of the State of New York.

I cite these merely as examples of important achievements in my administration in which Mrs. Moskowitz co-operated. Though she was tender, as only a fine woman could be, she was a fighter; she was a hater of insincerity and bunk and in the many struggles which I had against those who misrepresented my motives or those who tried to block my efforts, I always received encouragement and help from her. She hated sham and hypocrisy.

On the purely political side of the administration, she did not pretend to have expert judgment, but I usually found her political sense sound. She was my closest adviser in connection with my two national campaigns. She worked day and night to bring about my nomination.

It was a privilege to work with her and to live up to her high ideals of public service. The country has lost a great citizen, the Jewish people one of their most illustrious women and a real friend.

Notes

Introduction

1. Mary (Molly) Dewson, "An Aid to the End," unpublished memoir, p. 21, Dewson Papers; Matthew and Hannah Josephson, *Al Smith, Hero of the Cities* (1969), 457. In *Beyond Suffrage: Women in the New Deal* (1981), Susan Ware shows how female New Dealers shared the historical experience of Progressive Era social reform.

1: Guardian Angel of the Lower East Side

1. Jeffrey S. Gurock, *When Harlem Was Jewish 1870–1930* (1979); Wilson's *Business Directory of New York City* (1871), 576, lists Isidor Lindner's address as 109 Canal St.; U.S. Census, 1880; author interviews (1975–76) with Miriam Israels Gabo, Dorothy Lindner Omansky, Barbara Lindner Samuels, Nan Lindner Redell, and Esther Lindner Haas; correspondence of Grace H. Goodale to Miriam Gabo, 1933–46 (author's collection); article on Temple Israel, *New York Tribune,* 16 Mar. 1902.

2. On N.Y. school centralization controversy see David Hammack, *Power and Society. Greater New York at the Turn of the Century* (1982), ch. 9. On progressive education, see Lawrence A. Cremin, *The Transformation of the School. Progressivism in American Education, 1876–1957* (1961), 3–22, and *passim.* On Horace Mann and Teachers College, see "Horace Mann Alumni Register," "Horace Mann School Circular of Information," 1893–94, and "Teachers College Circular of Information, 1897–98" (all in TC Archives); also, James E. Russell, "The Horace Mann School," *Teachers College Record,* 3/1 (Jan. 1902), 1–3; and Lawrence A. Cremin, David A. Shannon, and Mary Evelyn Townsend, *A History of Teachers College* (1954). A mimeographed biography issued in 1930 by Moskowitz's firm, Publicity Associates, says she attended P.S. 39 and P.S. 68 (Moskowitz Papers), neither of which kept 19th-century records. Horace Mann H.S. for Girls left only the materials cited above. The same situation holds for Teachers College.

3. On Benfey (later, Ida Benfey-Judd), see George C. D. Odell, *Annals of the New York Stage (1891–94),* XV, 465, 468, and 751; reviews in the *New York Times* (hereafter, *NYT*), 18 Nov. 1894; 22 Jan., 5 Feb. 1896; 24 Jan. 1899; and *New York Tribune,* 25 Mar. 1895. On Conried, see Walter Rigdon, *The Biographical Encyclopedia & Who's Who of the American Theatre* (1966); Odell, XV, *passim;* Norman Hapgood, "Heinrich Conried and What He Stands For," *Outlook,* 77 (7 May 1904), 80–84; and "A Talk About Acting," *NYT,* 3 Dec. 1899. On Brighton Beach see Kathy Peiss, *Cheap Amusements: Working Women and Leisure in Turn-of-the-Century New York* (1986), 122–23. Bernays, sister of Edward Bernays and niece of Sigmund Freud, met Belle in 1902 in the Catskills. They remained lifelong friends, and Belle's children called her "Aunt Judith" (author interview, 1976).

4. On emotional bonding among Victorian women see Carroll Smith-Rosenberg, "The Female World of Love and Ritual: Relations Between Women

in Nineteenth-Century America," *Signs: A Journal of Women in Culture and Society,* 1 (Autumn, 1975), 1–29.

5. My portrait of Belle's late adolescence is based on her letters to Abbie Fridenberg (11, 21 Aug. 1896); Grace Goodale's letters to Carlos Israels (2 Jan. 1935) and Miriam Israels Gabo (8 July 1941; another, undated but written shortly afterward; and 15 Sept. 1935)—all in the Moskowitz Papers. Abbie's poem appeared in the *American Jewess,* II/5 (Feb., 1896), 244.

6. Goodale to Gabo, 20 April 1936.

7. Goodale's copy of the *Mortarboard* is in the Barnard College Archives. King's *Handbook of New York City* (Boston: Moses King, 1893), 551–52, briefly describes the Fidelio Club (110–12 East 59 St.).

8. Lindner to Fridenberg, *op. cit.*

9. See Victor Margolin, *American Poster Renaissance* (1975), Intro., 17–21. Martha Banta told me about the poster craze, and Agnes DeMille Prude sent me a photo of her mother Anna George DeMille's party. Grace Goodale attributes the name "Poster Room" to Anna.

10. See interview with Goodale in *Barnard Alumnae Monthly* (Mar., 1933), 9–10.

11. I am grateful to Martha Banta's insightful lectures delivered at Ind. Univ. in 1982–83 on late-19th-century women's photographs, ideals of beauty, and female transvestism. On turn-of-the-century women's choices between careers and domesticity, see Jessie Taft, *The Woman Movement From the Point of View of Social Consciousness* (1916).

12. See Charlotte Baum, Paula Hyman, and Sonya Michel, *The Jewish Woman in America* (1976), ch. 4. Women were barred from City College until WWI, but could attend Hunter College, tuition free.

13. See Allen F. Davis, *Spearheads for Reform: The Social Settlements and the Progressive Movement 1890–1914* (1967), chs. 1–2; Robert A. Woods and Albert J. Kennedy, *The Settlement Horizon: A National Estimate* (1922); Jane Addams, *Twenty Years at Hull House* (1910); and Lillian Wald, *The House on Henry Street* (1915).

14. See Moses Rischin, *The Promised City: New York's Jews, 1870–1914* (1962), and Irving Howe, *World of Our Fathers* (1976). Robert H. Bremner, *From the Depths: The Discovery of Poverty in the United States* (1956) discusses late-19th-century charity organization and social work movements.

15. Minute Books, Temple Israel Sisterhood (1891–93), American Jewish Archives. For Belle's remarks on Jewish women's charity, see her "The Jewish Woman's Opportunity for Service," *The American Hebrew* (hereafter, *AH*), 100 (6 April 1917), 739. On working girls' clubs and vacation societies, see Kathy Peiss, *Cheap Amusements,* 168–71.

16. On the Educational Alliance, see Rischin, 101; S. P. Rudens, "A Half Century of Community Service: The Story of the New York Educational Alliance," *American Jewish Year Book,* 46 (18 Sept. 1944–7 Sept. 1945), 73–86; "The Educational Alliance," *AH,* 64 (9 Dec. 1898), 208–17 (includes photographs of Alliance activities); and "Reformation of the Great East Side," *NYT,* 19 Nov. 1900. The Alliance's *Annual Reports* contain much information; Straus's statement comes from that of 1897. On patriotic assemblies, see, e.g., *N.Y. Trib.,* 29 Nov. 1900.

17. Rischin, 102; Rudens, 74–76, 82; "Plan and Scope," *Annual Report* (1896); and David Blaustein, "The Making of Americans," *Proceedings, N.Y.S. Conference*

of Charities and Corrections, 4 (1903), 231–39, and "From Oppression to Freedom," *Charities,* 10 (4 Apr. 1903), 337–43. Paul Abelson's term, "rational altruism," comes from a letter to E. R. A. Seligman, 12 July 1901 (Paul Abelson Papers, American Jewish Archives).

18. Gordin's play is mentioned in Irving Howe, *World of Our Fathers,* 233–34. Bonny Fetterman alerted me to *AH* coverage of the controversy over the play in vol. 72, 6, 13, 20, and 27 Mar. 1903; on the Educational League, see also vol. 67 (13 July 1900).

19. Gollomb, *Unquiet* (1935), 324–25. This novel of Lower East Side life is cited in David Hollinger, *Morris R. Cohen and the Scientific Ideal* (1975), which relies on Gollomb's description of the Alliance's famous "Breadwinners' College" where Cohen first taught. See also the Alliance's *Annual Report* for 1899, p. 11; and for 1902, "Activities in Communal Work." The controversy over Americanization and patronization at the Alliance can be followed in *AH,* 67 (12 Oct. 1900), 625–27; 71 (14 Nov. 1902), 724–25; and 72 (5 Dec. 1902), 107–09.

20. Educational Alliance, "Announcements," 1900–1902; *Annual Reports,* 1896–1903; Minutes, Alliance Board of Trustees, 1898–1903 (Microfilm, YIVO Institute); and the Alliance House Committee Minutes and Reports, 3 Oct. 1898–31 May 1901 (Educ. All. Papers, fol. 13, YIVO).

21. Alliance exhibits were described in the *N.Y. Trib.,* 16 Mar., 19 Apr. 1902; and in *AH,* 69 (31 May 1901), 47–48; the poster contest was announced in 71 (3 Oct. 1902), 553. Artists who benefited from Alliance art programs included Ben Shahn, Chaim Gross, Sir Jacob Epstein, and Raphael, Moses, and Isaac Soyer.

22. *AH* carried notices and descriptions of Alliance entertainments (70 [4 Apr. 1902], 607; 71 [31 Oct. 1902], 675). The comments by "One of the Submerged" appeared in vol. 71 (7 Nov. 1902), 705.

23. *Ibid.,* 72 (13 Feb. 1903), 429.

24. *Ibid.,* 72 (20 Mar. 1903), 598.

25. See Samuel Shipman's reminiscence of Belle Lindner, *N.Y. Evening Journal,* 12 Mar. 1932; Harry Roskolenko, *The Time That Was Then: The Lower East Side 1900–1914, An Intimate Chronicle* (1971), 210–13, mentions Alliance children who went on to theatrical or film careers.

Belle Israels opposed the increasing professionalization of Alliance children's theater. See her "Another Aspect of the Children's Theatre," *Charities and The Commons,* 19 (4 Jan. 1908), 1310–11, a response to J. Garfield Moses's "The Children's Theatre," *ibid.,* (18 [Apr. 1907], pp. 23–34). The debate continued in *AH* editorials, 87 (3 June 1910) and 90 (17 Nov. 1911).

26. For more on recreation theory, see ch. 3. Abelson later took history and law degrees, and became a labor conciliator. See Jesse Carpenter, *Competition and Collective Bargaining in the Needle Trades, 1910–1967* (1972). On Ethical Culture, see ch. 2, p. 26.

27. Educ. All., *Ann. Rep.,* 1901.

28. Belle Lindner, "Social Work Among Young Women," *AH,* 68 (22 Feb. 1901), 425–27. Her optimistic view of settlements gibed with 19th-century women's notion of the family circle radiating virtue out into society. See Barbara Welter, "The Cult of True Womanhood," *American Quarterly,* XVIII (Summer, 1966), 151–74.

29. Her salary is mentioned in her mimeographed biography in the Moskowitz papers; on her resignation, see Sidney Blumenthal to Louis Marshall, 18

Sept. 1903, Louis Marshall Papers; Benjamin Tuska to Henry Moskowitz, 3 Jan. 1933, Moskowitz Papers.

30. "Jewish Women as Settlement Workers" appeared in *The Hebrew Standard*, 50/11 (5 Apr. 1907), 9. Sue Elwell found it for me.

31. Condolence letters, Moskowitz Papers; author interview with Judith Bernays and Monroe Goldwater (1976); lists of Alliance club leaders in the Annual Reports, *op. cit.*; Jacob Billikopf to Josef Israels II, 16 Oct. 1943 (Jacob Billikopf Papers); "Memorabilia of Friends," Box 4, Abelson Papers, AJA.

2: A New Woman

1. *AH*, 72/19 (27 Mar. 1903), 636, reported the Educational Alliance dinner-dance. Other sources for this section include: Grace Goodale to Miriam Gabo, 15 Sept. 1935, 21 April 1939, and 13 Sept. 1944 (author's collection); author interview with Judith Bernays Heller.

2. See "The Death of Josef Israels. Sketch of the Life of the World-Renowned Painter," in *AH*, 89/16 (18 August 1911), 447; J. Ernst Phythian, *Jozef Israëls* (London, 1912); and C. L. Dake, *Jozef Israëls* (Paris, n.d.: series, "L'Art et le Beau").

3. On Lehman Israels, see the following letters in the Whitelaw Reid Papers: Lehman Israels to Reid, 24 May 1875; Reid to Shanks, n.d.; Israels to Reid, 6 Nov. 1876; and obituaries of Lehman Israels in the *NYT*, 21 Feb. 1896, and *N.Y. Trib.*, 22 Feb. 1896.

4. See Charles H. Israels to Edwin Blake Seely, 7 Feb. 1882 (Rice Family Papers).

Israels's chief buildings include: Apartments—the Devon (70–72 West 55th St.); Warrington (161 Madison Ave.); Arlington (18 West 25th St.); Walton (104 West 70th St.); Howard (66 West 46th St.); Holland (351 West 42nd St.); "bachelor apartments" at 22 East 31st St.; and "apartment house" at 78 Irving Place. Residences—Edward Thaw (4 East 89th St.); row houses built for developer Bernard S. Levy, 306–14 West 81st St. (1892) and 307–19 West 80th St. (1894), all part of a historic district now (see *NYT*, 13 Mar. 1985). Offices—Lord & Taylor; Silo Building; Physicians Building (East 41st St.). Civic—Hall of Records façade (Manhattan); Hahneman Monument (Scott Circle, Washington, D.C.). Religious—Temple Emanu-El, Yonkers (now, Ohab Zedek). Theater—the Hudson Theatre. See *American Architect and Building News*, 87 (3 June 1905), 180, on the Devon; 88 (8 July 1905), 16, on the Howard; 89 (28 April 1906), plate 1583, on the Hahneman Monument; and 87 (22 April 1905), 132, on the Thaw residence; *Architectural Record*, 19 (April, 1906), 308, on the Walton, and 20 (July, 1906), 7 on the Devon.

5. Charles published three articles in *Architectural Record:* "New York Apartment Houses," 11 (July, 1901), 477–508; "John Rogers, Sculptor," 16 (Nov., 1904), 483–87, and "Socialism and the Architect," 17 (April, 1905), 329–35.

6. Charles Israels, "The Busy Man's Bungalow," *Good Housekeeping*, 48 (April, 1909), 486–88; *Charities*, 19 (19 Oct. 1907), 928, and (16 Nov. 1907), 1078–81; *Biographical Directory of the State of New York*, 1900; *NYT* obit., 14 Nov. 1911; death notice, *Quarterly Bulletin of the American Institute of Architects*, 12/4 (June, 1912), 298–99; and *American Jewish Year Book* (1904), 123.

7. Goodale to "James's children," n.d. (post July 1941).

8. Charles H. Israels to *AH*, 83 (8 May 1908), 15. Belle Israels's role in the Ethical Culture Society's Women's Conference was mentioned in Charity Organization Society's *25th Annual Report* (1907), 246. Her presidency of a chapter of the Society for the Study of Child Nature is mentioned in a Madison House *Newsletter*, II/7 (May, 1905), Madison House Papers (Micro. ed.).

9. See interview of Belle Moskowitz by Denneen Hanlon, *N.Y. World*, 24 May 1925. See Allen F. Davis, *Spearheads for Reform*, 33–34, on female settlement workers: they averaged three years in residence, and those who left did so for marriage. Mary K. Simkhovitch, founder of Greenwich House and a career settlement worker who married, sent her two children to a farm in New Jersey, a decision they resented (author interview with Helena Simkhovitch, 1977); see Mary's autobiography, *Neighborhood* (1938). *Statistics of Women At Work*, based on the schedules of the 12th U.S. census (Dept. of Commerce and Labor, Bur. of the Census, Washington, D.C., 1907), Table 21, shows that 92.4% of native white females "in professional service" were single.

10. *The American Jewess*, I/1 (April, 1895), 40. On the so-called "domestic" or "social feminists," see William L. O'Neill, *Everyone Was Brave. The Rise and Fall of Feminism in America* (1969), ch. 3; on clubs and the "new woman," Karen Blair, *The Clubwoman as Feminist: True Womanhood Redefined, 1868–1914* (1980).

11. See Ellen Sue Levi Elwell, "The Founding and Early Programs of the National Council of Jewish Women: Study and Practice as Jewish Women's Religious Expressions," Ed.D. diss., Indiana Univ., 1982, chs. I–III; Deborah Grand Golomb, "The 1893 Congress of Jewish Women: Evolution or Revolution in American Jewish Women's History?" *American Jewish History*, 70/1 (Sept., 1980), 52–67; Hannah G. Solomon, *A Sheaf of Leaves* (1911) and *Fabric of My Life* (1946); Rebekah Kohut, *My Portion* (1925); Reena Sigman Friedman, "Their Sisters' Keepers: The Response of the National Council of Jewish Women to East European Jewish Women," M.A. thesis, Columbia Univ., 1978. Sue Elwell called my attention to this work. The Council did not keep its early papers, but published its triennial proceedings and plans of work; local sections also published yearbooks, and some have preserved their manuscript minute books. On women's organizations see Eleanor Flexner, *Century of Struggle: The Woman's Rights Movement in the United States* (1975), ch. XIII; on clubs, Blair, *The Clubwoman as Feminist.*

12. On the charity organization movement, see Bremner, *From the Depths*, 51 ff. On the Council's work, see its *Program of Work*, 1894–1895 (pamphlet); *Proceedings of the Triennial Conventions*, 1896: 64, 159 (Corr. Secy. report); 226–27 (philanthropy rep.); 387–88 (sabbath controversy). The sabbath controversy can be followed in the *Proceedings* for 1900, 139–49. The 1905 proceedings reveal an overwhelming stress on service. See also Elwell, ch. IV. The debate over the shift to philanthropy can be followed in many issues of *The American Hebrew* between 1911 and 1914.

13. Rebekah Kohut called Minnie Louis "sweet singer in Israel and southern aristocrat"; Louis founded the Hebrew Technical School for Girls (*My Portion*, 200–201). For Sulzberger's remarks, see National Council of Jewish Women, *Proceedings*, 1900, 58–60. N.Y. Section programs are described in its *Yearbook*, 1905–06, 27–41; and 1907–08, 51–54, and in ms minutes of board meetings. See also *Proceedings* for 1902, 14, 98, 200–201. Recreation rooms are described in *AH*, 65/9 (30 June 1899), 261; 66/14 (9 Feb. 1900), 438.

14. Marion L. Misch, "The Americanization of the Immigrant Girl," *The American Citizen,* I/1 (July, 1912), 30–31, credits Sadie American with developing the first international contacts to protect immigrant Jewish women. On American's career, see her extremely self-serving autobiog. (10 pp. typescript), National Association of Jewish Social Workers (I-88, Box 1), American Jewish Historical Society; [Mary S.] Logan, *The Part Taken by Women in American History* (Repr., 1972), 642–46; Council *Proceedings,* esp. for 1900, 141, 145; *AH,* 70/25 (9 May 1902), 753.

15. See Edward J. Bristow, *Prostitution and Prejudice. The Jewish Fight Against White Slavery 1870–1939* (1983); Mark Thomas Connelly, *The Response to Prostitution in the Progressive Era* (1980), esp. ch. 6 and bibliography; Ruth Rosen, *The Lost Sisterhood: Prostitution in America, 1900–1918* (1982), ch. 7.

16. Belle Lindner Israels, "Report of the Philanthropy Committee," Council of Jewish Women, New York Section, *Yearbook,* 1905–1906, 27–41.

17. On Lakeview, see *Charities and The Commons,* XVIII (24 Aug. 1907), 616; Section Yearbooks, 1906–1907 (29–33); 1907–1908 (17, 50–54); 1909–1910 (59); 1911–1912 (19–22). See also *AH,* 83/25 (23 Oct. 1908), 614, and 88/24 (14 Apr. 1911), 707–09.

18. See N.Y. Section, *Yearbook,* 1907–1908, 34. On the controversies Sadie American aroused in the National Council, see *Proceedings,* 1905 (241–65), 1908 (349 ff.); Board Minutes, N.Y. Section, 30 Dec. 1908; and papers of the National Council (Box 103), Library of Congress.

19. See the *N.Y. Evening Post,* 19, 26 Apr., 7 June 1907, on the Davis bill; on the conference, see Belle Israels to Louis Marshall, 2 Feb. 1908, Box 23, Louis Marshall Papers; and on her views in 1910, Gen. Corr. (1910), Committee of Fourteen Papers.

20. See the Section Minute Books, Board of Directors meetings for 28 Mar., 25 Apr., 2 May 1904; 27 Feb., 24 Apr., 8 May, 15, 19 Nov. 1905; 20 Mar., 23 Apr. 1906; 28 Oct. 1907; 27 Apr., 8 May, 9 Nov., 1908; 11 Jan., 8 Feb., 20 Apr., 11 Oct., 1909. Also, the *Yearbook,* 1905–1906 (41); 1906–1907 (31); 1907–1908 (34).

21. Belle Lindner Israels, "Jewish Women as Settlement Workers," *The Hebrew Standard,* 50/11 (5 Apr. 1907), 9. See Ruth Bordin, *Woman and Temperance. The Quest for Power and Liberty, 1873–1900* (1981), chs. 1–4. On the rhetoric of motherhood in women's reform movements, see Mary Ryan, *Womanhood in America, From Colonial Times to the Present* (1975), 225–26.

22. On the methods of progressive reformers, see Allen Davis, *Spearheads,* 96–98, and *passim,* for examples of "fact-gathering" among social reformers. See also *Charities* editor Edward Devine in *Charities and The Commons,* 19 (27 Oct. 1907), 867.

23. New York State Conference of Charities and Corrections, *Proceedings,* 1900 (2–5), 1906 (284–97; the quotation is on p. 289). On the national conferences, see Thomas L. Haskell, *The Emergence of Professional Social Science* (1977).

24. *Proceedings,* 1907, 271–74.

25. *Charities,* 19 (23 Nov. 1907), 1093.

26. *Proceedings,* 1908, iii.

27. "The New View," pamphlet published by *Charities,* Apr. 1907. On the history of this publication, see Clarke A. Chambers, *Paul U. Kellogg and The Survey: Voices for Social Welfare and Social Justice* (1971). Belle's financial arrange-

ments with the journal appear in a file, "Charities Publication, 1899–1907," Community Service Archives; *Survey* papers at the Social Welfare History Archives in Minnesota hold no further clues.

28. "Widowed Mothers," *Charities,* 22 (4 Sept. 1909), 741. See also her "A Contribution to Play," 22 (31 July 1909), 598, and "For Summer Reading," 22 (7 Aug. 1909), 622.

29. "Crime Among the Jews," *Charities,* 20 (19 Sept. 1908), 701–02. On the Bingham incident, see Arthur Goren, *New York Jews and the Quest for Community* (1970), 25–37. The article on *Landsmanschaften* was called "Imported Neighborhood Spirit," *Charities,* 18 (21 Sept. 1907), 720–21.

30. "Salvation Nell—A Lost Opportunity," *ibid.,* 21 (23 Jan. 1909), 705–06; "The Battle," 21 (27 Feb. 1909), 1023–24.

3: The Motherhood of the Commonwealth

1. Addams, "Some Reflections on the Failure of the Modern City to Provide Recreation for Young Girls," *Charities,* 21 (5 Dec. 1908), 365–68, and *The Spirit of Youth and the City Streets* (1909), which Israels called an "exquisite piece of work, full of beautiful meaning and spiritual essence of your own philosophy," "an inspiration" (BLI to JA, 14 Jan. 1910, Jane Addams Papers). See also Michael M. Davis, Jr., *The Exploitation of Pleasure: A Study of Commercial Recreations in New York City* (1911), which opposes "laissez-faire" policies on recreation.

2. Elite and popular cultures have often clashed, especially over dancing. On dancing, see Peter Burke, *Popular Culture in Early Modern Europe* (1978), ch. 8, p. 212; on music halls, Gareth Stedman Jones, "Working-Class Culture and Working-Class Politics in London, 1870–1900; Notes on the Remaking of a Working Class," *J. of Social Hist.,* 7 (Summer 1974), 460–508; on unsafe theaters and lewd films in America's fledgling moving picture industry, Robert Sklar, *Movie-Made America, A Social History of American Movies* (1975), 16–32; on Progressive Era recreation programs designed to wean workers from commercial establishments, Francis G. Couvares, "The Triumph of Commerce: Class Culture and Mass Culture in Pittsburgh," in Michael H. Frisch and Daniel J. Walkowitz, eds., *Working-Class America* (1983), 123–52.

3. On Progressive Era anti-prostitution, see Mark Connelly, *The Response to Prostitution* (1980); Ruth Rosen, *Lost Sisterhood* (1982); and on an earlier movement, David Pivar, *Purity Crusade: Sexual Morality and Social Control, 1868–1900* (1973). On temperance, see Ruth Bordin, *Woman and Temperance* (1981); James H. Timberlake, *Prohibition and the Progressive Movement, 1900–1920* (1963); and Norman H. Clarke, *Deliver Us From Evil: An Interpretation of American Prohibition* (1976). On recreation reform, see Richard F. Knapp and Charles F. Hartsoe, *Play for America. The National Recreation Association 1906–1965* (1979), 3–49; Lawrence A. Finfer, "Leisure as Social Work in the Urban Community: The Progressive Recreation Movement, 1890–1920" (Ph.D. diss., Michigan State Univ., 1974); and Dominick Cavallo, *Muscles and Morals: Organized Playgrounds and Urban Reform, 1880–1920* (1981). Both Finfer and Cavallo interpret recreation reform as attempts to curb American spontaneity and turn play over to state control, views I find exaggerated. See my "Recreation as Reform in the Progressive Era," *Hist. of Educ. Qtly.* (Summer 1984), 223–28.

4. On the popularity of dance halls, see Rheta Childe Dorr, *What Eight Million Women Want* (1910), chs. 7–8. On the link between dance halls, liquor, and

prostitution, see Henry Moskowitz's letter to the *NYT,* 7 Apr. 1909, and John Dillon, *From Dance Hall to White Slavery, Ten Dance Hall Tragedies* (1912). Also, Michael Davis, *Exploitation,* 3; George Kneeland, *Commercialized Prostitution in New York City* (1913), 68–70; Ruth S. True, *The Neglected Girl* (1914), 16–17, 68 ff.; Lillian Wald, *The House on Henry Street,* 174, 213–14, 225; and Mary R. Beard, *Woman's Work in Municipalities* (1915), 139–42. For a modern assessment, see Kathy Peiss, *Cheap Amusements.*

5. Either Israels did not save the papers of the Committee on Amusements, or they were discarded at her death. My information on its history comes from books cited in the note above; articles in *Charities* (esp. 27 Feb. 1909, 1019) and the *NYT,* 1909–1912 (esp. a *Sunday Magazine* feature on Israels, 10 Nov. 1912); the papers of the Committee of Fourteen, a citizen anti-prostitution group; Mayor William Gaynor; and Lillian Wald (both NYPL and Columbia Univ. holdings).

6. See studies by Pivar, Connelly, and Rosen, *op. cit.,* note 3. For a fuller discussion of the anti-prostitution debate in the early 1900s, see chap 4.

7. The 1896 Raines Law exempted saloons with hotels attached from the Sunday drinking ban. To qualify, saloons bought adjoining apartments that soon became brothels. In 1905, a Committee of Fourteen formed to repeal the Raines Law exemption. See its annual report for 1912; John P. Peters, "The Story of the Committee of Fourteen of New York," *J. of Social Hygiene,* 4/3 (July 1918), 347–88; and W. C. Waterman, *Prostitution and its Repression in New York City, 1900–1931* (1932).

8. Schoenfeld's report was summarized in Belle Israels, "The Way of the Girl," *The Survey,* 22 (3 July 1909), 486–96, and in the Committee of Fourteen, *The Social Evil in New York City* (1910), xxvii, 53–59, 65. On Schoenfeld, see Logan, *The Part Taken by Women* (1912), 648–49. Robb O. Bartholomew, dance hall inspector in Cleveland, Ohio, after 1911, reported similar conditions in his town (*Dance Hall Report,* 24 Aug. 1912).

9. The phrase comes from her speech before the Yonkers City Council, reported in the *Yonkers Herald,* 11 Feb. 1913.

10. Committee members eventually included William Dean Embree (Assistant District Attorney for New York), Edith Rich Isaacs (journalist, wife of Lewis Isaacs, Stanley Isaacs's brother and law partner), W. Frank Persons (Director, New York Charity Organization Society), Mrs. William H. Jackson, Carlotta Nicoll, Mrs. Josephine Redding, Mrs. Frederick R. Swift (her husband was a lawyer), Elizabeth Williams, and Mrs. Alfred Martin. Miss Parsons may have been related to Herbert Parsons, a prominent Republican; both gave generously to philanthropic projects. Gertrude Robinson-Smith founded a Vacation Savings Stamp Fund to provide vacation possibilities for Manhattan working girls, an organization that later became the American Women's Association; she also founded the Berkshire Music Festival at Tanglewood in Lenox, Mass.

11. See note 8, above. On Lewis Hine, see Judith Mara Gutman, *Lewis W. Hine and the American Social Conscience* (1967).

12. On "treating," see Kathy Peiss, *Cheap Amusements,* 108 ff.

13. Vacation homes for poor working girls evoked mixed reactions. See Rose Cohen, *Out of the Shadow* (1918); Elizabeth Hasanovitz, *One of Them. Chapters from a Passionate Autobiography* (1918), 228–229. See also Ruth Rosen and Sue Davidson, eds., *The Maimie Papers* (1977), 141.

14. See, e.g., Jane Addams, note 1, above. Joseph Lee at the 1912 National

Conference of Charities and Corrections argued that sexual attraction, the "great budding force of nature," should be bent to "its true task of producing strength and beauty, instead of permitting it to go to waste or worse" (*Proceedings,* 126).

15. See her speeches: "The Dance Problem," 9 June 1910; "Social Dancing," 12 May 1911; "Regulation of Dance Halls," 7 June 1912, all in *Playground,* the organ of the Playground and Recreation Assoc. It also published her "Recreation for Money" (1912–13), and reported her speech to the Recreation Institute for the New England States (Brookline, Mass., 16 Feb. 1912). To the Nat. Conf. on Charities and Corrections, she spoke on "Recreation in Rural Communities," 11 June 1911; and "The Dance Problem," 14 June 1912, both in the conference proceedings. "The Dance Hall and the Amusement Resorts" appeared in *Transactions* of the American Society of Sanitary and Moral Prophylaxis, 3 (1909–10), 46–50. "Regulation of Public Amusements," delivered to the Academy of Political Science, 19 Apr. 1912, appeared in its proceedings. Israels also spoke to the Philadelphia Consumers League, the Council of Jewish Women, the Society for Ethical Culture, the Association of Neighborhood Workers, and the Charity Organization Society (Israels to Whitin, 4 Nov. 1909, "Gen. Corr. 1909, Committee of Fourteen Papers; *N.Y. Sun,* 20 Apr. 1919). *Leslie's Weekly* published her "Diverting a Pastime," 27 July 1911, 94, 100. For Lee's remark, see National Conference of Charities and Corrections, *Proceedings,* 1912, 547.

16. *NYT,* 13 Jan. 1909. See letters from settlement workers, *NYT,* 5 Feb., 7 Apr., 1909. On separating dancing and liquor, see Bartholomew on Cleveland (note 8, above). On Graubard's bill, see *NYT,* 13, 20, 22 Apr. 1909; *N.Y. Sun,* 6 Oct. 1909. On Duryea's lawsuit, see *NYT,* 25 June, 28 Sept., 6 Oct., 1909; 23 Feb. 1910. Israels later pacified Duryea and won his cooperation on reforming dancing styles.

17. *The Survey,* 24 (2 Apr. 1910), 11; *NYT,* 22 Apr., 28 May, 7 June, 1910.

18. *NYT Sunday Mag.,* 10 Nov. 1912.

19. The Young Women's Christian Association, Charity Organization Society, Russell Sage Foundation, Downtown Ethical Society, Women's Trade Union League, Federation of Churches, New York Probation Association, Council of Jewish Women, and National Civic Federation all sent observers. Prominent citizens included Oswald G. Villard, Ruth Standish Baldwin, Walter Laidlaw, Percy Straus, and E. W. Bloomingdale. See *NYT,* 17 Feb. 1909; and *Charities,* 21 (27 Feb. 1909), 1018–19.

20. *NYT,* 3, 6 Feb. 1910.

21. *NYT,* 3 Feb., 26 Mar., 1, 11 Oct., 10 Nov. 1912. Settlements, such as Greenwich House, also experimented with dancing pavilions. See *The Survey,* 24 (26 Aug. 1911), 752–53.

22. See Tom Fletcher, *100 Years of the Negro in Show Business* (New York: Burdge, 1954), ch. 13. Also, Lewis Erenberg, *Steppin' Out: New York Nightlife and the Transformation of American Culture, 1890–1930* (1981); Frederick Lewis Allen, "When America Learned to Dance," *Scribner's,* 102 (Sept. 1937), 11–12; Marshall and Jean Stearns, *The Jazz Dance: The Story of American Vernacular Dance* (New York: Macmillan, 1968), 95–97; and A. H. Franks, *Social Dance—A Short History* (London: Routledge, and K. Paul, 1963), 178–82.

23. For articles on the new dances, see *NYT,* 15, 19 Jan., 19 Feb., 20 Sept., 15 Oct., 22 Dec. 1911; 8, 13, 24, 25 Jan., 2, 4, 7 Feb. 1912. The debate can also be

followed in smaller city newspapers, such as the *Yonkers Herald,* 8 Feb., 13 Oct., 1913. Mr. Dooley's complaint is in the *NYT,* 1 Mar. 1913.

24. Circular from the Committee on Amusements and Vacation Resources of Working Girls, "Parks & Playgrounds, Corr.," Lillian Wald Papers, Columbia Univ.

25. *NYT* and *N. Y. Sun,* 1 Jan. 1912.

26. *NYT,* 11 Oct., 10 Nov., 1912.

27. *NYT,* 5 Jan. 1914. See Erenberg and Allen (note 22, above) on the evolution of ballroom dancing. Also, Needham, "Virtues of the Trot," *Collier's,* 51 (2 Aug. 1913), 19; editorial, *Independent,* 77/51 (12 Jan. 1914); Brill, remarks made to the Vidonian Club, printed in the *N. Y. Medical Journal,* 99 (25 Apr. 1914).

28. In "Diverting a Pastime," Israels claimed that, by 1911, 133 cities had instituted dance hall licensing. See Louise de Koven Bowen's discussion of reform in Chicago and Denver in *The Survey,* 26 (3 June 1911), 383–87; 28 (28 Sept. 1912), 788–89, and Julia Schoenfeld, "The Regulation of Dance Halls," *Playground,* 6 (1912–13), 340–42, and "Commercial Recreation Legislation," *ibid.,* 7 (1913–14), 461–81.

29. Andrew Linn Bostick, "The Regulation of Public Dance Halls. Municipal Legislation," *St. Louis Public Library Monthly Bulletin* (July 1914), covers cities of over 300,000.

30. Committee of Fourteen, *Annual Report,* 1914, 11; Whitin to Israels, n.d. ("Gen. Corr., 1913–15," Comm. of Fourteen Papers); Israels to Gaynor, 30 July 1913, "Gen. Corr.," Gaynor Papers. Other halls forced to close included the Bal Tabarin, Kid McCoy's, and the Eldorado Cafe.

31. Gaynor to Israels, 22 Mar. 1910. Gaynor's acerbic comments about "pious reformers" are in *Some of Mayor Gaynor's Letters and Speeches* (1913). Biographies include Mortimer Smith, *William Jay Gaynor* (1951) and Lately Thomas, *The Mayor Who Mastered New York* (1969).

32. Israels to Gaynor, 30 June 1913.

33. Gaynor to Israels, 21 June 1912; Israels to Gaynor, 31 Mar. 1913; *NYT,* 10 Nov. 1912.

34. The *Yonkers Herald* covered the controversy almost on a daily basis between 28 Jan.–29 Apr., 1913.

35. See Frederick Rex, "Municipal Dance Halls," *National Municipal Review,* 4 (July 1915), 418; "Should Dancing be Municipally Encouraged?" *Literary Digest,* 45 (9 Nov. 1912), 847–48; and Bartholomew's plea for public dance facilities in Cleveland where licensing forced small neighborhood establishments out of business (*Report,* 16).

36. Bowen, "The Public Dance Halls of Chicago," Juvenile Protection Assoc., rev. ed., 1917.

37. Ella Gardner, "Public Dance Halls. Their Regulation and Place in the Recreation of Adolescents," U.S. Dept. of Labor, Children's Bureau Publication, No. 189 (1929), 9, 36–50. For other views on dance hall reform in the 1920s, see Maria Ward Lambin, "Report of the Advisory Dance Hall Committee of the Women's City Club and the City Recreation Committee," 1924, a study Belle Moskowitz inspired; Paul G. Cressy, *The Taxi-Dance Hall: A Sociological Study in Commercialized Recreation and City Life* (1932).

38. Cavallo, *Muscles and Morals,* 10, shows that the ideal "team player" of recreation theorists was a male adolescent. Most views of women's sports stressed

non-competitive ethics and the protection of reproductive organs (see, e.g., Katherine D. Blake, "General Health of Girls in Relation to Athletics," *American Physical Education Review,* 11 [1906], 171–74).

39. Mrs. Charles Henry Israels, "The Dance Problem," *Proceedings,* National Conference of Charities and Corrections (June 14, 1912), 145.

4: Beyond the Committee Stage

1. Grace Goodale to Miriam Gabo, 13 Jan. 1933; 19 June 1934; 13 Jan. 1944.

2. The house was at 21 (now 43) Edgecliff Terrace. For a description of the service on the Putnam railroad, see the *Yonkers Herald,* 8 Mar. 1913.

3. The prayer book was from *Hours of Devotion, A Book of Prayers and Meditations for the Use of the Daughters of Israel,* trans. from the German by M. Mayer (New York, 1866), 21 (copy owned by Miriam Israels Gabo, on whose memories this section on family life is based); see also Denneen Hanlon interview of Belle Moskowitz, *N.Y. World,* 24 May 1925, the essential facts of which Miriam has confirmed.

4. Charles Israels's will of 6 Aug. 1908 named William Poey de Luna as an executor; a codicil of 3 Feb. 1909 removed him, making it possible to date at least the diagnosis of Josie's illness. Grace Goodale's letter to Miriam Gabo, 13 Sept. 1944, corroborates the Josie story. On the status of health care for syphilitics, see James F. Gardner, Jr., "Microbes and Morality: The Social Hygiene Crusade in New York City, 1892–1917," Ph.D. diss., Indiana Univ., 1974, 177. Salvarsan, an imperfect cure, was not introduced until 1910–11, too late for Josie.

5. See Nancy Schrom Dye, "History of Childbirth in America," *Signs: J. of Women in Culture and Society,* 6/11 (1980), 97–108; "Donation of Articles," *Yearbook 1912–13,* Council of Jewish Women–New York Section.

6. Charles to Belle, 18 Aug. 1911 (author's collection); emphasis is Charles's.

7. Belle to Charles, 29 July 1907 (Moskowitz Papers); emphasis hers.

8. Goodale to Gabo, then Miriam Franklin, 13 Jan. 1933; an undated carbon of a letter written to "my James's dear children," sometime after July 1941; and to Miriam, 21 Apr. 1939. The Kipling quotes come from *The Story of the Gadsbys* (1895) and *The Light That Failed* (1891). Charles's death certificate confirms the cause of death. He was 45 years old. See *Yonkers Herald,* 15 Nov. 1911. The issues of 12 or 13 Nov. are missing from the microfilm. Charles's service on the Art Commission was covered on 9–10 Aug. 1910; 23 Feb. 1911.

9. Charles's will, filed in Westchester County Surrogate's Court; recollections of the funeral come from Goodale to Gabo, 13 Jan. 1933.

10. "A Certain Person," *New Yorker,* 9 Oct. 1926.

11. Perkins, Oral History, II, 522–23 (Columbia Univ.).

12. The story of Carlos's feeling neglected was told to me by Jean Ellis in 1976.

13. In a letter to Henry Moskowitz (3 Jan. 1933, Moskowitz Papers), William S. Bennet identified himself as part of a little group of friends who helped Belle plan a career in social welfare after Charles died; Alice Long Goldsmith to author, 7 Nov. 1976.

14. The 1916 edition of "The Child" makes no mention of midwives at all. The controversy may be followed in the Lee K. Frankel Papers.

15. "The Child," p. 27.

16. Metropolitan Life archives shed no light on editorial policy regarding the pamphlet. Frankel's file there indicates that the pamphlet was the second in a series designed for "housewives of the industrial families and for the children at school." It was "well received and established the success of the whole series." See Louis I. Dublin, *A Family of Thirty Million. The Story of the Metropolitan Life Insurance Company* (1943), 428–29; and Marquis James, *The Metropolitan Life, A Study in Business Growth* (1947), 188, who called the pamphlet the company's most widely read. In its various revisions, 31,872,000 copies had been distributed by 1945.

17. See Knapp and Hartsoe, *Play for America* (1979), 3–49, 56. Some of the organization's early papers are at the Social Welfare History Archives (Univ. of Minnesota), but contain no documents pertaining to Israels.

18. See the summary of the arguments against regulation in the Committee of Fifteen's *Social Evil* (2nd ed., 1912), 60–62, 76–78, 110–13.

19. See Rockefeller to Felix Warburg, 19 Aug. 1910, and Rockefeller to Allan Robinson, 23 Aug. 1912, "John D. Rockefeller, Jr., Boards," boxes 8 and 9, Rockefeller Archives.

20. Comm. of Fifteen, *The Social Evil* (1912), 218 (describing the Comm. of Fourteen).

21. See John C. Burnham, "The Progressive Era Evolution in American Attitudes toward Sex," *J. of American History*, 59/4 (Mar. 1973), 885–908.

22. See, e.g., the Committee's *Annual Report, October 1912*, 28, which used the terms "improper dancing" and "disorderly" without defining them. Mayor Gaynor thought the Committee's chief purpose was to "exploit themselves in the newspapers" (Gaynor to Rockefeller, 4 Jan. 1911, Box 8, Rockefeller Archives). For a discussion of prostitutes' lives, see Rosen (ed.), *The Maimie Papers* (1977).

23. Comm. of Fourteen, *The Social Evil* (1910), "Preface."

24. On the Raines Law, see Richard L. McCormick, *From Realignment to Reform: Political Change in New York State, 1893–1910* (1981), 94–98. On the origins of the Committee of Fifteen, see below, chapter 6, and Jeremy P. Felt, "Vice Reform as a Political Technique: The Committee of Fifteen in New York, 1900–1901," *N.Y. History*, 54 (Jan. 1973), 24–51.

25. Comm. of Fourteen, *Annual Report, January 1912*, "Summary of Results"; *October 1912*, 1–5; and Rev. John P. Peters, "The Story of the Committee of Fourteen of New York," *J. of Social Hygiene*, 4/3 (July 1918), 367.

26. Israels to Frederick Whitin, 23 Nov. 1908; Michael Davis, Jr. to Whitin, 27 Apr., 27 July 1909; Israels to Whitin, 4 Nov. 1909; Whitin to Henry Moskowitz, 25 Mar. 1912; Israels to Whitin, 26 Sept. 1912, 21 July 1913 (Comm. of Fourteen Papers).

27. Comm. of Fourteen, *Annual Report, October 1912*, 29–31; *Annual Report for 1921;* and Peters, "The Story of the Committee of Fourteen," *passim*.

28. See exchanges between Israels and Whitin, June–July 1913. Henry Moskowitz, another link to the Jewish community, also joined the committee but later, over-burdened with the "social worker's disease of Committeeitis," resigned (Moskowitz to Whitin, 4 Mar. 1912).

29. Andy Logan, *Against the Evidence: The Becker-Rosenthal Affair* (1970); Smith, *William Jay Gaynor*, 126–28; and Herbert Asbury, *The Gangs of New York* (1927), 340–43. None of these works are documented. The Rosenthal affair is best followed in the newspapers.

30. Examples were the victory of a Tammany candidate in 1897, and the defeat of Mayor Low in 1901. Mayor Mitchel would also be a one-term reformer. See James F. Richardson, *The New York Police, Colonial Times to 1901* (1970).

31. [Jerome D.] Greene to John D. Rockefeller, Jr., Memo, 20 Aug. 1912 (Rockefeller Archives, Box 9). On the Bureau of Municipal Research, see Robert Caro, *The Power Broker. Robert Moses and the Fall of New York* (1974), 60–62, and Jane S. Dahlberg, *The New York Bureau of Municipal Research* (1966).

32. *NYT* and *N.Y. Eve. Post,* 15 Aug. 1912.

33. "Report of the Citizen's Committee Appointed at the Cooper Union Mass Meeting August 14, 1912," in New York (City), Board of Alderman, *Report of the Special Committee to Investigate the Police Department,* 10 June 1913 (Repr. 1971, Arno Pr.). There is an original copy in the Henry L. Stimson Papers, ser. IV, box 201, fol. 11.

34. The following signed the call for the meeting: Eugene H. Outerbridge, Jacob H. Schiff, Eugene A. Philbin, Henry Moskowitz, F. S. Tomlin, Raymond V. Ingersoll, Mrs. Charles H. Israels, George B. Agnew, all businessmen, lawyers or settlement workers. The other two women on the committee were Ruth Baldwin (Comm. of Fourteen), and Helen Harley Jenkins (philanthropist and benefactor of Columbia Univ. and Teachers Coll.). The Executive Committee consisted of Robert S. Binkerd, Joseph Cotton (ex-officio), Charles P. Howland, Belle Israels, Samuel A. Lewisohn, Henry Moskowitz, Outerbridge, and Allan Robinson.

35. See Israels-Rockefeller exchanges, 4–17 Dec. 1912, Rockefeller Archives, Boards, Box 9.

36. "Report of the Citizen's Committee," 6–16. An earlier draft called the Board of Social Welfare the "Commission of Public Safety" ("Confidential draft report of the Executive Committee," Comm. of Fourteen Papers). The debate can be followed in the newspapers, esp. the *N.Y. Eve. Post,* which ran a series on it from 21–30 Dec. 1912, and the *N.Y. Eve. Mail,* 20 Feb.–7 Mar. 1913.

37. Buckner to Jerome D. Greene, 25 Mar. 1913; Joseph P. Cotton to Allan Robinson, 22 Oct. 1912 (Rockefeller Papers, Boards, Box 9). Felix Frankfurter greatly respected Cotton, "one of the most trenchant, original of minds," "capacious-minded," as "effective a man of law as anyone I know of in my time," a "powerful person, wise as well as shrewd" (Harlan B. Phillips, ed., *Felix Frankfurter Reminisces* [1960], 108, 218). Cotton became undersecretary of state under Stimson during the Hoover administration.

38. See the Board of Alderman report (note 34, above), 138; on the state law, Clement Driscoll, "The New York Police Situation," *Nat. Munic. Rev.,* II (Apr. 1913), 279–83, (July 1913), 401–07. Driscoll supervised the study of the police done by the Bureau of Municipal Research for the Aldermanic investigation.

39. *Yonkers Herald,* 12 Aug. 1912. The *Yonkers Statesman* (14 Aug.) and *Herald* (15 Aug.) featured Israels as the first Yonkers woman to achieve political prominence.

40. Speech, Conference of the Women's Organization for National Prohibition Reform, 12 Apr. 1932, Moskowitz Papers.

41. *Statesman,* 4 Sept. 1912. There was one other woman delegate from Yonkers, Mrs. Leo H. Baekeland, a suffragist and wife of a prominent doctor.

42. See Herbert Hillel Rosenthal, "The Progressive Movement in New York State," Ph.D. diss., Harvard Univ., 1955, 402 ff. Also, Thomas J. Kerr, IV, "New

York Factory Investigating Commission and the Progressives," Ph.D. diss., Syracuse Univ., 1965, ch. III.

43. *Herald,* 6 Sept. 1912; *N.Y. Eve. Post,* 4, 6 Sept. 1912. Henry Moskowitz withdrew either because someone else was more favored, or because he decided to run for Congress instead.

44. See Rosenthal, "The Progressive Movement"; Richard S. Skolnik, "The Crystallization of Reform in New York City, 1890–1917," Ph.D. diss., Yale Univ., 1964, and his "Civic Group Progressivism in New York City," *N.Y. History,* 51 (July 1970), 411–39; Augustus Cerillo, "The Reform of Municipal Government in New York City: from Seth Low to John Purroy Mitchel," *N.Y. Hist. Soc. Qtly.,* 57 (Jan. 1973), 51–71; and Robert Crosby Eager, "Governing New York State: Republicans and Reform, 1894–1900," Ph.D. diss., Stanford Univ., 1977.

45. Rosenthal, "The Progressive Movement," 155 ff.

46. The Straus nomination can be followed in New York newspapers for September 1912; in Naomi W. Cohen, *A Dual Heritage: The Public Career of Oscar S. Straus* (1969); in Box 41, Oscar Straus Papers. The *N.Y. Press* (7 Sept. 1912) reported that Moskowitz had originally supported Hotchkiss because of his closeness to ex-governor Hughes.

47. "The Buffalo Progressive," 12 Sept. 1912, Microfilm ed., Buffalo and Erie County Historical Society.

48. *Yonkers Herald,* 28 Oct. 1912.

49. See Cohen, *Straus,* 218 ff.; and Clarke A. Chambers, *Paul U. Kellogg and the Survey: Voices for Social Welfare and Social Justice* (Minneapolis, 1971), 47–48, writing on the impact of social workers on Progressive party platforms. See also *Annotated Edition of the Platform of the National Progressive Party of the State of New York,* Adopted by the State Convention, Syracuse, N. Y., 5 Sept. 1912, 3–53 (Wald Papers, NYPL).

50. Cohen, *Straus,* 53–55; William L. Ransom to Oscar Straus, 5 Nov. 1912, Straus Papers: the other three women were Mrs. Carlos Alden of Buffalo, Mrs. Charles M. Down of Jamestown, and Mary E. Dreier of Brooklyn.

5: Apostle of Industrial Peace

1. The most gripping account is Leon Stein, *The Triangle Fire* (1962).

2. For garment industry conditions, see Charles H. Winslow, *Industrial Court of the Cloak, Suit, and Skirt Industry of New York City* and *Conciliation, Arbitration, and Sanitation in the Dress and Waist Industry of New York City* (1914); *Final Report and Testimony submitted to Congress by the Commission on Industrial Relations created by the Act of August 23, 1912* (1916), I, 575 ff. and 587 ff., and II, 1027–1161; Louis Levine (Lorwin), *The Women's Garment Workers, A History of the International Ladies' Garment Workers' Union* (1924); Hyman Berman, "Era of the Protocol: A Chapter in the History of the International Ladies' Garment Workers' Union, 1910–1916," Ph.D. diss., Columbia Univ., 1956; Melvyn Dubofsky, *When Workers Organize. New York City in the Progressive Era* (1968), 40–101; Jesse Carpenter, *Competition and Collective Bargaining in the Needle Trades, 1910–1967* (1972). For women's treatment, see Baum, et al., *The Jewish Woman in America* (1976), ch. 5; Leslie Woodcock Tentler, *Wage-Earning Women. Industrial Work and Family Life in the United States, 1900–1930* (1979); Alice Kessler-Harris, *Out to Work. A History*

of Wage-Earning Women in the United States (1982). On the WTUL, see Nancy Schrom Dye, *As Equals and As Sisters: Feminism, the Labor Movement, and the Women's Trade Union League of New York* (1980).

3. See Winslow, *Conciliation,* 11–12, and Berman, ch. 1.

4. Berman, 83–85.

5. *Ibid.,* 112–153; Carpenter, ch. 2.

6. Winslow, *Industrial Court,* App. A., 56–58.

7. Berman, 154 ff., and Winslow, *Conciliation,* 13 ff.

8. *NYT,* 17 Jan. 1913; *Women's Wear Daily* [hereafter, *WWD*], 15, 16, 20 Jan. 1913; Berman 168–72; and Winslow, *Conciliation,* 38. The DWMA's house organ, *Dress and Waist Bulletin* (published from Feb.–April, 1913), predicted the association's membership would reach 500, 5/7ths of the entire trade. The fourth bulletin (15 Apr.) reported a membership of only 317. By early 1916, membership had dropped to 211 (see *NYT,* 10 Feb. 1916).

9. Abraham Baroff, "Splendid Results of a Bloodless Struggle. . . ," *The Ladies' Garment Worker,* IV, 3, 1–3; *N.Y. Eve. Post,* 22 Jan. 1913; *WWD,* 20–25 Jan. 1913; and Berman, 171–175.

10. See Julius Henry Cohen, *Law and Order in Industry: Five Years' Experience* (New York, 1916), and "The Revised Protocol in the Dress and Waist Industry," *Annals of the American Academy of Political and Social Science,* 69 (Whole No. 158, Jan. 1917), 183–96. See also Samuel Haber, *Efficiency and Uplift. Scientific Management in the Progressive Era, 1890–1920* (1964), a study of other "modernizing" industrial reformers.

11. Stein, *Triangle Fire;* Baum, et al., *Jewish Woman in Amer.,* 148–53. The final count was 147. The events following the fire received detailed coverage in the *NYT.*

12. Minutes, General Meeting, Council of Jewish Women—N.Y. Sec., 21 Dec. 1909.

13. Kerr, "The New York Factory Investigating Commission and the Progressives," 9–21. Belle Israels was named as an organizer of the Committee of Safety in *Challenging Years: The Autobiography of Stephen Wise* (1919), 64.

14. Winslow, *Conciliation,* 12–13. The union hired some women clerks, but none as an executive in the protocol system. No historian of the protocol points out the uniqueness of Israels's appointments. Levine (303) praises her "broad vision," then relegates her to a footnote; Berman (352) comments only that she became chief clerk and was a "staunch friend of industrial democracy" (453); Winslow (*Conciliation,* 77), writing in March 1914, refers to the chief clerks as "expert men." Only Julius Henry Cohen, who engineered the appointment, saw its significance. See his *They Builded Better Than They Knew* (1946), 246.

15. "Trade Unions From the Employer's Point of View," 19–20, in Papers of the Bureau of Vocational Information (Lecture 7, 23 Nov. 1915, part 2), Schlesinger Library. Nancy Cott found this unindexed transcript. See also a brief interview with Israels in *The Survey,* 30 (12 July 1913), 506.

16. "Trade Unions . . . ," 25–26, 31–32 on unions; 22–23, 29–30, and 33 on the association; and 42, where she justified her own role. Her friend Mary Dreier shared the platform, speaking about unions from the workers' viewpoint.

17. On Henry's resignation from the clerkship, see H. Moskowitz to Brandeis, 26 Nov. 1915; Brandeis to Moskowitz, 30 Nov. 1915, Brandeis Papers.

18. For statistics on grievance settlements, see Winslow, *Conciliation,* 48 ff., and 68. Transcripts of the ILGWU-DWMA board meetings are in the Research

Division, ILGWU headquarters in N.Y.C. The DWMA records have vanished, but the papers of Louis Brandeis and Paul Abelson provide much useful information. Neither the Meyer London nor Morris Hillquit Papers offered much that was useful. Cohen's papers have disappeared. Bartholomew's plaint is in his confidential memo to Brandeis, 18 Nov. 1913, Brandeis Papers.

19. Cohen to Felix Adler, 11 Aug. 1915 (copy in the Charles L. Bernheimer Papers, N.Y. Historical Society).

20. H. Moskowitz to Filene, 12 Apr. 1913, Brandeis Papers; Winslow (*Conciliation,* 65–68); Bartholomew's memo to Brandeis, cited in note 18, above. For the testimony of the union clerks, see *Transcript,* Arbitration Proceedings Between the DWMA and the ILGWU, 9 Nov. 1913, 310 and 328.

21. See H. Moskowitz to Brandeis, 10 July 1913, Brandeis papers.

22. Briefs filed by each side are summarized in Winslow, *Conciliation, 62–64.*

23. On her return to the DWMA, see H. Moskowitz to Brandeis, 12 Nov. 1913, and Bartholomew to Brandeis (note 18, above); in a letter to Brandeis (29 Nov.) Bartholomew described his negotiations with Cohen; *WWD,* 7 Jan. 1914, announced Israels's appointment, and she described her tasks in the grievance office in *Transcript,* Arb. Proc. DWMA-ILGWU, 17 May 1914, 47.

24. Cohen to Brandeis, 2 Mar. 1914, Brandeis Papers. Although upset by Cohen's high fees, Henry Moskowitz and Abelson supported his work.

25. Cohen to Brandeis, 27 Mar. 1914. Selig Perlman studied the dress and waist protocol for the federal government. For his view of the bad politics within both union and association, see his "Industrial Relations in the Dress and Waist Industry" (Records of the U.S. Commission on Industrial Relations, R. G. 174, National Archives). He interviewed Israels, who told him about the conference in Washington with Brandeis, but his notes on the interview have vanished. Perlman's "Preliminary Report," 14 Mar. 1914, slightly different from the final version, is at the Wisconsin Historical Society (U. S. Commission on Industrial Relations Papers, micr. ed.) or at the National Archives.

26. *Transcript,* Arb. Proc. DWMA-Local 10 (Cutters' Union)-ILGWU, 6 Nov. 1914, 99–172.

27. *Transcript,* Conferences on the Preferential Clause held at the direction of the Board of Arbitration, 11, 18, 19 Nov. 1914; *Minutes,* Wage-Scale Board, 11 Jan. 1915, 17; Ordway Tead, "Trade Unions and Efficiency" (*Amer. J. of Sociology,* 22/1 [July 1916], 34). For the conclusions of the November conferences, see B. Moskowitz to Brandeis, 24 Dec. 1914 (Brandeis Papers). Two years later, the timely exchange of membership lists was still an issue.

28. See Nahum I. Stone, *Wages and Regularity of Employment and Standarization of Piece Rates in the Dress and Waist Industry of New York City* (1914), 7; *Transcript,* Arb. Proc., 8 Nov. 1913, 111 ff. (testimony of Floersheimer and Margulies); Stone to H. Moskowitz, 3 June 1915, Brandeis Papers; *Minutes,* Wage-Scale Board, 11 Jan. 1915, 6–7. Cohen's attribution is in *They Builded* . . . , 247; articles in *Munsey's Magazine,* 49/4 (July, 1913), 527, and *WWD,* 24 Feb. 1914, attributed the idea to Bartholomew.

29. Levine, 304–309; *WWD,* 4 Oct. 1916; *The Survey,* 35 (11 Mar. 1916), 694; Mack to Brandeis, 29 Sept. 1916, Brandeis Papers; and R. G. Valentine, "Cooperating in Industrial Research," *The Survey,* 35 (9 Sept. 1916), 586–88. See also Milton J. Nadworny, *Scientific Management and the Unions, 1900–1932* (1955), esp. 98–101, 116; Robert F. Hoxie, *Scientific Management and Labor* (1915), 3; Haber, *Efficiency,* 33–34. Shortly after Valentine submitted his

proposals to the DWMA, he died of heart failure; see *WWD,* 14 Nov. 1916, and his friend Felix Frankfurter's eulogy, *Harvard Alumni Bulletin,* 19/12 (14 Dec. 1916).

30. *Minutes,* Conferences Between the DWMA and Local 10, 19 Sept., 5, 10 Oct. 1916.

31. *Transcript,* Arb. Proc. DWMA-ILGWU, 5, 7 Feb. 1916, 290–323.

32. *Transcript,* Conferences Between the DWMA and Local 10, 19 Nov. 1916, 1262, for Polakoff's comment; *Transcript,* Arb. Proc. DWMA-ILGWU, 13 May 1916, for Moskowitz's (19), Cohen's (37), and Hillquit's (47).

33. The union charged her with inattention to critical cases in *Transcript,* Arbitration Proceedings, 8 Apr. 1916, 1776 ff. See Mannheimer to Rabbi Judah Magnes, 26 Mar. 1915, Box 4, fol. 2, Abelson Papers, AJA, and Moskowitz's defense in the arbitration proceedings of 13 May 1916, pp. 15–22.

34. The best source for the 1916 dress and waist crisis is *Women's Wear Daily (WWD).* Hereafter in this chapter, all dates refer to *WWD* articles unless otherwise noted. See 3–4 Jan. 1916, and the Mannheimer memo cited above.

35. 19, 28 Jan.; 4 Feb. The DWMA also demanded an "experimental test shop," and the "schedule method of settling prices."

36. 7, 8 Feb.

37. 9 Feb.

38. 15, 16, 18, 24, 26, 28 Feb. The number of insurgents varied from 26 to 52, depending on the enumerator (see George Lewy's statement, 29 Feb.).

39. 1 Mar.; *WWD,* 4 May 1915, clipping in Brandeis's scrapbook, "Garment Trades," Brandeis Papers.

40. 3, 6, 10, 14 Mar.; letters of 9, 11 Mar. from the committee of ten waist manufacturers, to Floersheimer, Abelson Papers (Cornell Univ.).

41. 23 Mar. On the failure of the cloak protocol, see Berman, ch. 8, and Dubofsky, 92–97.

42. 28 Apr.

43. 1–3 May; 2 Sept.

44. 31 Aug., 5 Sept.

45. 21, 25–27 Sept.; 5 Oct.

46. 30 Oct., 2, 8, 9, 20, 21 Nov. At the Nov. 2 elections, nominees for seats included Max Blanck of the Triangle and Bruno Stern of Aero Waist Companies, both targets of numerous union complaints and at one time expelled from the DWMA. Stern was elected.

47. 23, 24 Nov.; *The Survey,* 37 (9 Dec. 1916), 288 and (23 Dec. 1916), 385; and *NYT,* 27 Nov. 1916.

48. See Mannheimer's memo, "Dress and Waist Industry Memorandum re. Joseph Rosenberg," 3 Nov. 1915, Box 4, fol. 2, Abelson Papers, AJA.

49. 29 Nov.

50. *NYT,* 30 Jan. 1917; 28 Dec. 1918; 12, 21 Jan. 1919; and 8 Apr. 1919. See also, "Statement issued by Harry A. Gordon, April 8, 1919," Abelson Papers (Cornell Univ.). Copies of letters of 9, 16 Jan. 1919 between Benjamin Schlesinger, ILGWU president, the DWMA, and Mayor John F. Hylan in the Samuel Untermyer Papers review the history and future prospects of the protocol.

51. 29 Nov. This statement revealed that her salary had grown to $5,500: the DWMA gave her $500 severance pay for "five-weeks vacation."

52. *Special Investigation,* Rosenberg Company, 8 Feb. 1915.
53. *Hearing,* The Bonwit Case, 25 July 1914. Cohen had wanted to call the protocol a "treaty of peace"; London had suggested "collective agreement"; but Louis Marshall, "with Talmudic shrewdness," offered "protocol" because "nobody would know what that meant." His prophecy, Cohen concluded wryly, "was only too true." (Cohen to Hillquit, 23 Nov. 1923, Hillquit Papers, Microfilm Ed.)
54. "Industrial Service" brochure, Moskowitz Papers.

6: From Social Reform to Politics

1. Henry's natal town was variously called Huesche, Husse (German), or Yassi (English). Personal information comes from Henry's will (Abelson Papers, AJA) and the recollections of Miriam Israels Gabo and Irma Commanday Bauman.
2. See Moskowitz to Wald, 27 May 1901, 25 Aug. 1902, 6 July and 28 Nov. 1903, Wald Papers, Columbia Univ.; *Madison House Newsletter,* reel Two, Hamilton-Madison House Papers.
3. See Holt, "Henry Moskowitz: A Useful Citizen," *Independent,* 77 (12 Jan. 1914), 67; "The Garment Trade and the Minimum Wage, An Interview with Dr. Henry Moskowitz," *Outlook,* 10 May 1916, 66; club memorabilia in Abelson's Papers, "University Settlement Club" folder; reminiscences of former club members, reel 22, University Settlement Society Papers. The Settlement's "Year Books" and Annual Reports also contain information on settlement club life. There is a brief biographical section on HM in Simkhovitch, *Neighborhood,* 74–77. In *When Harlem Was Jewish,* 89–90, Geoffrey Gurock describes the Harlem S.E.I. club's efforts to promote "an almost totally uncritical acceptance of the virtues of Americanization."
4. See Poole, *The Bridge* (1940), 68–73; Jack Meyer Stuart, "William English Walling: A Study in Politics and Ideas," Ph.D. diss., Columbia Univ., 1968, 35; Wald to Abelson, 31 Mar. 1937, Box 5, fol. 6, Abelson Papers, AJA.
5. "Founders Day Journal" and "Forty Years of Madison House," Abelson Papers, AJA; Dr. Henry Neumann, "80th Birthday Dinner," 20 Jan. 1962, Ethical Cult. Archives. See also Benny Kraut, *From Reform Judaism to Ethical Culture* (1979), 184–85. Kraut, who calls the settlement the "East Side Ethical Club," and locates it on Madison Avenue, claims the uptown society kept its association with the downtown group secret. But the society's name revealed its inspiration.
6. Those who identified Moskowitz as the impetus behind the Comm. of Fifteen included Henry Neumann, leader of the Brooklyn Ethical Society (interviewed by Howard Radest, 8 Aug. 1961, Ethical Cult. Archives); Lillian Wald, *The House on Henry Street,* 174; Woods and Kennedy, *The Settlement Horizon,* 261. Abelson's papers contain memorabilia of the Downtown Ethical Society's involvement in the Committee, e.g. a flyer inviting the public to bi-monthly discussions of district vice. See the Committee of Fifteen, *The Social Evil* (1902); and Jeremy P. Felt, "Vice Reform as a Political Technique: The Committee of Fifteen in New York, 1900–1901," *N.Y. History,* 54 (Jan. 1973), 24–51.
7. Some of this group's activities are described in the *AH,* 68/14 (22 Feb. 1901), 430–31; 68/24 (3 May 1901), 717; 70/11 (31 Jan. 1902), 345; 70/16 (7 Mar. 1902), 493. In its second year, "Boys'" was dropped from the name.

8. The *AH,* 71/12 (8 Aug. 1902), 330 and 71/13 (15 Aug. 1902), 358. Clippings and other items on the club are in Box 2, fol. 5, Abelson Papers (AJA). The petition campaign is described briefly in Simkhovitch, *Neighborhood,* 76.

9. Moskowitz to Wald, 20 May 1906, Wald Papers, Columbia Univ. Woods and Kennedy, *Handbook of Settlements* (1911), gives the following Madison Street locations for the Downtown Ethical: No. 232 (1898–1900), No. 310 (1900–1904), and No. 300 (1904–1910). See also, Society for Ethical Culture, *Newsletter,* I/1 (Nov. 1903); II/2 (Dec. 1904); III/5 (Mar. 1906); IV/3 (Jan. 1907); V/2 (Dec. 1907); and J. Salwyn Schapiro, "Henry Moskowitz: A Social Reformer in Politics," *Outlook,* 26 Oct. 1912, 446–47. Henry's diss., *Das Moralische Beurteilungsvermögen in der englischen Ethik von Hobbes bis John Stuart Mill. Eine historische-kritische Studie* (Erlangen, 1906), was a "History of English Ethics with special reference to the Theory of Conscience."

10. *Newsletters of the Down-Town Ethical Society,* 1909–11, Reel Two, Hamilton-Madison House Papers.

11. Speech reproduced in the "Independent Progressive," published by the Independent Henry Moskowitz Campaigners of the 12th Congressional District, Wald Papers, NYPL.

12. See ch. 5, and Hyman Berman, "Era of the Protocol," ch. 3. Henry's many speeches on the protocol include "The Power for Constructive Reform in the Trade Union Movement" (to the Boston Women's Trade Union League, June 1911; see *Life and Labor,* 2/1 [Jan. 1912], 10–15); "The Moral Challenge of the Ethical Struggle," *Ethical Addresses,* 18/9 (May 1911), 223–41; "The Joint Board of Sanitary Control in the Cloak, Suit and Skirt Industry of New York City," *Annals of the American Academy of Arts & Sciences,* 44 (Nov. 1912), 39–58; "Public Responsibility for Working Conditions," 26th New York State Conference of Charities and Corrections (1925), *Proceedings,* 148–162.

13. See Allen Davis, *Spearheads for Reform,* 208 ff.; *Preliminary Report of the Factory Investigating Commission* (1912), II, 98–107; 127–35; *Second Report* (1913), V, 1210–15; and Jesse Carpenter, *Competition and Collective Bargaining,* 574, 588–89.

14. On New York civic reform, see Richard Skolnick, "Civic Group Progressivism in New York City," *N.Y. History* 51 (July 1970), 411–39, and Augustus Cerillo, "The Reform of Municipal Government in New York City: from Seth Low to John Purroy Mitchel," *N.Y. Historical Soc. Quarterly* 57 (Jan. 1973), 51–71. On the NAACP, see Mary White Ovington, "How the National Association for the Advancement of Colored People Began," NAACP, *Annual Report,* No. 7, 1916, and *The Walls Came Tumbling Down* (1947).

15. Goldstein lived at the University Settlement while executive secretary of the ESNA, and later became a judge, Court of General Sessions, New York City. See his 70th birthday tribute, *Congressional Record,* 11 Apr. 1956, and the Goldstein Papers, AJHS. For the Goldstein-Magnes affair, see Arthur Goren, *New York Jews and the Quest for Community,* 164–69.

16. Advertising circular for the Association; Goldstein to Charlotte Waterbury, 18 Nov. 1913; "General Resume of Work Done by the E.S.N.A.," 5 Dec. 1912–5 Jan. 1913; "Conference of the Heads of Settlements . . . called by the E.S.N.A.," 21 Mar. 1913; "Summary" [of activities], March 1913, Jacob A. Riis Neighborhood Settlement Papers; corr. between Wald and Lawrence Veiller, April 1913 (Wald Papers, Columbia Univ.), on the "Schlapp bill," developed by Goldstein when he found that a boy was twice committed to the House of Refuge before being discovered as a "defective."

17. Moskowitz to Warburg, 12 Sept. 1913; Karl de Schweinitz (Charity Org. Soc.) to Ida Clemons (Schiff's secretary), 17 July 1913; Magnes to Warburg, 1 and 18 Sept. 1913; Mr. Newburger (3rd Deputy Police Commissioner), memo to Magnes, n.d. (but after 12 Sept. 1913), Felix Warburg Papers; Henrietta M. Voorsanger to author, 12 Feb. 1978. Miriam Israels Gabo recalls Goldstein's visits to the Park Hill house during this period. As evidence of Moskowitz's continuing support for Goldstein, see his letter recommending the ESNA and Goldstein to Arthur Woods, Mayor Mitchel's secretary, 7 Feb. 1914, Box 208, J. P. Mitchel Papers (Munic. Arch.).

18. See Warburg, Schiff, and Marshall collections, American Jewish Archives, *passim,* for Moskowitz corr. on Jewish matters; Binkerd, Columbia Univ. Oral History; Adler, "Notes on the Meetings of the American Ethical Union, May 9–12, 1907," Adler Papers.

19. Joseph M. Levine to Wald, 3 Jan. 1914; Wald to Hughes, 3 Mar. 1909, in which she praises the work of "our Settlement young men" in organizing the League of Independent Voters (Wald Papers, NYPL); Stimson's correspondence (boxes 17–18, Stimson Papers) refers often to Moskowitz and the 1909 fusion movement. See also Schapiro, "Henry Moskowitz"; Moskowitz's obituary, *N.Y. Her. Trib.,* 18 Dec. 1936; Allen Davis, *Spearheads of Reform,* 183–85.

20. On the 1910 Campaign, see esp. "Speeches and Statements," Series III, Box 190; General Corr., Boxes 21–23, Stimson Papers; *NYT,* 22 Oct. 1910; Moskowitz, "Henry L. Stimson," *The Independent,* 69 (27 Oct. 1910), 904–07.

21. Moskowitz to Wald, 22 Aug. 1912; Price to Wald, 11 Dec. 1912, Corr. files, Wald Papers, NYPL; H. Moskowitz campaign material and platform of the state Progressive party, *ibid.,* Collateral files. See also H. H. Rosenthal, "The Progressive Movement in New York State," 307, 330, 412, 415, 417, 422; Schapiro, "Henry Moskowitz"; John Allen Gable, "The Bull Moose Years: Theodore Roosevelt and the Progressive Party, 1912–1916" (Ph.D. diss., Brown University, 1972), 40, 65, 165, and ch. 3 for discussion of "lily-white" progressivism; *NYT,* 6 Aug. 1912; Allen Davis, *Spearheads,* 151, 197, 201, 206–07; corr. between Moskowitz and Theodore Roosevelt, e.g. TR to HM, 22 July 1912, Theodore Roosevelt Papers (Microf. ed.); and Ethical Culture Society, *Newsletter,* X/2 (Nov. 1912). Henry's campaign committee held a "Monster Mass Meeting" for him on 28 Oct. 1912 at the Lafayette Casino on Avenue D, with speeches by politicians Leon Sanders and James J. Fitzgerald, recreation reformer Luther Gulick, and others (Box 4, Abelson Papers). See also ch. 4.

22. On Moskowitz's role in the 1913 fusion movement, see his letters to Theodore Roosevelt, 21 April-5 Sept. 1913; Series Four, Box 201, fol. 13, and Gen. Corr., Boxes 39–41, *passim,* Stimson Papers; and the J. P. Mitchel Papers (Munic. Arch.), esp. Scrapbook No. IV, 107–19: news release (14 Sept. 1914) on Moskowitz's Civil Service Commission reforms and Mitchel's response; No. V, 2–24: Mitchel's Address to the City Club, 2 Oct. 1914; and corr., boxes 162, 186, 208, 219, 229, 240. On Mitchel, see Edwin R. Lewinson, *John Purroy Mitchel. The Boy Mayor of New York* (1965). Moskowitz discusses views of civil service in "Sound Business Principles of Civil Service," *7th Annual Conf. of the National Assembly of Civil Service Commissions, Los Angeles, Calif., June 15–19, 1915* (1915); "Old and New Problems of Civil Service," *Annals,* 64 (Mar. 1916), 153–67. See also, Caro, *The Power Broker,* ch. 5.

23. In 1904, 629 children passed through Camp Felicia, the summer program of the downtown society (*Newsletter,* II/1 [Nov. 1904]); during the 1909 cam-

paign, Henry cancelled his uptown office hours (VII/1 [Oct. 1909]); the debate was summarized in VII/8 (May 1910). Alvin Johnson, *Pioneer's Progress. An Autobiography* (1952), 142–43, says Adler held that every "movement for political reform was bound to issue in half-baked radicalism. Only in the advance of ethics could there be any hope of a humane and healthy society."

24. *Ethical Record,* IV/2 (Jan. 1903); *Newsletter,* VI/1 (Nov. 1908).

25. *The Yonkers Herald,* 23 Nov. 1914; author interviews with Edward Bernays and Judith Bernays Heller, 1975–76.

26. The following is based primarily on Denneen Hanlon, "Career and Home—Can Women Have Both?", *N. Y. World,* Editorial Sec., 24 May 1925, profile of Belle Moskowitz.

27. The Moskowitzes treated Atlantic City as the ideal vacation spot. Miriam's letters, for the most part undated, are now in the Moskowitz papers.

28. Joe's novels were *Rebecca The Wise* (1930) and *The Sea and The Land* (1931). *Rebecca,* about a young Jewish woman who had moved from Oklahoma to New York City, was withdrawn from circulation after her family protested. The second novel was about Ethiopia.

29. Author interviews with Aline MacMahon, Doris Fleischman Bernays, Warren Moscow, Dorothy Omansky, Clara Rabinowitz, Seymour Hammel, Dorothy Rosenman, 1975–76; Perkins, Oral History, III, 486–87; Ivor Montagu, *With Eisenstein in Hollywood* (1968), 37–38.

30. See corr. files, 1914–15; collateral file, "Women's Constitutional Convention Committee, N.Y.S.," Wald Papers, NYPL. The Committee published two pamphlets: "Memorandum on the Eligibility of Women to Act as Delegates to the Constitutional Convention to be Held in New York in 1915"; "The Service of Women to New York State: A Memorandum on the work women are doing in the State, entitling them to sit as Delegates in the Constitutional Convention of 1915." The State Attorney General confirmed that women could run for delegate, but said the convention was the final arbiter of who might serve (*NYT,* 29 July 1914).

31. Three letters to Wald detail these events: Frances Kellor and Anne Rhodes, 2 Sept.; Harriet B. Lowenstein, 8 Sept. 1914; see also, "Minutes of meeting of Campaign and Executive Committees," 14 Sept. 1914. For Israels's progress toward nomination, see the *Yonkers Statesman,* 29 Sept., 4 Nov., 1914. Dr. Davis received 87,273 votes, a respectable showing. See *NYT,* 6 Dec. 1914, for tallies.

32. See fol. 145, Harriet Burton Laidlaw Papers; Box 2, fol. 53, Jane Norman Smith Papers, Schlesinger Library. Moskowitz's reelection to the Council of Jewish Women board was reported in the *AH,* 101 (5 Oct. 1917), 614.

33. On the founding of the Women's City Club, see *NYT,* 27 Dec. 1915; 23, 25 Jan. 1916; and Minutes, Executive Board, WCC, 23 Feb. 1916.

34. BLM to Theodore Rousseau (the mayor's secretary), 12 Apr. 1917, Box 263; BLM to Mitchel, 20 June 1917; "Progress Report of the Committee of Women," 25 Aug. 1917, Box 237; Scrapbook No. XVI, p. 59, Mitchel Papers (Munic. Archives). The 72 women on the Committee included many known to Moskowitz from other projects and members of the Women's City Club or Council of Jewish Women.

35. "Semi-Annual Report, Mayor's Committee of Women on National Defense," Box 239, Mitchel Papers (Munic. Archives).

36. Anna Moscowitz Kross was among those who visited Sheriff Smith

(Voorsanger to author, 12 Feb. 1978). See also Belle Moskowitz to Frederick Whitin, 5 Aug. 1918 (Comm. of Fourteen Papers).

7: Building the Partnership

1. I have relied heavily on Matthew and Hannah Josephson, *Al Smith: Hero of the Cities* (1969), chs. 2–9. See also Henry Moskowitz, *Alfred E. Smith, An American Career* (1924); Norman Hapgood and Henry Moskowitz, *Up From the City Streets: Alfred E. Smith* (1927); Henry F. Pringle, *Alfred E. Smith: A Critical Study* (1927); Alfred E. Smith, *Up To Now, An Autobiography* (1929); Emily Smith Warner, *The Happy Warrior. The Story of My Father, Alfred E. Smith* (1956); Oscar Handlin, *Al Smith and His America* (1958).

2. Frederick M. Davenport, "Human Nature in Politics. Al Smith and the Human Side of Tammany," *The Outlook,* 31 July 1918, 522–24. Warner, *The Happy Warrior,* 33, has a sympathetic description of Tammany. On urban machines and immigrant welfare, see J. Joseph Huthmacher, "Urban Liberalism and the Age of Reform," *Mississippi Valley Historical Review,* 44 (Sept. 1962), 231–41, and John D. Buenker's *Urban Liberalism and Progressive Reform* (1973).

3. For a detailed study of the commission, see Thomas J. Kerr, IV, "New York Factory Investigating Commission and the Progressives" (1965). New York State published the commission's reports.

4. Paula Eldot, *Governor Alfred E. Smith, The Politician As Reformer* (1983), 6–7; Robert Wesser, "The Impeachment of a Governor: William Sulzer and the Politics of Excess," *N. Y. History,* 60/4 (Oct. 1979), 407–38.

5. See Thomas Schick, *The New York State Constitutional Convention of 1915 and the Modern State Governor* (1978); on Smith's fight against the "Barnes' amendment," which would have barred the state from passing social legislation, see Eldot, 194–96.

6. See the *NYT* report of Abram Elkus speech, 17 Oct. 1918; for the full committee membership, 18 Oct.; Kerr, "Factory Invest. Comm.," 22–30, for the membership of the 1911–14 "coalition" of Tammany politicians, social workers, "good government" reformers, labor leaders, and Socialists; Josephson, *Al Smith,* 184, 190–95; Frances Perkins, *The Roosevelt I Knew* (1946), 50.

7. Other women on the list included Mrs. Thomas L. Chadbourne, Jr., Mrs. Walter B. Chambers, Sara Conboy, Mrs. William Temple Emmet, Mrs. Charles L. Guy, Mrs. Frederic Kernochan, Anne Lee, Miriam Sutro Price, Mrs. Albert de Roode, Mrs. William Spinney, Ethel Stebbins, Mrs. Laurence A. Tanzer, Helen Todd, Blanche Wylie Welzmiller, and Mrs. William G. Willcox. Women registered in political parties voted in the May 1918 primaries, but the non-aligned voted for the first time in November.

8. Perkins, Columbia Univ. Oral History, I, 412–13; Henry Moskowitz, "New York's East Side as a Political Barometer," *The Outlook,* 27 Feb. 1918, 325–27; *NYT,* 23 Oct. 1918. John Buenker (*Urban Liberalism and Progressive Reform*) argues that machine politicians had discovered the vote-getting results of backing social programs.

9. Howard Shiebler, "Mrs. Moskowitz Wanted to be a Doctor But Fate Made her People's Spokesman," *Brooklyn Daily Eagle,* 26 July 1925.

10. *NYT* and *N.Y. Herald,* 1, 2, 3 Nov. 1918.

11. Publicity Associates biog., Dec. 1932 (Moskowitz Papers); Josephson, 194; Pringle, 67–68; Henry Moskowitz, 49, gives Smith's words as: "If I do wrong,

you may be sure that it will not be from ignorance and you can hold me responsible."

12. HM to Wald, n.d., *ca.* 1917–18, Wald Papers, NYPL.

13. For references to national reconstruction, see *NYT,* 4, 17, 21 Nov. 1918. For a wistful portrayal of the "bright dreams" of reconstruction, see Lewis Mumford, *Sketches From My Life* (1982), 218–19.

14. Perkins, Oral History, II, 207–25. Perkins recalled conversations in the most authoritative manner, but her memory could be faulty, e.g. she confused the identity and politics of Belle Moskowitz's two husbands (I, 255–58), and claimed Belle had influenced Smith to support Wendell Willkie for president in 1940 (II, 80–81, 335–37), though Belle had been dead for seven years; Perkins had spoken at a memorial service for her in 1934.

15. Caro, 95–96, differs from Perkins, who seems more plausible here. There is no other corroboration that Smith invited Belle Moskowitz to attend a Biltmore Hotel conference to map out a program of administrative reform, much less that Belle, described by Caro as a "plump little Jewish matron," approached Smith's "bluff old Irish" friends. Caro's analysis of why Smith would like the idea is on the mark.

16. For a full list of the Commission members with their professions and locations, see H. Moskowitz, *Alfred E. Smith,* 54–55.

17. Perkins, Oral History, II, 218–19.

18. Caro, chs. 5–6, *passim.* Moses was chief of staff only for the retrenchment committee, not for the entire commission.

19. *NYT,* 25 Jan. 1919; *The Nation,* 108 (11 Jan. 1919), 38. Reconstruction Commission papers were not saved in Smith's papers in Albany; some might be found in Robert Moses's, which are still closed to the public. I have consulted newspapers, the commission's reports (published separately and then in Smith's public papers for 1919 and 1920), and Smith's executive correspondence.

20. See *NYT,* 2 Jan. 1919; Smith, *Pub. Pap.* (1919), 30–31. The commission would address permanent tax needs, the high cost of living, health and welfare legislation, and economic issues.

21. Sweet's speech is in the N.Y. State *Journal of the Assembly,* 142nd Session (1919), 9–10. See *NYT,* 5, 10 Feb., 2 Mar. 1919, for hints of Republican action to come. The Assembly journal provides substantial portions of the defeated bill during the amendment process (see pp. 208–09).

22. *NYT,* 19 Mar. 1919; Smith, *Pub. Pap.* (1919), 670–71.

23. Kerr, "Factory Investigating Commission," 20–22.

24. See Caro, *Power Broker,* 103–04; *NYT,* 2 Mar. 1919; and "A History of the Reconstruction Commission," 18 Dec. 1920, in Smith, *Pub. Pap.* (1920), 569–79 [hereafter "RC History"]. I am assuming here that Belle Moskowitz wrote the "RC History," written a month after Smith's 1920 defeat in a style similar to other Moskowitz documents.

25. Caro, 99–101; Perkins, Oral History, II, 221–22.

26. Caro, 99.

27. "RC History," 570–71, and Belle Moskowitz, "Report of the Committee on Reconstruction Problems," *Proceedings of the New York State Conference on Charities and Corrections,* XX (1919), 138–46 [hereafter "RC Speech"]. On the film projects, see Smith [AES], Exec. Corr., ser. 260, fol. 182, exchanges between Moskowitz and Smith's secretary, George Van Namee.

28. "Demobilization, War Department Regulations and Unemployment in

New York City," 7 Apr. 1919, in Smith, *Pub. Pap.* (1919), 614–15; "RC History," 572; "RC Speech," 140–41. On Moskowitz's lobbying efforts, see "World War I Reconstruction," Wald Papers, Columbia Univ.

29. "Business Readjustment and Unemployment," 14 Apr. 1919, in Smith, *Pub. Pap.* (1919), 616–26. An app., "Public Improvements in Progress, Not Started, and Contemplated," showed figures for highways; for engineering, conservation, and architectural projects at the federal, county, and city levels; and for the development of water supply and rapid transit systems.

30. "A Permanent Employment Program," 17 June 1919, in Smith, *Pub. Pap.* (1919), 636–43. John Agar chaired this committee. See also Eldot, *Governor Alfred E. Smith,* 201.

31. Smith, *Pub. Pap.* (1919), 553; see Elkus to Smith, 20 Aug. 1919, Exec. Corr., ser. 260, fol. 172, for the original on which Moskowitz signed Elkus's name.

32. Moskowitz to Smith, 27 Aug. 1919 (Exec. Corr., ser. 260, fol. 172).

33. Moskowitz to Van Namee, 2, 9 Sept. 1919; invitation lists, with corrections; Moskowitz to Smith, 4 Sept. 1919 (ser. 260, fol. 172).

34. "Proceedings of the Conference on Industrial Conditions in the State of New York," 16 Sept. 1919, in Smith, *Pub. Pap.* (1919), 553–83; Smith quotation, 572. A stenographic record was kept of this conference (see Moskowitz to Van Namee, 9 Sept.). Moskowitz ran the Reconstruction Labor Board out of her Hall of Records office. For the text of the law, see *Laws of the State of New York* (1920), ch. 894.

35. "Terminal Markets in New York City," 14 June 1919, in Smith, *Pub. Pap.* (1919), 629–36; "A State Endorses the Rural Motor Express," 28 June 1919, in AES, Exec. Corr., ser. 260, fol. 171; see also Moskowitz to Van Namee, 22 Aug. 1919, in fol. 172. Fol. 171 shows Moskowitz engineering this minor episode in the decline of America's railroad system.

36. "Food Production and Distribution," 30 Jan. 1920, in *Pub. Pap.* (1920), 137–72. See AES, Exec. Corr., ser. 260, fol. 148, and *NYT,* 15 Apr. 1920.

37. For a thorough discussion of housing legislation, see Paula Eldot, *Governor Alfred E. Smith,* ch. 5.

38. For background on housing reform, see Roy Lubove, *The Progressives and the Slums. Tenement House Reform in New York City, 1890–1917* (1962) and *The Urban Community: Housing and Planning in the Progressive Era* (1967); Anthony Jackson, *A Place Called Home. A History of Low-Cost Housing in Manhattan* (1976).

39. Hamilton, "Housing," N. Y. S. Conference on Charities and Corrections, *Proceedings* (1919), 161–76.

40. See "Report of the Reconstruction Commission on the Housing Emergency," 10 May 1919, in Smith, *Pub. Pap.* (1919), 672–73, and the final report, "Report of the Housing Committee of the Reconstruction Commission of the State of New York," 22 Mar. 1920, in *Pub. Pap.* (1920), 207–41; also, housing discussions in the *NYT,* 5, 16–19, 22 Apr. 1919. Moskowitz did not sign the final report, but certainly wrote it. In 1920 she alone did commission work, and the report's style differs from Hamilton's. There is proof that she prepared Smith's speech on housing to the 1920 Governors' Conference (AES, *Pub. Pap.,* 1920, 676–84); see Eldot, 168.

41. See the press release announcing the report; a Moskowitz addendum called for a conference in the Governor's name, Exec. Corr., ser. 260, fol. 46.

42. *NYT,* 17, 22, 28, 31 May 1919; speech by Smith, *Pub. Pap.* (1919), 751–56.

43. "Report of the Housing Committee" (1920), 228–29. See also "Competition for the Remodeling of a New York City Tenement Block," a joint Lockwood-Reconstruction Comm. project, in Smith, *Pub. Pap.* (1920), 241–51.

44. See *Laws of the State of N.Y.* (1919), chs. 647–50; and Lockwood, "Report of the Joint Legislative Committee on Housing," 16 June 1919 (Smith's *Pub. Pap.* [1919], 676–79), acknowledged the cooperative effort among the Lockwood, Mayoral, and Reconstruction Commission housing committees.

45. *NYT,* 17, 18, 22, 31 May; 11 June 1919.

46. "Public Health and Reconstruction," in Smith, *Pub. Pap.* (1919), 644–70. Frances Perkins singled out this report for its impact on rural health services (Oral Hist., II, 223).

47. Smith, *Pub. Pap.* (1920), 35–37; (1923), 176–78, 523–24. The commission's reports on education proposed a more intense "Americanization" program and an end to the compulsory military training of adolescent boys. See "Americanization," 14 May 1919, in Smith, *Pub. Pap.* (1919), 617–19; and "Military Training," 7 Apr. 1919, typescript of Legislative Document No. 78. Smith wanted the commission's advice as he planned to veto any appropriation for the program. The American Union Against Militarism reprinted the report as "New York's Second Sober Thoughts: 'No Compulsory Military Drill,'" (Washington, D.C., 1919). On Americanization, see John Higham, *Strangers in the Land* (1965), ch. 9, esp. 254–57.

48. See Schick's study of the 1915 State Constitutional Convention; Eldot's discussion of administrative reorganization (ch. 2); Arthur E. Buck, *The Reorganization of State Governments in the United States* (1938); Joseph McGoldrick, "Governor Smith Introduces the Cabinet in New York State," *National Municipal Review,* XVI (Apr. 1927), 226–29; Alfred E. Smith, "How We Ruin Our Governors," in *ibid.,* X (May 1921), 277–80.

49. *NYT,* 18 Apr. 1920, and "RC Speech," 143.

50. "Summary of Report of Reconstruction Commission to Governor Alfred E. Smith on Retrenchment and Reorganization in the State Government," 10 Oct. 1919, in *Pub. Pap.* (1919), 585–614.

51. "Summary," ch. 2, secs. IX, X, XI, and XV; full report (Albany, 1919), Part V, chs. 1–2.

52. *NYT,* 22 Sept. 1919; the survey was probably prepared by Buck; see also Caro, 106.

53. See Eldot, ch. 2. In the end, New York consolidated into 19 departments with four elected officials, including the attorney general.

54. Women's City Club, Minutes, Exec. Bd., 6, 20 Jan., 16 Mar. 1920; *N. Y. Eve. Post,* 15 Dec. 1919.

55. "RC Speech," 143–44.

56. *NYT,* 13, 29, 24 Oct., 22 Dec. 1919, and 13 Jan., 28 Apr. 1920.

57. *NYT,* 20, 21 Apr. 1920.

58. Caro, 108–10; "RC History," 578; *NYT,* 11, 12, 14 Apr. 1919.

8: A Power for Good

1. Edward L. Bernays, *Propaganda* (1928), 9, and *Public Relations* (1952); Harwood L. Childs, *Public Opinion: Nature, Formation, and Role* (1965), 273 ff.

2. Author interview with Miriam Israels Gabo, 1975. See also Bernays, *Propaganda,* 78 ff.

3. Henry Clay Gipson, *Films in Business and Industry* (1947), ch. 1.
4. *NYT,* 5 Mar. 1920.
5. See *NYT,* 23 May 1920 (VI, 7:1); 5 Nov. 1921 (VI, 6:1); 1 Jan. 1922 (II, pp. 1–3); also, *Report of the Port of New York Authority to the Governors of the States of New York and New Jersey,* 21 Dec. 1921, p. 9; and Julius Henry Cohen, *They Builded,* 248.
6. See *NYT,* 19 Nov. 1920; 24, 29 Mar., 30 Apr., 24 Aug. 1921; 11, 21, 24 Jan. 1922.
7. *NYT,* 13 Feb. 1921 (VII, 1:2).
8. *NYT,* 29 July 1921. An article on 8 Jan. 1922 (II, 1:3) explains automated railroad cars.
9. *NYT,* 29 Apr., 25 Aug., 11 Sept., 20 Nov., 20, 21 Dec. 1921. See also *Report of the Port,* 7–8, and Cohen, *They Builded,* 249.
10. Five women's organizations attended: the Women's City Club, the League of Women Voters, the New York State Federation of Women's Clubs, the Women's Municipal League, and the National Council of Jewish Women. See *NYT,* 31 Jan. 1922.
11. See *NYT,* 1, 2, 11 Feb. 1922.
12. On women's political organizations in the 1920s, see J. Stanley Lemons, *The Woman Citizen* (1973). See the Marion Dickerman Papers for ms material on the Women's Joint Legislative Conference, esp. her eulogy of Mary Dreier (1963).
13. *NYT,* 11 Nov. 1917; 16 June 1918; 9 May 1920. On the club's earliest years, see above, ch. 6.
14. Minutes, Exec. Board., WCC, 12 Dec. 1916; 20 Mar. 1917; 3 Apr. 1917; 8, 22 Jan. 1918.
15. *Ibid.,* 10 Mar., 14 Apr. 1919; 6, 20 Jan., 16 Mar. 1920; and Minutes, Monthly Club Meetings, 14 Apr. 1919.
16. *Ibid.,* Monthly Meeting, 3 Jan. 1921; Board Meeting, 7 Dec. 1921; 25 Apr., 3 May 1922; 7th Annual Meeting, 15 May 1922. See also *NYT,* 13 Feb. 1921; 19 May 1925.
17. See my "Training for Public Life: ER and Women's Political Networks in the 1920s," 28–45; William Chafe, "Biographical Sketch," 3–27; and Susan Ware, "ER and Democratic Politics: Women in the Postsuffrage Era," 46–60, in Joan Hoff-Wilson and Marjorie Lightman, eds., *Without Precedent: The Life and Career of Eleanor Roosevelt* (Bloomington: Indiana Univ. Press, 1984). See also Joseph P. Lash, *Eleanor and Franklin* (1971).
18. The *Women's Democratic News* [hereafter, *WDN*] began as a newsletter, became a printed monthly in May 1925, and was eventually absorbed by the Women's Division of the Democratic National Committee, its name changed to the *Democratic Digest.* On relations among the members of the Women's Division, see Kenneth Davis, *Invincible Summer. An Intimate Portrait of the Roosevelts based on the recollections of Marion Dickerman* (1974), *passim.*
19. Hilda R. Watrous, "In League With Eleanor Roosevelt," and my "The Political Apprenticeship of Eleanor Roosevelt," Centennial Conference on Eleanor Roosevelt, Vassar College, 1984; author interview with Aline MacMahon (1975). On ER's anti-Semitism, see Joseph Lash, *Eleanor and Franklin,* 135, 214, 323.
20. For a detailed narrative of Smith's relations with Hearst, see Eldot, *Governor Alfred E. Smith,* ch. 8.

21. Draft, Manifesto of the Citizens' Fair Play Committee, 22 Oct. 1919, Scrapbook No. Two, George Van Namee Papers, on which Moskowitz's handwriting appears. For the Smith-Hearst feud see Josephson, ch. 11, Smith (*Up to Now*), and New York newspapers.

22. *NYT,* 27 May 1922. The founding officers of the Democratic Union included Ethel Stebbins, Pres.; Ida Blair, Mrs. Montgomery Hare, and Mrs. Charles Dana Gibson, Vice-Presidents; Anne O'Hagan Shinn, Sec'y; and Mrs. Abram I. Elkus, Treas.

23. *Ibid.,* 28, 29 May 1922. Moskowitz did not head the Union, but was its acknowledged leader. Of course, in 1913 Belle Israels was not yet married to Henry Moskowitz.

24. *Ibid.,* 28, 29 May, 11, 21, 28 June; 7, 8 July 1922.

25. *Ibid.,* 10, 11, 12, 22 July 1922.

26. *Ibid.,* 19 Aug., 29 Sept. 1922. In the vote for Dickerman's resolution, 54 out of 62 representatives favored Smith.

27. On Smith's election in 1922, see Eldot, *Alfred E. Smith,* 15–20; Josephson, 259–60; Smith, *Up to Now,* 228–45, none mentions women's roles in Smith's victory; but see K. Davis, *Invincible Summer,* p. 15. On Democratic Union activity, see *NYT,* 17 Nov., 5 Dec. 1922; 1, 22 Jan., 4 Feb., 6 Mar. 1923. In June, the Union held elections, and Moskowitz became one of several vice-chairs.

28. Smith, *Pub. Pap.,* 1923, pp. 47–82 (Annual Message to the Legislature, 3 Jan. 1923).

29. The first (1921) federally-funded health care program, Sheppard-Towner provided matching funds to states for the establishment of health centers. Its renewal was defeated in 1929. See Lemons, *The Woman Citizen,* ch. 6; and Sheila Rothman, *Woman's Proper Place* (1978), 136 ff.

30. Moskowitz to Wald, 7 Aug. 1923, box 3, Wald Papers, NYPL. See the Appendix for the text of this letter.

31. Author interview with Robert Moses, 1976. Perkins also remembered that Belle often knitted, Oral History, III, 16.

32. Barrie's *What Every Woman Knows* was first performed 3 Sept. 1908 in London and ran for a year. See the *Dictionary of Literary Biography,* vol. 10, pt. I, 33.

33. Perkins, Oral History, III, 13–15.

34. Ibid., II, 221–22.

35. See the New York State Conference on Charities and Corrections, *Proceedings,* 1923, p. 138; Max Meltzer to Alfred E. Smith, 4 Jan. 1933 (Moskowitz Papers).

36. Oliver H. P. Garrett, "A Certain Person," *The New Yorker,* 9 Oct. 1926.

37. Author interview with Aline MacMahon, 1975.

38. Perkins, Oral History, II, 225. Moskowitz occasionally used her influence to help personal friends. She arranged a rereading of Regents examinations of the daughters of two friends, and secured a confidential medical report on a young man, once enamored of her daughter and then confined to the state hospital (Moskowitz to Graves, 28 July 1925; Lewis Farrington to Moskowitz, 16 June 1925; AES, Exec. Corr., ser. 200, fols. 13, 29, 30, 35).

39. Eleanor Booth Simmons, "Women Lacking in Patience in Politics," *N.Y. Herald,* 20 May 1923, interview with Moskowitz, who repeated the point in an interview in the *Brooklyn Eagle,* 26 July 1925.

40. *NYT,* 28 Apr. 1926. The next day the *Times* editorialized that women

would soon develop intellectual power. Of all the speakers, Moskowitz received the most press coverage.

41. E. Roosevelt, *It's Up to the Women* (1933), 197, 202.

42. *NYT,* 28 Apr. 1926.

43. *It's Up to the Women,* 198.

44. Smith, "My Closest Advisor: Belle Moskowitz as I Knew Her," *The Jewish Chronicle,* 17 Feb. 1933. See the Appendix for the text of this article.

45. See above, ch. 1. Admittedly, there is a ten year gap between the two documents I am comparing, and they were written under much different circumstances. Still, they are the only surviving documents in which each spoke of the other in personal terms.

46. Author interview with Robert Moses; Garrett, "A Certain Person," *New Yorker,* 9 Oct. 1926, p. 26.

47. See Smith's speech at a memorial for Moskowitz, reported in the *NYT* and *New York Daily News,* 13 Mar. 1934.

48. Garrett, "A Certain Person," 27.

49. Author interview with Miriam Israels Gabo, 1975.

50. Garrett, 28; profiles in the *New York Sun,* 30 Oct. 1926 and 1 July 1928.

51. Moskowitz to Walter Lippmann, 17 May 1928, Lippmann Papers.

52. Perkins, Oral History, III, 15–16.

53. *Ibid.,* 16.

54. *Ibid.,* 488.

9: Policy and Publicity

1. Garrett, "A Certain Person," *New Yorker,* 9 Oct. 1926, 26; the $4,000 figure also appears in Henry F. Pringle, *Alfred E. Smith,* 69, and Josephson, *Hero of the Cities,* 302. According to Garrett, a Public Service Commissioner would have earned $15,000. Smith's comment on Moskowitz and the news is from AES to Proskauer, 9 Oct. 1926, George Graves Papers.

2. On the publicity bureau see *Hearings Before a Special Committee Investigating Presidential Campaign Expenditures,* U.S. Senate, 70th Congress, 83–85, 333–41, and a ms article by Moskowitz [Moskowitz Papers] about the 1928 presidential nomination [hereafter BLM, "AES Ms."], pp. 11–12 (see below, ch. 10). In 1925, Herbert C. Pell, then chair of the Democratic State Committee, tried to raise the bureau's budget to $18,000 and to add an "Information Bureau" for another $3,000. See Pell to AES, 20 Oct. 1925, Graves Papers.

3. Mrs. Henry Moskowitz, "A Department of Public Relations for the Plan of New York and Its Environs," Regional Planning Association Papers, Box 32, No. 2688. Daniel Bluestone found this reference, Ann Polisar located the document at Cornell Univ.

4. My model for this pattern is the industrial conference she developed for Smith in 1919 (see above, ch. 7). See BLM to Smith, 27 Aug. 1919: the plural "we"—"It seems to me that you should set the date for the industrial conference which we have suggested. . . . The constructive thing we hope to accomplish. . . . What we must attain by the conference is an agreement. . . ."; references to voter appeal—"This State of all the others could take the lead in this matter. . . . If we can accomplish this result, every State in the Union will follow our example, and you would really be leading a tremendous movement. . . . I hope that you will

agree with me that this conference will be most popular throughout the State. . . ." See also BLM to AES, 3 Sept. 1919: assuring him he is in control—"Enclosed is a list of names suggested for the conference. . . . I send you also copy of the proposed invitation, which I have already sent to Mr. VanNamee [*sic*]. Of course it may require changes and is merely a suggestion to you" [AES, Exec. Corr., ser. 260, fol. 172].

5. Pringle, 70; Josephson, 345–46.

6. Eldot, *Alfred E. Smith,* 41 ff.; Caro, *The Power Broker,* 108 ff.

7. Eldot, 50–54; Hapgood & Moskowitz, *Up From the City Streets,* 229–31.

8. I have claimed Moskowitz involvement only where I found evidence. On reorganization, see drafts of the legislative message of 26 Feb. 1923 with Moskowitz handwriting, and press release on Machold, 9 Apr. 1924; AES to McGinnies, 14 July 1925 (AES Papers, Exec. Corr., ser. 200, fols. 241, 241–1, and 246). See also Moskowitz to Charles Bennett Smith of Buffalo, 16 Apr. 1923, and her copy of the press release on McGinnies, filed in AES's Private Papers, fol. 25.

9. See AES, Priv. Pap., fol. 42 on the state publicity campaign of 1925; and AES to BLM, 22 Aug. 1925; BLM to AES, 10 Sept. 1925, carbon copy, fol. 224, sec. 4.

10. See *NYT,* 7 Apr. 1924, 8 Mar. 1925, and a note in Mary W. Dewson Papers, FDR Library (misidentifying the topic of the Smith-Lowman debate as the 48-hour bill for women workers).

11. *WDN,* I/1 (May 1925), 8.

12. On the conference of women, see Mack to Graves, 25 June 1925; Graves to Mack, 3 July 1925, AES, Exec. Corr., ser. 200, fol. 504. Mack claimed that some Buffalo women were never sure what the conference had been about. Graves replied, "While I would very much dislike to have any of the ladies know of the reference, I do not possibly see how anything could have been explained to them more fully than the simple, straight talk which the Governor made. What's the use?"

13. See my essay, "Training for Public Life: ER and Women's Political Networks in the 1920s," in Hoff-Wilson and Lightman, *Without Precedent,* 37–38.

14. See Eldot, ch. 3, for an explanation of fiscal issues in the Smith administration.

15. See AES, Exec. Corr., ser. 200, fol. 351: Mrs. George E. Owens (President, The Government Club) to AES, 23 Oct. 1923, answered in the Governor's name by Moskowitz, 29 Oct.; Women's City Club: Minutes, Monthly Meetings, 1 Oct. 1923; Minutes, Executive Board, 3 Oct. 1923; see also *NYT,* 14 Oct. 1923. Also, see BLM to AES, 2 Oct. 1923, in which she relays a request from the Citizens' Committee (Moskowitz fol., Graves Papers).

16. "Steps to take in Organizing Support for Constitutional Amendment Authorizing $10,000,000 Bond Issue for Ten Years," carbon copy in AES, Priv. Pap., fol. 42. The document was not signed by Moskowitz, but in my view the following establishes her authorship: the presence of many similar carbons and originals in Smith's private papers; later corr. from her discussing issues raised in the publicity plan; the similarity between this and her other public relations plans. See folder 617, for a "History of Governor Smith's Bond Issue Fight ($100,000,000)," a scrapbook on every event of this well-organized campaign that could have issued only from Moskowitz's office.

17. For more on the bond issues, see AES, Exec. Corr., ser. 200, fols. 504, 504-1 (Pts. I, II), and Eldot, 92–99.

18. AES, Priv. Pap., fol. 228, again, an unsigned carbon, but accompanied by a draft of the memo with notations in Moskowitz's hand. The pamphlet is in AES, Exec. Corr., ser. 200, fols. 504, 505.

19. Eldot, ch. 5, esp. 168–88, shows that the legislation of the 1930s "included elements of Smith's program such as government subsidies, local authorities, and public corporations." On Cohen, see Gerald Fetner, "Public Power and Professional Responsibility: Julius Henry Cohen and the Origins of the Public Authority," *American J. of Legal History,* 21 (Jan. 1977), 15–39. Cohen's private papers have disappeared; his communications with Smith are scattered in the Smith and Graves papers. On Stein, see Mumford, *Sketches,* 336ff.

20. AES, Priv. Pap., fol. 98, Pts. I, II.

21. *Ibid.,* fol. 228, primarily concerns water power, but also contains carbons of Moskowitz letters on housing: BLM to AES, 15 Jan. 1925; AES to BLM, 6 Jan. 1926; BLM to AES, 30 Apr. 1926. Her nominees were John Halkett (Chair, Building Trades Council); Oliver Cabana, Jr. (Buffalo financier); Louis Pink (Brooklyn lawyer and social worker); and Rabinowitz. On 21 May 1926 she suggested Leo Bing, who was not chosen. AES, Exec. Corr., ser. 200, fol. 51-1, Pt. II, contains Moskowitz-Graves exchanges that show her liaison role in housing matters in 1924.

22. *NYT,* 18 Feb., 11 May 1924; Minutes, Exec. Board, WCC, 7 May 1924, 1 Apr. 1925, 6 Jan. 1926, 2 Mar. and 6 Apr. 1927; *WDN,* I/9 (Jan. 1926), 1.

23. Eldot, 237–68; also, an 18-pp. unsigned memo on the history of Smith's water power policies, AES, Priv. Pap., fol. 228. See Josephson, 340, on Untermyer's role.

24. Untermyer's investigation of the transit situation upset lawyer Thomas L. Chadbourne, a Smith backer and a heavy investor in New York subways. When Smith failed to intercede with Untermyer on behalf of a fair raise, Chadbourne felt betrayed. See *NYT,* 22 May 1985, on the suppression of a chapter of Chadbourne's autobiography that details these events and claims to have provided Smith with major financial support throughout the 1920s.

25. AES, Priv. Pap., fol. 228, all corr. from 1927: Untermyer-Moskowitz exchanges of 29 July, 1, 8 Aug., 31 Oct., 2, 16 Nov.; also, Cohen to Untermyer, undated carbon (probably Oct.); undated Cohen carbons sent to Moskowitz of his letters to Smith containing extracts from the 7th Annual Report of the Federal Power Commission.

26. *Ibid.,* Moskowitz's 1927 letters to Walter Laidler of the League for Industrial Democracy, Warren Bishop from the *Nation's Business,* writer Norman Hapgood, and William Hodson of the Welfare Council.

27. *WDN,* III/10 (Feb. 1927), 6–7.

28. *Ibid.,* IV/1 (May 1928), 12–13. For a discussion of Progressives and the Muscle Shoals controversy, see Richard Lowitt, *George W. Norris. The Persistence of a Progressive, 1913–1933* (1971), ch. 14. After Smith left office, Moskowitz served on the advisory board of a "Public Committee on Power in New York State," which she had helped organize in 1927. See AES, Priv. Pap., fol. 228, Margaret Norrie to Moskowitz, 29 Jan. 1927; H. S. Raushenbush to Moskowitz, 18 Jan. 1929.

29. See Eldot, 105–08, and AES, Exec. Corr., ser. 200, fol. 245, for examples of Moskowitz's planning of the conference on education and the commission that resulted.

30. "Report of the Governor's Commission on School Finance and Administration," in AES, *Public Papers,* 1926, 250–351; *WDN,* II/12 (Apr. 1927), 5, and IV/1 (May 1928): "The Champion of the Public School," unsigned lead article, probably written by Moskowitz or her staff.

31. The following account is based primarily on AES, Exec. Corr., ser. 200, fol. 276, and *The Power Broker,* chs. 9–11, by Robert Caro, the only writer who has treated the parks controversies thoroughly.

32. Caro, 187–88; this version of the "rabble" story came from Carlos Israels and Joe Proskauer, who heard it from Moskowitz and Smith the day after it happened. For a slightly different version see Eldot, 116 ff., and 429, n18.

33. Caro, 202–03, and Eldot, ch. 4.

34. Caro, 237, 245–56, 287–91; Eldot, 146–50; the many Smith-Moses exchanges in the Exec. Corr.

35. *WDN,* I/2 (June 1925), and I/3 (July 1925).

36. Caro, 257–59, on Emily Smith Warner's recollections of how Smith and Moses enjoyed one another's company; Moses missed Moskowitz deeply after her death (author interview, 1976; Dorothy Omansky, Dorothy Rosenman, 1975).

37. Eldot, 189–96, argues that Smith's labor policies were closer to those of Progressive Era reformers than of labor unions, which preferred collective bargaining.

38. Eldot, 204–06; AES, Exec. Corr., ser. 200, fol. 444, esp. Battle to AES, 8 July 1924, 12 Nov. 1925; fol. "Garments' industry, 1927," Lindsay Rogers Papers, Columbia Univ.

39. In summer 1924, Moskowitz hired Carlos Israels to type Smith's correspondence; see also BLM to AES, 8 June 1925, Priv. Pap., fol. 228.

40. On protectionism for women workers, see Kessler-Harris, *Out to Work,* ch. 7; Susan Estabrook Kennedy, *If All We Did Was to Weep at Home: A History of White Working-Class Women in America* (1979), ch. 7; Lemons, *The Woman Citizen,* ch. 7; and Nancy Cott, "Feminist Politics in the 1920s: The National Woman's Party," *J. of Amer. Hist.,* 71 (1984), 43–68.

41. Jane Norman Smith to Alice Paul, 7 Jan. 1923, fol. 111, Jane Smith Papers, Schlesinger Library. Nancy Cott alerted me to this reference. On Cotillo, a Tammany Democrat who that year became a state supreme court judge, see Nat J. Ferber, *A New American* (1938).

42. JNS to Miss Woodson, 13 Mar. 1924, fol. 61.

43. BLM to AES, 2 Oct. 1923, Graves Papers.

44. *WDN,* I/9 (Jan. 1926).

45. *Ibid.,* IV/1 (May 1928), "Why New York State Leads in Labor Legislation," an abstract of Smith's annual message on labor, and probably a Moskowitz product. See also Eldot, 212, 223–26, and *NYT,* 31 Jan. 1924 and 14 Mar. 1925.

46. See Smith to Simeon D. Fess, 26 Mar. 1923, Exec. Corr., ser. 200, fol. 4, later released to the press; and Eldot, 343–46.

47. Eldot, 356–65; Frankfurter to Henry Moskowitz, 22 May 1923 (AES Papers, Misc. box); the memo on repeal was printed in Smith's *Public Papers,* 1923, 293–303.

48. Late in May, Belle Moskowitz and four other Women's City Club board members petitioned for a special meeting on Mullan-Gage, which led to an overwhelming vote for a veto, alerting Smith to how women would react. See Minutes, Annual Meeting, WCC, 28 May 1923; and "Meet Mrs. Warwick!", *New Yorker,* 14 Mar. 1925.

49. Scrapbook No. 3, Van Namee Papers. See also AES, Priv. Pap., fol. 137, for three drafts of the message, all in Moskowitz's hand. Louis Hacker and Mark Hirsch, *Proskauer: His Life and Times* (1978), 61, claim that Proskauer "prepared a 4,500 word statement, setting forth the legal (and personal) reasons for his decision, and Smith affixed his signature to it." Van Namee and fol. 137 tell a different story.

50. "Mrs. Moskowitz, Al Smith's Confidante and Adviser, is Power in State Affairs," *Eagle,* 8 June 1923.

51. "An Address Made by Mrs. Charles H. Sabin, at a meeting of the Massachusetts Branch of the National Civic Federation, in Boston, November 18th, 1929," in AES, Priv. Pap., fol. 154. See Grace C. Root, *Women and Repeal. The Story of the Women's Organization for National Prohibition Reform* (1934), and David Kyvig, *Repealing National Prohibition* (1979), 118–27, and *passim.*

52. "Speech by Mrs. Henry Moskowitz at Conference of [WONPR], April 12, 1932," Moskowitz Papers. See also *N. Y. Eve. Post,* 10 Nov. 1928, in which she called herself a "former dry."

53. Smith's opponents in these races were: Judge Nathan L. Miller (1920, 1922); Colonel Theodore Roosevelt, Jr. (1924); Congressman Ogden Livingston Mills (1926).

54. Eldot, 38; AES, Misc. papers, Pamphlet of Speeches from the City Club, 8 Dec. 1919.

55. See Eldot, 316–17; BLM to AES, 30 July 1920, AES Exec. Corr., ser. 260, fol. 271; Shiebler profile of Moskowitz, *Brooklyn Daily Eagle,* 26 July 1925.

56. Kenneth Davis, *Invincible Summer,* 15. The tours were described in articles entitled "Trooping for Democracy" published in the *Women's Democratic News.*

57. See Lawrence H. Madaras, "Theodore Roosevelt, Jr. Versus Al Smith: The New York Gubernatorial Election of 1924," *N.Y. History,* 47 (1966), 372–90.

58. News release on the "Singing Teapot," in "Dem. State Comm.—Women's Division," Collateral Files, Lillian Wald Papers, NYPL. Also described in Lash, *Eleanor and Franklin,* 291.

59. *WDN,* I/3 (July 1925), 11.

60. See, e.g., "Our Governor's Message," I/9 (Jan. 1926); "Claiming Everything as Usual," II/1 (May 1926); "Calm After Storm," II/2 (June 1926). Moskowitz also kept up her role in the Women's Democratic Union, speaking on state issues. See Ida Blair to Wald, 26 May 1923, "Corr. Rec'd.," Wald Papers, NYPL.

61. The platform drafts in AES, Priv. Pap., fol. 147, show extensive work in Moskowitz's hand.

62. "Democratic State Platform, 1926," 2–5. The accomplishments included a motor vehicle licensing law, soldier's bonus, home rule for cities, progress in eliminating tubercular cattle, strengthening forest conservation laws, and separating transit regulation from transit construction in New York City. Smith's papers do not show how or to what extent she was involved in some or all of these projects.

63. *Ibid.,* 6–9. Robert Moses provided the wording for the parks section; see his memo to Moskowitz ("Dear Lady"), AES, Priv. Pap., fol. 147.

64. "Platform," 9–12.

65. *WDN,* II/6 (Oct. 1926).

66. *Ibid.,* "The Elections," II/8 (Dec. 1926).

67. Her involvement stands out in rural health care (AES, Exec. Corr., ser.

200, fol. 188; Priv. Pap., fols. 157–58); the N.Y.-N.J. vehicular tunnel link (Exec. Corr., ser. 260, fol. 4; ser. 200, fol. 230); a Port Authority dispute with the Dept. of War over the purchase of the Hoboken railroad (ser. 200, fol. 57).

10: The Selling of Al Smith

1. Allen J. Lichtman, *Prejudice and the Old Politics: The Presidential Election of 1928* (1979), ch. 2; he uses political surveys made after the 1928 election to challenge David Burner's classic interpretation of 1928 as the turning-point in America's swing from a rural to urban base (*The Politics of Provincialism: The Democratic Party in Transition, 1918–1932* [1975]). See also William E. Leuchtenburg, *The Perils of Prosperity, 1914–32* (1958), and Robert K. Murray, *The 103rd Ballot: Democrats and the Disaster in Madison Square Garden* (1976), prologue, and ch. I.

2. See Burner, *The Politics of Provincialism,* 160, for a table showing party strengths in statehouses and Congress, 1920–30; Murray, 24, 57, 61; Emily Warner, *The Happy Warrior,* 125; Josephson, *Hero of the Cities;* Richard O'Connor, *The First Hurrah: A Biography of Alfred E. Smith* (1970).

3. Murray, 38–83; Denis Tilden Lynch, "Friends of the Governor," *North American Review,* Oct. 1928, 420. Other candidates included Senator Oscar W. Underwood and lawyer John W. Davis. The "Great Commoner," William Jennings Bryan, was not a candidate, but intended to "keep the Democratic party straight." Asked why, as the former leader of "the progressive forces of the country," he now opposed Smith, Bryan replied, "coldly and curtly, 'I am opposed to Smith, not because he is a progressive, but because he is a wet,' and turned his back," *Up From the City Streets,* 305. See also Nancy Weiss, *Charles Francis Murphy, 1858–1924: Respectability and Responsibility in Tammany Politics* (1968).

4. BLM, "AES Ms.," 3, 14; *The Happy Warrior,* 152; Kenneth Davis, *FDR. The Beckoning of Destiny, 1882–1928* (1971), 735.

5. Warner, 151-2; Murray, 68–71; *Hearings Before a Special Committee Investigating Presidential Campaign Expenditures,* U.S. Senate, 70th Congress (May, 1928), 84, 336; BLM, "AES Ms.," 15–17. A copy of the pamphlet "What Everybody Wants to Know," 1927–28 version, is in Box 102, Franklin Roosevelt Family, Business, and Personal Papers. Robert Wesser located it for me.

6. Murray, 117–18 on the women, 169 on McAdoo.

7. BLM, "AES Ms.," 18–22.

8. On the speech's authorship, see Hacker and Hirsch, *Proskauer,* 84–86.

9. Murray, 126–34, 143–218: the issues were prohibition, the League of Nations, and the Klan. The first was settled with a bland statement, the second and third went to a floor fight. Newton D. Baker pleaded for immediate League approval without a referendum, but lost. A strong condemnation of the Klan mentioning the organization by name also lost, by one vote.

10. Murray, 209–10; O'Connor, 153. Moskowitz and Hapgood, *Up From the City Streets,* 153, 159, describe Smith's speaking style: "His voice is resonant and large but rough. His gestures are free and vigorous, but not especially graceful. His arms are likely to swing away from somewhere near his heart. He often bends forward like a perspiring evangelist." These authors admitted that the 1924 convention speech failed; listeners thought him "egotistic" (312).

11. Daniel C. Roper, *Fifty Years of Public Life* (1941), 223–24.

12. BLM, "AES Ms.," 24–28.

13. In May 1928, Smith, Van Namee, and Moskowitz were among thirty witnesses from both political parties who testified before a special United States Senate committee investigating corruption in presidential campaign funding. See *Hearings Before a Special Committee Investigating Presidential Campaign Expenditures,* U.S. Senate, 70th Congress, 76–81, 83–109, 332–41; see also, BLM, "AES Ms.," 61–67.

14. *Hearings,* 93, 335–37; Scrapbook No. 4, unidentified news clippings from April 1928, George Van Namee Papers; BLM, "AES Ms.," 2–3, 50.

15. Josephson, *Hero of the Cities,* 358–66; Hacker and Hirsch, *Proskauer,* 88–92; Proskauer, Columbia Univ. Oral History, 29; *Hearings,* 84, 337. Moskowitz's hand appears on drafts of Smith's reply in AES, Priv. Pap., fol. 121.

16. BLM, "AES Ms.," 31–32, 44.

17. BLM, "AES Ms.," 10, 32–35.

18. BLM, "AES Ms.," 37–41. See John Sword Hunter Smith, "Al Smith and the 1928 Presidential Campaign in Idaho, Nevada, Utah, and Wyoming: A Media Perspective," Ph.D. diss., Univ. of Utah, 1976.

19. This section was based on BLM, "AES Ms.," 43–70. About 40 exchanges between Lippmann and Moskowitz are in the Lippmann papers, most from 1927–28.

20. Neither the Hapgood family papers at Yale Univ. nor Norman's at the Library of Congress discuss the co-authored book, although the latter collection contains warm exchanges between Henry and Norman. After Belle died, Henry hoped Norman would write a book about her, but nothing came of his request. On *Progressive Democracy,* see HM to AES, 4 April 1928, Box 50, George Graves Papers.

21. *Hearings,* 106, 335; BLM, "AES Ms.," 66.

22. *Ibid.,* 70. On the selection of Houston, see Roper, *Fifty Years,* 232, and Tom Connally, *My Name is Tom Connally* (1954), 133. The best analysis of the convention and campaign is David Burner, "The Brown Derby Campaign," in *Politics of Provincialism,* 179–216.

23. BLM, "AES Ms.," 73–74; Josephson, 367–69.

24. BLM, "AES Ms.," 79.

25. Bowers, *My Life* (1962), 228. Harrison was senator from Mississippi. See also Burner, *Politics of Provincialism,* 200; Roper, *Fifty Years,* 232–33; Josephson, *Hero of the Cities,* 369.

26. Eunice Fuller Barnard, "Governor Smith's 'Kitchen Cabinet,'" *NYT,* 23 Sept. 1928. By the time FDR spoke to Bowers, his hostility to BLM had deepened. Drafts of the wire of acceptance, with Moskowitz handwriting, are in AES, Priv. Pap., fol. 175.

27. "Smith . . . to be plain 'Al' for Day," *New York World,* 1 July 1928; Josephson, 356–57, 370–72; Burner, 198–200.

28. Burner, 213–14; blueprint of DNC power structure, Van Namee Papers; Frank Freidel, *Franklin D. Roosevelt: The Ordeal* (1954), 243, 247, and *The Triumph* (1956), 17; Elliott Roosevelt, ed., *F.D.R. His Personal Letters 1928–1945* (1950), II, 772; Democratic National Committee, *The Campaign Book of the Democratic Party: Candidates and Issues in 1928* (1928), 386–92. Bowers thought his own speech should have been put out in pamphlet form; Moskowitz printed it only in the campaign handbook (374–86). See Bowers, *My Life,* 197.

29. BLM, "AES Ms.," 61: "If it had rested with me, I would never have moved out of [the Biltmore] even for the national campaign." See also, *N.Y. Her. Trib.,* 19 July 1928.

30. Marbury to AES, 12 July 1928, George Graves Papers; Smith's reply (14 July) was evasive and asked Marbury to "do what you can to keep them [the women] happy." For a vivid portrait of Marbury, see Bowers, *My Life,* 181–84; novelist Fanny Hurst made her the protagonist of *Lonely Parade* (1942). Marbury was a chronic complainer and gossip; see Marbury to Mack, 18 Jan. 1927 (box 40, fol. 6, Norman E. Mack Papers): she warned that choosing Herbert Lehman as chair of the State Committee "would be a grievous mistake. Jews are not any more popular upstate than are the Catholics. There is an idea here that the Governor is being closed in by the Semitic race due to certain influences. A great deal of criticism is going around here in high quarters which will not help him in the presidential campaign. Surely there is someone of wealth and position for the state position who is not a Jew!" Eleanor Roosevelt's letter is quoted in Lash, *Eleanor and Franklin,* 315.

31. Robert Moses and Joseph Proskauer thought Roosevelt an intellectual "lightweight." See Moses, *Public Works: A Dangerous Trade* (1970), 141; Hacker and Hirsch, *Proskauer,* 84; Josephson, 372–75; Lash, 316–18; Frances Perkins, *The Roosevelt I Knew,* 41–43; K. S. Davis, *FDR. The Beckoning of Destiny,* ch. 25; Elliott Roosevelt, ed., *F.D.R. His Personal Letters 1928–1945,* II, 772.

32. "Rival Political Machines Geared for Drives to Win Presidency," *N.Y. Her. Trib.,* 16 Sept. 1928, III/1–2. The DNC also maintained a Washington, D.C. publicity office, a regional campaign center in St. Louis, Mo., and scattered speakers' bureaus. For the publicity side of the Hoover campaign, see Edward Anthony, *This is Where I Came In* (1960), ch. 16. Some members of Hoover's staff considered Anthony a "hothead" and "troublemaker": see A. H. Kirchhofer to Hoover, 24 Aug. 1928; Anna Richardson to Lawrence Richey, 23 Feb. 1929; Hoover to Allen, 13 Sept. 1928 (Campaign & Transition [C&T], General Correspondence, Herbert Hoover Library [HHL]). At one point in the campaign, Allen fired Anthony, but was persuaded by Hoover to keep him on.

33. Davis, *FDR. The Beckoning of Destiny,* 831; Blueprint, Van Namee Papers. In a letter (11 Sept. 1928) to Hubert Work, RNC chair, Hoover suggested that Work broadcast that "99% of the expenditure of the National Committee" went to educating the public through speakers, printed matter, and radio (C & T, "Gen. Corres.-Work, Hubert S.," HHL); see also, Frank R. Kent, "Charley Michelson," *Scribner's Magazine,* 88 (July–Dec. 1930), 291, on the amounts each party spent on publicity. On FDR's attitude toward Moskowitz as publicity director, see Alfred B. Rollins, Jr., *Roosevelt and Howe,* (1962), 229.

34. Lash, *Eleanor and Franklin,* 314–15; see above, ch. 9, on "Touring for Democracy"; *N.Y. Her. Trib.,* 16 Sept. 1928. Gaston was Amos Pinchot's daughter and an actress.

35. Eleanor Roosevelt, *This I Remember* (1949), 39. When she retired in November, coworkers at the *Women's Democratic News* praised her "brilliant political sense," "unfailing energy and blythe courage," "understanding of human nature," and "sense of humor that has helped us over many a rough place." Calling her "distinguished in statesmanship," they concluded: "we love her most because of her unswerving loyalty to women and to the high ideals of women from which expediency has never swayed her" (IV/7 [Nov., 1928], 6).

36. See Vincent Sheean, *Personal History* (c. 1934, repr. 1969), 322–23. The

Republican Publicity Bureau saved some 80 DNC press releases (C & T, "DNC—Press Releases," HHL). On the authorship of the acceptance speech, see Proskauer, Oral Hist., 133.

37. The *Elmira Star-Gazette* (28 Sept. 1928) urged both sides to "lay off" Tammany and the oil scandals. Other newspapers mentioned the liability for Smith of "Coolidge prosperity" (e.g. the *N.Y. Sun,* 22 Sept.). Hoover's Clippings Files are a more convenient resource than Smith's scrapbooks, at Georgetown Univ.

38. Democratic National Committee, *The Campaign Book of the Democratic Party. Candidates and Issues in 1928* (1928), 5. The Smith chapter (17 pp.) closely follows *Up From the City Streets* and was probably supervised by Belle Moskowitz with Henry's participation.

39. A later chapter, "The Bogey of Tammany" (311–20), pointed out that Republicans also depend on city machines, that Smith had never "Tammany-ized" Albany, and that the Republican-dominated State Senate had always confirmed Smith's appointments.

40. *Campaign Book,* 20.

41. William Irwin to Hoover, 13 Sept. 1928 (C&T, "Gen. Corr.," HHL).

42. Burner, 216; Lichtman, 241–42; Josephson, 387; Nathan Strauss, Columbia Oral History, 119–20, on Smith's deliberate mispronunciation of select words. Roy V. Peel and Thomas C. Donnelly, *The 1928 Campaign, An Analysis* (1931), in their chapter on party organization, never mention Belle Moskowitz.

43. Frankfurter to B. Moskowitz, 11 Aug. 1928, Frankfurter Papers. Former Progressives were split in 1928. Many endorsed Hoover (Jane Addams, Julia Lathrop, Edith Abbott, Elizabeth Christman, Sophonisba Breckinridge, Raymond Robins, Lewis Hine, N. I. Stone, Ernest Poole, Luther Gulick, Paul Monroe, Charles Sabin); some endorsed Smith (Rudolph Spreckels, Alice Brandeis, Morris Cohen, Paul Kellogg, George W. Norris), and a smaller number were for Norman Thomas (Freda Kirchwey, W. E. B. Du Bois). For many former social worker colleagues of the Moskowitzes, the chief issues were prohibition, which they still supported, and Smith's associations with Tammany Hall. See C & T, "Endorsements" and "Polls, Surveys," HHL. For a brief introduction to the Progressive split, see LeRoy Ashby, *The Spearless Leader. Senator Borah and the Progressive Movement in the 1920s* (1972), 273–77.

44. Frankfurter to Moskowitz, 11 Aug. 1928; for Perkins's experiences campaigning for Smith, see Josephson, 388–92.

45. Perkins, Oral Hist., III, 83; II, 576.

46. "C & T," "Press Statements," 21 Sept. 1928. The Republicans vowed to run a "clean fight." See Henry J. Allen to George Akerson (Hoover's assistant), 26 Sept. 1928 (C&T, "Gen. Corres.-Allen, Henry J."), which includes a repr. from *The Protestant* (8/6 [August 1928]) attacking papal power. This "transgresses the policy we have been carrying out so carefully in the campaign," Allen wrote, and asked Hoover to get it stopped. Allen had fewer compunctions about using Tammany against Smith. See Allen to Akerson, 25 Aug.: "Heaven knows I don't want to make an abusive campaign, but I would like to have it strong enough to let in the facts that relate to the myth they have created about the New Tammany and the new leadership in New York." See also A. H. Kirchhofer (RNC Assoc. Pub. Dir.) to Hoover, 24 Aug., suggesting a book of anti-Tammany cartoons; Allen to Akerson, 28 Aug., reporting on instructions to the cartoonist Satterfield to "lay off on Tammany"; H. A. Mann to Hoover, 4 Sept., suggesting

that six southern Democrats sponsor an anti-Tammany document; Hoover to Mr. Brown, n.d. ("Repub. Nat. Comm."), instructing an aide to look up Civil War Tammany statements that criticized President Lincoln; and Edgar Rickard to Hoover, 6 Sept. ("Gen. Corr."), arranging for a writer to collect anti-Tammany material for church bulletins. On the rabid anti-Catholicism as well as snobbery and anti-Semitism in the campaign, see Edmund A. Moore, *A Catholic Runs for President* (1956), esp. 158–59.

47. BLM to AES, 21 May 1926 (AES, Priv. Pap., fols. 224–30) reports a visit from White during which she had given him "the material he wanted. I know you must have enjoyed your visit with him. He is a fine old liberal progressive." Also, fol. 233, "William Allen White," with drafts of Smith's reply to White, corrections in various hands, including Moskowitz's. The reply (18 Aug.) was repr. in the DNC *Campaign Book,* 320–26. On the Van Nostrand affair, see Vernie L. Hatch to BLM, 6 Sept.; BLM to Hatch, 11 Sept.; memo from Mr. [Joe] Canavan for BLM (n.d., résumé of phone conversation with Mrs. Medillia Cox of Valparaiso, Ind., about Van Nostrand's alleged charges); BLM to Cox, 1 Oct. 1928, Box 81, George Graves Papers.

48. *N.Y. World,* 24 Sept. 1928. In addition to the Anti-Saloon League, the Klan actively spread stories about Smith in "The Fellowship Forum."

49. Lippmann to BLM, 2 Aug. and 13 Sept. 1928, Walter Lippmann Papers.

50. Donald J. Lisio, *Hoover, Blacks, & Lily-Whites* (1985), untangles Hoover's racial policies; author interview with Eva Levy Marshall, 1975; Walter White, *A Man Called White* (1969), 99–101, for which reference I thank Don Lisio.

51. BLM, "AES Ms.," 73; Warner, *Happy Warrior,* 219.

52. BLM to Raskob, 14 Nov. 1928, 473/596, Raskob Papers; Perkins, Oral Hist., II, 527 ff.

53. Proskauer, Oral Hist., 133; Moskowitz to Frankfurter, 10 Nov. 1928. In the south, Smith lost Florida, North Carolina, Tennessee, Texas, and Virginia; the southern states he won were by small margins.

54. *N.Y. Telegram,* 31 Dec. 1928.

55. This account is based substantially on Perkins's Oral History, III, 11–14, 24–38; see also Rexford Tugwell, *The Democratic Roosevelt* (1957), 170–72; Eleanor Roosevelt, *This I Remember,* 48–51; Josephson, 408–12; Lela Stiles, *The Man Behind Roosevelt* (1954), 117–18.

56. Harlan B. Phillips, *Felix Frankfurter Reminisces* (1960), 199; Perkins, Oral Hist., III, 35–36, 82; Josephson, 414; Moskowitz to Frankfurter, 10 Nov. 1928, Frankfurter Papers. According to Edward Bernays, Smith called himself "janitor of the Empire State" (*Biography of an Idea,* 652). Though FDR rejected Moskowitz, he earlier, at her suggestion, hired lawyer Sam Rosenman to advise him on state legislative affairs (Rosenman, *Working With Roosevelt* [1952], 13), and made him his personal counsel and member of his "brain trust."

57. Evidence of the extent to which Smith relied on Moskowitz from 1928 until her death is ample in Smith's Private Papers, e.g.: her draft of an outline for an article on veterans' benefits, fol. 219; draft of his radio speech on the Depression, Boston, 1930, with her handwriting, fol. 335; BLM to Herbert Lefkowitz, 2 Feb. 1929, fol. 1; BLM to Grace Reavy, 20 Jan. 1931, fol. 6; Shouse to BLM, 11 May 1932, fol. 45; exchanges with Frederick Greene, State Superintendent of Public Works (Jan.–Feb., 1931), on railroad crossings and widening highways, fols. 85 and 87; with George Gove, Secretary of the State

Board of Housing (Feb. 1931), fol. 98, etc. See also fols. 501 ff. (espec. fols. 517, 539, and 540) for records of the Moskowitz-Smith literary and financial arrangements. Raskob's survey is in fol. 162. On Moskowitz hiring Hine to photograph construction of the Empire State Building, see Judith Mara Gutman, *Lewis Hine.*

58. Frank Kent, "Charley Michelson," 290–96; see also Michelson, *The Ghost Talks* (1944).

59. Perkins, Oral Hist., III, 75–77.

60. Moskowitz and Hapgood, *Up From the City Streets,* 243. On reforestation, see AES, Priv. Pap., fols. 3 and 165. Belle Moskowitz assisted in Smith's campaign against the referendum. See also Josephson, ch. XV; Samuel B. Hand, "Al Smith, Franklin D. Roosevelt, and the New Deal: Some Comments on Perspective," *The Historian,* 27 (1964–65), 366–81, argues that Smith was always a cautious constitutionalist.

61. Moskowitz's role in the renomination attempt is heavily documented in her state-by-state correspondence, alphabetically arranged in Smith's Private Papers.

62. BLM to Fred Johnson, 30 June 1931; 13 Jan. 1932; and Johnson to BLM, 2 Mar. 1932, fol. 591.

63. BLM to Frankfurter, 1 Mar. 1932. See also Frankfurter to W. C. Henson, 10 Nov. 1928; and, from 1932, Frankfurter to Proskauer, 27 Feb.; Proskauer to FF, 2 Mar.; FF to BLM, 5 Mar.; BLM to FF, 16 Mar.; and FF to BLM, 17 Mar.: "As for staying friends—My dear Belle, I hope I belong to the English school of politics, who are so much more civilized about these matters than we are. Differences of opinion, for me, do not make for differences in friendship. Otherwise life would be immeasurably more dreary than it is. . . . [T]he recent course of events . . . has not in the slightest changed my feeling for Smith or weakened my old ties with you you may be assured." Copies of this exchange are in AES, Priv. Pap., 1932 campaign, fol. "Massachusetts." Only the copies in FF's papers in the Library of Congress contain his marginalia.

64. BLM to Philip Perlman, 2 May 1932, fol. 568, AES Priv. Pap.; see also "Autobiographical Memorandum on Ghosting for Two Candidates," 20 April 1932, Lindsay Rogers Papers.

65. Perkins, Oral Hist., III, 82 ff, 335; Roper, *Fifty Years,* 259–60; Josephson, 439–41.

66. Author interview with Dorothy Rosenman, 1975; David Burner, *Herbert Hoover, A Public Life* (1979), 312.

67. Caro, *The Power Broker,* 322; a copy of the menu was in Smith's scrapbooks, in Emily Warner's possession when I consulted them.

68. On AES's 1932 campaign speeches, see Moses and Moskowitz traces in fols. 354 and 358, Priv. Pap.; Perkins, Oral Hist., III, 481–83, 500–01.

69. On Henry Moskowitz's walk from Smith, see *NYT,* 28 May and 8 Aug. 1936. See Jordan A. Schwarz, "Al Smith in the Thirties," *N.Y. History,* 45 (1964), 316–30, an article not based on primary sources; and Graham, *An Encore for Reform: The Old Progressives and the New Deal* (1967).

70. Lippmann to Frankfurter, 26 Sept. 1932, Frankfurter Papers; author interview with Miriam Gabo, 1986.

71. Margaret Streeter, Secretary, D.C. Women's Democratic Educational Council, to BLM, 4 Apr. 1932, fol. 559; BLM to Alexander McCrae, 21 July 1932, fol. 557; BLM to R. J. Dunham, 7 July 1932, fol. 562, AES Priv. Pap.

Epilogue

1. Author interview with Aline MacMahon, 1975.

2. Grace Montgomery to Smith, 3 Jan. 1933, Moskowitz Papers.

3. The following account of Belle's death, funeral, and estate is based on family letters in the Moskowitz Papers: Joe Israels to Miriam Franklin, 7, 13 Jan., 7 Feb. 1933; Carlos Israels to "Schatzie" (Miriam), 4 Jan. 1933; and on letters in Miriam Gabo's possession: Henry Moskowitz to Mr. & Mrs. [Ernest] Franklin, 7 Jan. 1933; A. J. Warner to Mrs. [Ernest] Franklin, 13 Jan. 1933; author interview with Eva Levy Marshall (1975). The condolences are also in the Moskowitz Papers. The final appraisal of her estate is described in the *NYT,* 10 July 1934.

Select Bibliography

Note: references to journal articles, newspapers, and to unpublished dissertations, theses, and oral histories appear only in the endnotes.

Unpublished Sources

Abelson, Paul. Papers. American Jewish Archives, Cincinnati, Ohio; and Martin P. Catherwood Library, New York State School of Industrial and Labor Relations, Cornell University.

Adler, Felix. Papers. Columbia University Library.

Baldwin Family. Papers. Yale University Library.

Barondess, Joseph. Papers. New York Public Library.

Baruch, Bernard. Papers. Princeton University Library.

Bernheimer, Charles L. Papers. New York Historical Society.

Brandeis, Louis Dembitz. Papers. University of Louisville Law School Library.

Butler, Nicholas Murray. Papers. Columbia University Library.

Colvin, Addison. Scrapbooks. Crandall Library, Glens Falls, N. Y.

Committee of Fourteen. Papers. New York Public Library.

Community Service Archives. New York, N. Y.

Corning, Edwin. Papers. Albany Institute of History and Art, Albany, N. Y.

Dewson, Mary (Molly). Papers. Schlesinger Library, Cambridge, Mass., and Franklin D. Roosevelt Library, Hyde Park, N.Y.

Dickerman, Marion. Papers. Franklin D. Roosevelt Library.

Ethical Culture Society of New York. Archives. New York, N.Y.

Frankel, Lee K. Papers. American Jewish Historical Society, Waltham, Mass.

Frankfurter, Felix. Papers. Library of Congress.

Gaus, John. Papers. State Historical Society of Wisconsin.

Gaynor, William. Papers. Municipal Archives of the City of New York.

Goldstein, Jonah J. Papers. American Jewish Historical Society.

Graves, George. Papers. Franklin D. Roosevelt Library.

Hamilton-Madison House. Papers (Microfilm ed.). Minnesota Social Welfare Archives.

Hapgood, Norman. Papers. Library of Congress.

Hapgood Family. Papers. Beinecke Library, Yale University.

Hillquit, Morris. Papers (Microfilm ed.). State Historical Society of Wisconsin.

Hoover, Herbert. Papers. Herbert Hoover Library, West Branch, Iowa.

Institute for Public Service. Papers. State Historical Society of Wisconsin.

International Ladies' Garment Workers Union. Arbitration Proceedings Between the Dress & Waist Manufacturers Association of New York and the I.L.G.W.U., 1913–1917; Joint Conference of the Dress and Waist Manufacturers' Association and the Waist and Dress Makers' Union, Minutes, 3 Sept. 1914. Research Division, I.L.G.W.U., New York, N.Y.

Jacob A. Riis Neighborhood Settlement. Papers. New York Public Library.

Kingsbury, John A. Papers. Library of Congress.

Laidlaw, Harriet Burton. Papers. Schlesinger Library.

Lehman, Herbert H. Papers. Columbia University Library.
Lippmann, Walter. Papers. Yale University Library.
Lusk Committee. Papers. Archives, New York State Library, Albany, N.Y.
Mack, Norman E. Papers. Buffalo & Erie County Historical Society.
Mahoney, James J. Papers. Manuscripts, New York State Library.
Marshall, Louis. Papers. American Jewish Archives.
McAneny, George. Papers. Columbia University and Princeton University Libraries.
Menken, Alice D. Papers. American Jewish Historical Society.
Miller, Emma Guffey. Papers. Schlesinger Library.
Mitchel, John Purroy. Papers. Municipal Archives of the City of New York and Library of Congress.
Morgenthau, Henry. Morgenthau, Sr., Correspondence; Morgenthau, Jr., Diaries. Franklin D. Roosevelt Library.
Moskowitz, Belle. Papers. Connecticut College, New London, Conn.
Nathan, Maud. Papers. Schlesinger Library.
National Association for the Advancement of Colored People. Papers. Library of Congress.
National Association of Jewish Social Workers. Papers. American Jewish Historical Society.
National Civic Federation. Papers. New York Public Library.
National Council of Jewish Women. Papers. Library of Congress.
National Council of Jewish Women–New York Section. Minute Books. New York, N.Y.
National Progressive Party (Theodore Roosevelt). Papers. Harvard University.
National Women's Trade Union League of America. Papers (Microfilm ed.). Library of Congress.
New York Child Labor Committee. Papers. Manuscripts, New York State Library.
New York Woman Suffrage Collection. Papers. Columbia University Library.
Peabody, George Foster. Papers. Library of Congress.
Perkins, Frances. Papers (Microfilm ed.). Columbia University Library.
Pinchot, Amos R. E. Papers. Library of Congress.
Raskob, John J. Eleutherian Mills Historical Library, Wilmington, Del.
Regional Planning Association. Papers. Manuscripts and University Archives, Cornell University.
Reid, Whitelaw. Papers (Microfilm ed.). Library of Congress.
Rice Family. Papers. Yale University Library.
Robins, Raymond. Papers. State Historical Society of Wisconsin.
Rockefeller, Jr., John D. Papers. Rockefeller Archives, New York, N.Y.
Rogers, Lindsay. Papers. Columbia University Library.
Roosevelt, Eleanor. Papers. Franklin D. Roosevelt Library.
Roosevelt, Franklin D. Political Papers, 1920–28; Family, Business, & Personal Affairs, 1920–1928; 1924 Campaign Correspondence; 1928 Campaign Correspondence; Governor's Personal Correspondence, 1929–32. Franklin D. Roosevelt Library.
Roosevelt, Theodore. Papers (Microfilm ed.). Library of Congress.
Schlesinger, Benjamin. Papers. Archives, International Ladies' Garment Workers' Union, New York, N.Y.
Simkhovitch, Mary K. Papers. Schlesinger Library.

Smith, Alfred E. Papers. Executive Papers (series 260: 1919–21; series 200: 1923–29) and Private Papers. Manuscripts and Archives, New York State Library.
Smith, Jane Norman. Papers. Schlesinger Library.
Stimson, Henry L. Papers. Yale University Library.
Straus, Oscar S. Papers. Library of Congress.
Temple Israel. Papers. American Jewish Archives.
University Settlement Society of New York City. Papers. American Jewish Historical Society and Microfilm ed., State Historical Society of Wisconsin.
Untermyer, Samuel. Papers. American Jewish Archives.
Van Namee, George R. Papers. Archives of the Diocese of Fresno, Fresno, California.
Wald, Lillian. Papers. Columbia University and New York Public Libraries.
Walling, William English. Papers. State Historical Society of Wisconsin.
Warburg, Felix. Papers. American Jewish Archives.
White, Sue Shelton. Papers. Schlesinger Library.
Wilson, Woodrow. Papers (Microfilm ed.). Library of Congress.
Wise, Stephen S. Papers. American Jewish Historical Society.

Contemporary Sources and Memoirs

Addams, Jane. *A New Conscience and an Ancient Evil.* New York: MacMillan, 1912.
——. *The Spirit of Youth and the City Streets.* New York: 1909; repr. Urbana, Ill.: 1972.
——. *Twenty Years at Hull House.* New York: Macmillan, 1910.
Anthony, Edward. *This is Where I Came In.* Garden City, N.Y.: Doubleday & Co., 1960.
Beard, Mary R. *Woman's Work in Municipalities.* New York: D. Appleton & Co., 1915.
Bernays, Edward. *Biography of An Idea: Memoirs of Public Relations Counsel Edward L. Bernays.* New York: Simon and Schuster, 1965.
Blaustein, Miriam, ed. *Memoirs of David Blaustein, Educator and Communal Worker.* New York: McBridge, Nast & Co., 1913.
Bowers, Claude. *My Life. The Memoirs of Claude Bowers.* New York: Simon and Schuster, 1962.
Cohen, Julius Henry. *Law and Order in Industry: Five Years' Experience.* New York: MacMillan, 1916.
——. *They Builded Better Than They Knew.* New York: J. Messner, Inc., 1946.
Cohen, Rose. *Out of the Shadow.* New York: George H. Doran, 1918.
Curran, Henry H. *Pillar to Post.* New York: Charles Scribner's Sons, 1941.
Davis, Michael M., Jr. *The Exploitation of Pleasure: A Study of Commercial Recreations in New York City.* New York: Russell Sage Foundation, 1911.
Democratic National Committee. *The Campaign Book of the Democratic Party: Candidates and Issues in 1928.* Albany, N.Y.: J. B. Lyon, 1928.
Dreier, Mary E. *Margaret Dreier Robins, Her Life, Letters, and Work.* New York: Island Press Cooperative, 1950.
Dillon, John. *From Dance Hall to White Slavery, Ten Dance Hall Tragedies.* New York: Willey Book Co., 1912.
Dorr, Rheta Childe. *What Eight Million Women Want.* 1910; Klaus repr., 1971.

Fleischman, Doris E., ed. *An Outline of Careers for Women: A Practical Guide to Achievement.* Garden City, N.Y.: Doubleday, Doran & Co., Inc., 1928.

Flynn, Edward J. *You're The Boss.* New York: Viking Pr., 1947.

Gollomb, Joseph. *Unquiet.* New York: Dodd, Mead, Co., 1935.

Harriman, Forence Jaffray. *From Pinafore to Politics.* New York: Henry Holt & Co., 1923.

Hasanovitz, Elizabeth. *One of Them. Chapters from a Passionate Autobiography.* Boston & New York: Houghton Mifflin Co., 1918.

Howe, Frederic C. *The Confessions of a Reformer.* New York: Charles Scribner's Sons, 1925.

Hoxie, Robert Franklin. *Scientific Management and Labor.* New York: D. Appleton & Co., 1915.

Kneeland, George. *Commercialized Prostitution in New York City.* New York: The Century Co., 1913.

Kohut, Rebekah. *My Portion (An Autobiography).* New York: Albert & Charles Boni, 1925.

_____. *More Yesterdays (An Autobiography, 1925–49).* New York: Bloch Publ. Co., 1950.

Lash, Joseph, ed. *From the Diaries of Felix Frankfurter.* New York: W. W. Norton & Co., 1975.

MacLeish, Archibald and E. F. Prichard, Jr., eds. *Law and Politics: Occasional Papers of Felix Frankfurter, 1913–1938.* Gloucester, Mass.: Peter Smith, 1971.

Moses, Robert. *Public Works: A Dangerous Trade.* New York: McGraw-Hill, 1970.

Mumford, Lewis. *Sketches From My Life.* New York: The Dial Pr., 1982.

Ovington, Mary White. *The Walls Came Tumbling Down.* New York: Harcourt, Brace & Co., 1947.

Perkins, Frances. *The Roosevelt I Knew.* New York: Viking Pr., 1946.

Phillips, Harlan B., ed. *Felix Frankfurter Reminisces.* New York: Reynal & Co., 1960.

Richardson, Dorothy. *The Long Day. The Story of a New York Working Girl. As Told by Herself.* New York: The Century Co., 1905; repr. 1972.

Roosevelt, Eleanor. *It's Up to the Women.* New York: Frederick A. Stokes, 1933.

_____. *This I Remember.* New York: Harper & Bros., 1949.

Roosevelt, Elliott, ed. *F.D.R. His Personal Letters 1928–1945.* New York: Duell, Sloan and Pearce, 1950.

Roper, Daniel C. *Fifty Years of Public Life.* Durham, N.C.: Duke Univ. Pr., 1941.

Rosen, Ruth and Sue Davidson, eds. *The Maimie Papers.* Old Westbury, N.Y.: Feminist Pr., 1977.

Rosenman, Samuel. *Working With Roosevelt.* New York: Harper & Bros., 1952.

Roskolenko, Harry. *The Time That Was Then: The Lower East Side 1900–1914, An Intimate Chronicle.* New York: Dial Pr., 1971.

Schneiderman, Rose. *All For One.* New York: Paul Eriksson, Inc., 1967.

Sheean, Vincent. *Personal History.* Boston: Houghton Mifflin, c. 1934, repr. 1969.

Simkhovitch, Mary K. *Neighborhood. My Story of Greenwich House.* New York: W. W. Norton, 1938.

Smith, Alfred E. *Public Papers.* 8 vols. Albany: J. B. Lyon, 1919–20; 1923–28.

_____. *Up To Now, An Autobiography.* New York: Viking Pr., 1929.

Solomon, Hannah G. *A Sheaf of Leaves.* Chicago: Print. priv., 1911.

———. *Fabric of My Life, the Autobiography of Hannah Greenebaum Solomon.* New York: Bloch Publishing Co., 1946.

Stiles, Lela. *The Man Behind Roosevelt. The Story of Louis McHenry Howe.* Cleveland, N. Y.: World Publ. Co., 1954.

Straus, Oscar S. *Under Four Administrations: from Cleveland to Taft.* Boston: Houghton Mifflin Co., 1922.

Taft, Jessie. *The Woman Movement From the Point of View of Social Consciousness.* Chicago: Univ. of Chicago Pr., 1916.

Tead, Ordway. *Instincts in Industry: A Study of Working-Class Psychology.* Boston: Houghton Mifflin Co., 1918.

True, Ruth S. *The Neglected Girl.* New York: Survey Associates, 1914.

Urofsky, Melvin I. and David W. Levy, eds. *Letters of Louis D. Brandeis.* 4 vols. Albany: State Univ. of New York Pr., 1971+.

Wald, Lillian. *The House on Henry Street.* New York: Henry Holt, 1915.

Warner, Emily Smith. *The Happy Warrior. The Story of My Father, Alfred E. Smith.* New York: Doubleday, 1956.

White, Walter. *A Man Called White.* New York: Arno Pr., 1969.

Wise, Stephen. *Challenging Years: The Autobiography of Stephen Wise.* New York: G. P. Putnam's Sons, 1919.

Woods, Robert A., and Albert J. Kennedy. *The Settlement Horizon: A National Estimate.* New York: Russell Sage Foundation, 1922.

Reports, Bulletins, and Proceedings

Association to Promote Proper Housing for Girls, Inc. (New York). *Ten Tales.* Celebration of its 10th Anniversary, 1913–24. Pamphlet.

Charity Organization Society. *Annual Reports.* 1903–10.

Committee of Fifteen. *The Social Evil, with Special Reference to Conditions Existing in the City of New York.* 1902; rev. 2nd ed., New York & London: G. P. Putnam's Sons, 1912.

Committee of Fourteen. *Annual Reports.* 1909–28.

———. *Department Store Investigation. Report of the Sub-Committee.* 1915.

———. *The Social Evil in New York City. A Study of Law Enforcement by the Research Committee of the Committee of Fourteen.* New York: Andrew H. Kellogg Co., 1910.

Dress and Waist Manufacturers' Association of New York City, *Bulletin* (Feb.–Apr. 15, 1913), nos. 1–4.

Educational Alliance. *Announcement of the Departments and Courses of Study, 1900–1901, 1901–02.*

———. *Annual Reports.* 1894–1908.

Empire State. *A History.* Publicity Associates: Empire State Building, N.Y.C., 1 May 1931.

Gardner, Ella. *Public Dance Halls. Their Regulation and Place in the Recreation of Adolescents.* Washington, D.C.: Children's Bureau Publication No. 189, U.S. Dept. of Labor, 1929.

Kansas City, Mo., Board of Public Welfare, Recreation Department. *Annual Report.* 1911–12.

National Association for the Advancement of Colored People. *Annual Reports.* 1911–26.

National Conference of Charities and Corrections. *Proceedings.* 1911–23.
National Conference of Jewish Charities in the United States. *Biennial Proceedings.* 1912–23.
National Council of Jewish Women. *Program of Work.* 1894+.
_____. *Triennial Proceedings.* 1897+.
_____. *The First Fifty Years: A History of the National Council of Jewish Women, 1893–1943.* Monroe Campbell, Jr. and Willem Wirtz, eds. 1943.
_____, New York Section. *Yearbooks.* 1905+.
New York (City) Board of Aldermen. *Report of the Special Committee to Investigate the Police Department.* 10 June 1913 (Arno repr., 1971).
New York City Conference of Charities and Corrections. *Proceedings.* 1910–26.
New York State Conference of Charities and Corrections. *Proceedings.* 1905–23.
New York State Factory Investigating Commission. *Reports.* 1912–14.
New York State Reconstruction Commission. *Reports.* 1919–20 (printed separately and in Smith, *Public Papers,* 1919–20).
Playground and Recreation Association. *Proceedings.* 1910–12.
Port of New York Authority. *Annual Reports.* 1924–26.
Stone, Nahum I. *Wages and Regularity of Employment and Standarization of Piece Rates in the Dress and Waist Industry of New York City.* Bull. of the U.S. Dept. of Labor, Bur. of Labor Stat., No. 146, 1914.
United States Commission on Industrial Relations. *Final Report and Testimony submitted to Congress,* August 23, 1912. Washington, D.C.: 1915.
United States Senate. *Hearings Before a Special Committee Investigating Presidential Campaign Expenditures,* 70th Congress. May, 1928.
Winslow, Charles H. *Conciliation, Arbitration, and Sanitation in the Dress and Waist Industry of New York City.* Bull. of the U. S. Dept. of Labor, Bur. of Labor Statistics, No. 145, 1914.
_____. *Industrial Court of the Cloak, Suit, and Skirt Industry of New York City.* Bull. of the U. S. Dept. of Labor, Bur. of Labor Statistics, No. 144, 1914.

Monographs, Biographies and other Studies

Ashby, LeRoy. *The Spearless Leader. Senator Borah and the Progressive Movement in the 1920s.* Urbana: Univ. of Illinois Pr., 1972.
Baum, Charlotte, Paula Hyman, and Sonya Michel. *The Jewish Woman in America.* New York: Dial Pr., 1976.
Bernays, Edward L. *Public Relations.* Norman: Univ. of Oklahoma Pr., 1952.
Bordin, Ruth. *Woman and Temperance. The Quest for Power and Liberty, 1873–1900.* Philadelphia: Temple Univ. Pr., 1981.
Blair, Karen. *The Clubwoman as Feminist: True Womanhood Redefined, 1868–1914.* New York & London: Holmes & Meier, 1980.
Bonnett, Clarence E. *History of Employers' Associations in the United States.* New York: Vantage Pr., 1956.
Bremner, Robert H. *From the Depths: The Discovery of Poverty in the United States.* New York: New York Univ. Pr., 1956.
Bristow, Edward J. *Prostitution and Prejudice. The Jewish Fight Against White Slavery 1870–1939.* New York: Schocken, 1983.
Buck, Arthur E. *The Reorganization of State Governments in the United States.* New York: Columbia Univ. Pr., 1938.

Buenker, John D. *Urban Liberalism and Progressive Reform.* New York: W. W. Norton, 1978.

Burner, David. *The Politics of Provincialism: The Democratic Party in Transition, 1918–1932.* New York: Norton Library ed., 1975.

———. *Herbert Hoover, A Public Life.* New York: Knopf, 1979.

Caro, Robert. *The Power Broker. Robert Moses and the Fall of New York.* New York: Knopf, 1974.

Carpenter, Jesse. *Competition and Collective Bargaining in the Needle Trades, 1910–1967.* Ithaca, N.Y.: Cornell Univ. Pr., 1972.

Cavallo, Dominick. *Muscles and Morals: Organized Playgrounds and Urban Reform, 1880–1920.* Philadelphia: Univ. of Pennsylvania Pr., 1981.

Chafe, William H. *The American Woman. Her Changing Social, Economic, and Political Roles, 1920–1970.* New York: Oxford Univ. Pr., 1972.

Chambers, Clarke A. *Paul U. Kellogg and The Survey: Voices for Social Welfare and Social Justice.* Minneapolis: Univ. of Minnesota Pr., 1971.

———. *Seedtime of Reform. American Social Service and Social Action, 1918–1933.* Minneapolis: Univ. of Minnesota Pr., 1963.

Childs, Harwood L. *Public Opinion: Nature, Formation, and Role.* Princeton, N.J.: D. Van Nostrand Co., 1965.

Clarke, Norman H. *Deliver Us From Evil: An Interpretation of American Prohibition.* New York: Norton, 1976.

Cohen, Naomi W. *A Dual Heritage: The Public Career of Oscar S. Straus.* Philadelphia: The Jewish Publication Society of America, 1969.

Commons, John R., et al. *History of Labor in the United States, 1896–1932.* 4 vols. New York: MacMillan, 1935.

Connelly, Mark Thomas. *The Response to Prostitution in the Progressive Era.* Chapel Hill: Univ. of N. Carolina Pr., 1980.

Cremin, Lawrence A. *The Transformation of the School. Progressivism in American Education, 1876–1957.* New York: Knopf, 1961.

Cremin, Lawrence A., David A. Shannon, and Mary Evelyn Townsend. *A History of Teachers College.* New York: Columbia Univ. Pr., 1954.

Cressy, Paul G. *The Taxi-Dance Hall: A Sociological Study in Commercialized Recreation and City Life.* Chicago: Univ. of Chicago Pr., 1932; repr., Greenwood Pr., 1968.

Dahlberg, Jane S. *The New York Bureau of Municipal Research.* New York: New York Univ. Pr., 1966.

Davis, Allen F. *Spearheads for Reform: The Social Settlements and the Progressive Movement 1890–1914.* New York: Oxford Univ. Pr., 1967.

Davis, Kenneth S. *FDR. The Beckoning of Destiny, 1882–1928. A History.* New York: G.P. Putnam's Sons, 1971.

———. *Invincible Summer. An Intimate Portrait of the Roosevelts based on the recollections of Marion Dickerman.* New York: Atheneum, 1974.

Dubofsky, Melvyn. *When Workers Organize. New York City in the Progressive Era.* Amherst: Univ. of Mass. Pr., 1968.

Duffus, R. L. *Lillian Wald, Neighbor and Crusader.* New York: MacMillan, 1938.

Dye, Nancy Schrom. *As Equals and As Sisters: Feminism, the Labor Movement, and the Women's Trade Union League of New York.* Columbia: Univ. of Missouri Pr., 1980.

Eldot, Paula. *Governor Alfred E. Smith, The Politician As Reformer.* New York & London: Garland Publ. Co., 1983.

Erenberg, Lewis. *Steppin' Out: New York Nightlife and the Transformation of American Culture, 1890–1930.* Westport: Greenwood Pr., 1981.

Ferber, Nat J. *A New American. From the Life Story of Salvatore A. Cotillo, Supreme Court Justice, State of New York.* New York: Farrar & Rhinehart, 1938.

Flexner, Eleanor. *Century of Struggle: The Woman's Rights Movement in the United States.* Cambridge: Harvard Univ. Pr., c. 1959, 1975.

Freidel, Frank. *Franklin D. Roosevelt.* Vol. 1: *The Ordeal;* Vol. 2: *The Triumph.* Boston: Little, Brown and Co., 1954, 1956.

Gaylin, Willard, Ira Glasser, Steven Marcus, and David J. Rothman. *Doing Good. The Limits of Benevolence.* New York: Pantheon Books, 1978.

Goren, Arthur. *New York Jews and the Quest for Community.* New York: Columbia Univ. Pr., 1970.

Graham, Otis. *An Encore for Reform. The Old Progressives and The New Deal.* New York: Oxford Univ. Pr., 1967.

Gruberg, Martin. *Women in American Politics: An Assessment and Sourcebook.* Oshkosh, Wis.: Academia Pr., 1968.

Gurock, Jeffrey S. *When Harlem was Jewish, 1870–1930.* New York: Columbia Univ. Pr., 1979.

Gutman, Judith Mara. *Lewis W. Hine and the American Social Conscience.* New York: Walker & Co., 1967.

Haber, Samuel. *Efficiency and Uplift. Scientific Management in the Progressive Era, 1890–1920.* Chicago: Chicago Univ. Pr., 1964.

Hacker, Louis M. and Mark D. Hirsch. *Proskauer: His Life and Times.* University, Ala.: Univ. of Alabama Pr., 1978.

Hammack, David. *Power and Society. Greater New York at the Turn of the Century.* New York: Russell Sage Foundation, 1982.

Handlin, Oscar. *Al Smith and His America.* Boston: Little, Brown & Co., 1958.

Hapgood, Hutchins. *The Spirit of the Ghetto.* New York & London: Funk & Wagnalls, 1902, 1909.

Hapgood, Norman and Henry Moskowitz. *Up From the City Streets: Alfred E. Smith, A Biographical Study in Contemporary Politics.* New York: Harcourt, Brace & Co., 1927.

Haskell, Thomas L. *The Emergence of Professional Social Science.* Urbana: Univ. of Illinois, 1977.

Higham, John. *Strangers in the Land.* New York: Atheneum, 1965.

Hoff-Wilson, Joan and Marjorie Lightman, eds. *Without Precedent: The Life and Career of Eleanor Roosevelt.* Bloomington, Ind.: Indiana Univ. Press, 1984.

Hollinger, David. *Morris R. Cohen and the Scientific Ideal.* Cambridge, Mass.: MIT Press, 1975.

Howe, Irving. *World of Our Fathers.* New York: Harcourt Brace Jovanovich, 1976.

Irwin, Inez Haynes. *Angels and Amazons: A Hundred Years of American Women.* Garden City, N.Y.: Doubleday, Doran & Co., Inc., 1933.

Jackson, Anthony. *A Place Called Home. A History of Low-Cost Housing in Manhattan.* Cambridge, Mass.: The MIT Pr., 1976.

Josephson, Matthew and Hannah. *Al Smith: Hero of the Cities. A Political Portrait Drawing on the Papers of Frances Perkins.* Boston: Houghton Mifflin, 1969.

Kennedy, Susan Estabrook. *If All We Did Was to Weep at Home: A History of White Working-Class Women in America.* Bloomington, Ind.: Indiana Univ. Pr., 1979.

Kessler-Harris, Alice. *Out to Work. A History of Wage-Earning Women in the United States.* New York: Oxford Univ. Pr., 1982.

Knapp, Richard F. and Charles F. Hartsoe. *Play for America. The National Recreation Association 1906–1965.* Arlington, Va.: National Recreation Assoc., 1979.

Kraut, Benny. *From Reform Judaism to Ethical Culture: The Religious Evolution of Felix Adler.* Cincinnati: Hebrew Union College Pr., 1979.

Kurland, Gerald. *Seth Low. The Reformer in an Urban and Industrial Age.* New York: Twayne Publishers, 1971.

Kyvig, David. *Repealing National Prohibition.* Chicago: Univ. of Chicago Pr., 1979.

Lash, Joseph P. *Eleanor and Franklin.* New York: W. W. Norton, 1971.

Lemons, J. Stanley. *The Woman Citizen: Social Feminism in the 1920s.* Urbana: Univ. of Illinois Pr., 1973.

Leuchtenburg, William E. *The Perils of Prosperity, 1914–32.* Chicago: Univ. of Chicago Pr., 1958.

Levine (Lorwin), Louis. *The Women's Garment Workers, A History of the International Ladies' Garment Workers' Union.* New York: B. W. Huebsch, 1924.

Lewinson, Edwin R. *John Purroy Mitchel. The Boy Mayor of New York.* New York: Astra Books, 1965.

Lichtman, Allen J. *Prejudice and the Old Politics: The Presidential Election of 1928.* Chapel Hill: Univ. of North Carolina Pr., 1979.

Lisio, Donald J. *Hoover, Blacks, & Lily-Whites.* Chapel Hill: Univ. of North Carolina Pr., 1985.

Logan, Mary S. *The Part Taken by Women in American History.* 1912; repr., Arno Pr., 1972.

Lowitt, Richard. *George W. Norris. The Persistence of a Progressive, 1913–1933.* Urbana: Univ. of Illinois Pr., 1971.

Lubove, Roy. *The Professional Altruist. The Emergence of Social Work as a Career, 1880–1930.* New York: Atheneum, 1969.

———. *The Progressives and the Slums. Tenement House Reform in New York City, 1890–1917.* Pittsburgh: Univ. of Pittsburgh Pr., 1962.

———. *The Urban Community: Housing and Planning in the Progressive Era.* Englewood Cliffs, N. J.: Prentice-Hall, 1967.

Manners, Ande. *Poor Cousins.* New York: Coward, McCann & Geoghegan, Inc., 1972.

Marcaccio, Michael D. *The Hapgoods: Three Earnest Brothers.* Charlottesville: Univ. Pr. of Virginia, 1977.

Margolin, Victor. *American Poster Renaissance.* New York: Watson-Guptill, 1975.

McCormick, Richard L. *From Realignment to Reform: Political Change in New York State, 1893–1910.* Ithaca & London: Cornell Univ. Pr., 1981.

Moore, Edmund A. *A Catholic Runs for President.* New York: The Ronald Press Co., 1956.

Moskowitz, Henry. *Alfred E. Smith, An American Career.* New York: Thomas Seltzer, 1924.

Mowry, George E. *Theodore Roosevelt and the Progressive Movement.* New York: Hill & Wang, c. 1946, 1963.

Murray, Robert K. *The 103rd Ballot: Democrats and the Disaster in Madison Square Garden.* New York: Harper & Row, 1976.

Nadworny, Milton J. *Scientific Management and the Unions, 1900–1932: A Historical Analysis.* Cambridge: Harvard Univ. Pr., 1955.

Nelson, Daniel. *Frederick W. Taylor and the Rise of Scientific Management.* Madison: Univ. of Wisconsin Pr., 1980.

Noble, David W. *The Paradox of Progressive Thought.* Minneapolis: Univ. of Minnesota Pr., 1958.

O'Connor, Richard. *The First Hurrah: A Biography of Alfred E. Smith.* New York: G. P. Putnam's Sons, 1970.

O'Neill, William L. *Everyone Was Brave. The Rise and Fall of Feminism in America.* Chicago: Quadrangle Books, 1969.

Peel, Roy V. and Thomas C. Donnelly. *The 1928 Campaign, An Analysis.* New York: Richard R. Smith, Inc., 1931.

Peiss, Kathy. *Cheap Amusements: Working Women and Leisure in Turn-of-the-Century New York.* Philadelphia: Temple Univ. Pr., 1986.

Pivar, David. *Purity Crusade: Sexual Morality and Social Control, 1868–1900.* Westport: Greenwood Pr., 1973.

Pringle, Henry F. *Alfred E. Smith: A Critical Study.* New York: Macy-Masius, 1927.

Rischin, Moses. *The Promised City: New York's Jews, 1870–1914.* Cambridge: Harvard Univ. Pr., 1962.

Rollins, Jr., Alfred B. *Roosevelt and Howe.* New York: Knopf, 1962.

Rosen, Ruth. *The Lost Sisterhood: Prostitution in America, 1900–1918.* Baltimore: Johns Hopkins Univ. Pr., 1982.

Rothman, Sheila. *Woman's Proper Place: A History of Changing Ideals and Practices, 1870 to the Present.* New York: Basic Books, 1978.

Richardson, James F. *The New York Police, Colonial Times to 1901.* New York: Oxford Univ. Pr., 1970.

Root, Grace C. *Women and Repeal. The Story of the Women's Organization for National Prohibition Reform.* New York: Harper & Bros., 1934.

Ryan, Mary. *Womanhood in America, From Colonial Times to the Present.* New York: New Viewpoints, 1975.

Schick, Thomas. *The New York State Constitutional Convention of 1915 and the Modern State Governor.* Lebanon, Pa.: National Municipal League, 1978.

Sklar, Robert. *Movie-Made America, A Social History of American Movies.* New York: Random House, 1975.

Stein, Leon. *The Triangle Fire.* New York: Lippincott, 1962.

Tentler, Leslie Woodcock. *Wage-Earning Women. Industrial Work and Family Life in the United States, 1900–1930.* New York: Oxford Univ. Pr., 1979.

Timberlake, James H. *Prohibition and the Progressive Movement, 1900–1920.* Cambridge: Harvard Univ. Pr., 1963.

Tugwell, Rexford. *The Democratic Roosevelt.* Garden City, N.Y.: Doubleday & Co., Inc., 1957.

Ware, Susan. *Beyond Suffrage: Women in the New Deal.* Cambridge: Harvard Univ. Pr., 1981.

Waterman, Willoughby Cyrus. *Prostitution and its Repression in New York City, 1900–1931.* New York: AMS Pr., 1932.

Wiebe, Robert H. *Businessmen and Reform: A Study of the Progressive Movement.* Chicago: Quadrangle Books, c. 1962, 1968.

Weiss, Nancy. *Charles Francis Murphy, 1858–1924: Respectability and Responsibility in Tammany Politics.* Northampton: Smith College, 1968.

Index